D0927829

Wood Technology: Chemical Aspects

Wood Technology: Chemical Aspects

Irving S. Goldstein, EDITOR
North Carolina State University

A symposium sponsored by the Cellulose, Paper and Textile Division at the 172nd Meeting of the American Chemical Society, San Francisco, Calif., Aug. 31–Sept. 2, 1976.

ACS SYMPOSIUM SERIES **43**

AMERICAN CHEMICAL SOCIETY
WASHINGTON, D. C. 1977

Library of Congress CIP Data

Wood technology, chemical aspects
(ACS symposium series; 43)

Includes bibliographies and index.

1. Wood-using industries—Congresses. 2. Wood—Chemistry—Congresses.
I. Goldstein, Irving S., 1921– . II. American Chemical Society. Cellulose, Paper, and Textile Division. III. Series: American Chemical Society. ACS symposium series; 43.

TS801.W66 674'.8 77-2368
ISBN 0-8412-0373-3 ACSMC8 43 1-372

PRINTED IN THE UNITED STATES OF AMERICA

ACS Symposium Series

Robert F. Gould, *Editor*

FOREWORD

The ACS Symposium Series was founded in 1974 to provide a medium for publishing symposia quickly in book form. The format of the Series parallels that of the continuing Advances in Chemistry Series except that in order to save time the papers are not typeset but are reproduced as they are submitted by the authors in camera-ready form. As a further means of saving time, the papers are not edited or reviewed except by the symposium chairman, who becomes editor of the book. Papers published in the ACS Symposium Series are original contributions not published elsewhere in whole or major part and include reports of research as well as reviews since symposia may embrace both types of presentation.

CONTENTS

PREFACE

The role of chemistry in the forest products industries has always been secondary to mechanical influences. Most chemical effort has been expended in the pulp and paper sector to the virtual neglect of solid wood and board products even though the latter comprise 60% of the volume.

As pointed out in a recent National Research Council report ("Renewable Resources for Industrial Materials," National Academy of Sciences, Washington, D.C., 1976, 267 pp.), renewable resources are a great and underused national resource whose substitutability for dwindling coal and petroleum resources can reduce U.S. dependence on foreign energy and materials sources. Wood is our most abundant renewable resource and has substantial advantages, particularly from the standpoint of energy requirements, over alternative materials.

It was in this context that the symposium presented here was planned. The intent is to summarize the chemical aspects of solid wood technology, provide this background in a useful reference volume, and hopefully to stimulate additional chemical activity in this area. Most of the papers presented are reviews, but a few current research reports have been included to indicate the pioneering work being done.

The book opens with a paper on the structure and composition of wood to define the material under discussion and then considers molds, permeability, wood preservation, thermal deterioration and fire retardance, dimensional stability, adhesion, reconstituted wood boards such as fiberboard and particleboard, plywood, laminated beams, wood finishes, wood–polymer composites, and wood softening and forming. A final paper treats the common theme of wastewater management. Only one of the papers presented at the meeting is not included in this volume, and its subject of conventional wood preservation methods is adequately treated in detail elsewhere (e.g., Nicholas, D. D., *Ed.*, "Wood Deterioration and Its Prevention by Preservative Treatments," 2 vols., Syracuse University Press, 1973).

As is inevitable with so many topics and so many authors, there is great diversity in style, depth, and quantitative or qualitative degree of treatment. There might be more than a reader would want to know about a particular subject and not enough about another. However, I hope that in toto this collection of papers by knowledgeable individuals will to some extent meet the symposium objectives.

Raleigh, N.C.
January 3, 1977

IRVING S. GOLDSTEIN

1

Wood: Structure and Chemical Composition

R. J. THOMAS

Department of Wood and Paper Science, North Carolina State University,
Raleigh, N.C. 27607

Within living trees, wood is produced to perform the roles of support, conduction and storage. The support role enables the tree stem to remain erect despite the heights to which a tree grows. Because of these heights, wood also must perform the role of conduction, that is, transport water from the ground to the upper parts of the tree. Finally, food is stored in certain parts of the wood until required by the living tree. The wood cells which perform the role of conduction and/or support make up 60 to 90 percent of the wood volume. Within the living tree these cells are dead, that is, the cytoplasm is absent leaving a hollow cell with rigid walls. The only living cells within the wood portion of a tree are the food-storing cells. The close relationship between "form and function" simplifies the study of wood anatomy if the role of the cells in the living tree is kept constantly in view.

Gross Anatomical Features

The end view of a log (Figure 1) exposes the wood and bark portion of a tree trunk. Each year a growing center located between the wood and bark inserts a new layer of wood adjacent to the existing wood. In addition, new bark is deposited next to the pre-existing bark. Wood occupies the largest volume of a tree stem because more wood cells are produced than bark cells and also because the wood cells are retained and thus accumulate while the outermost bark cells are continually discarded.

The central wood portion of the log depicted in Figure 1 is considerably darker in color than the part adjacent to the bark. The dark-colored wood is termed heartwood and the light-colored wood is termed sapwood. The discoloration is due to the production and secretion of substances which are a by-product of the death of food-storage cells. As new wood, that is sapwood, is formed to the outside of the tree stem, additional interior sapwood adjacent to the heartwood zone is converted to heartwood. Some trees do not form discolored heartwood upon the death of

Figure 1. End view of a log showing both the wood and bark. Note the small total volume occupied by the bark. The central, dark portion of the wood is heartwood, and the outer ring of light colored wood is sapwood.

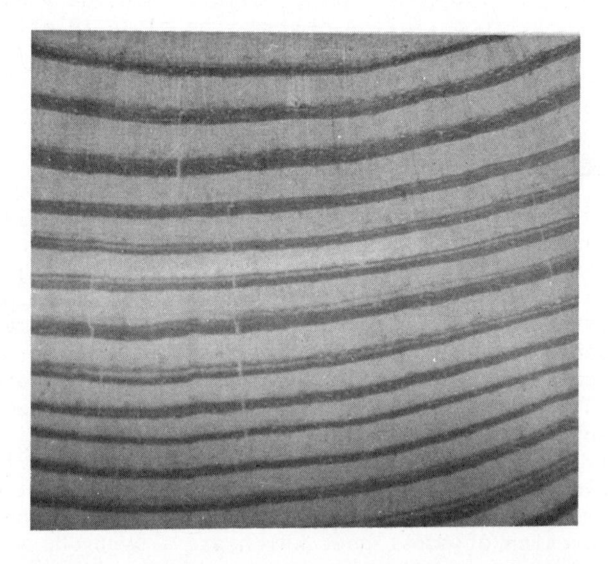

Figure 2. End view of a softwood showing growth rings. Each growth ring consists of a light and dark area. The light portion is called springwood or earlywood, and the dark area is termed summerwood or latewood. 15×

food-storage cells, thus recognition of heartwood based on color is not always possible. It should be apparent that the vast majority of cells which constitute the wood portion of a living tree are dead. The sapwood zone contains the only living cells found in mature wood and they constitute, depending upon species, from 10 to 40 percent of the sapwood volume. As heartwood is formed these living, food-storage cells die, thus the heartwood contains only dead cells.

Trees are classified into two major groups termed softwoods (gymnosperms) and hardwoods (angiosperms). The botanical basis for classification is whether or not the tree seed is naked as in softwoods or covered as in hardwoods. A more familiar classification, which with some exceptions is valid, is based on the retention of leaves by softwoods or the shedding of leaves by hardwoods. Thus the softwoods are often referred to as evergreen trees and hardwood as deciduous trees. The major difference with regard to wood anatomy is the presence of vessels in hardwoods. Vessels are structures composed of cells created exclusively for the conduction of water. Softwoods lack vessels but have cells termed longitudinal tracheids which perform a dual role of conduction and support.

The terms softwood and hardwood are not to be taken as a measure of hardness since some hardwoods are softer than many softwoods. For the commercially important domestic woods, the average specific gravity for softwoods is .41 with a range of .29 to .60. Hardwoods average .50 and vary from .32 to .81.

Growth rings, or annual increments, as seen on the end of a softwood log are depicted in Figure 2. Growth rings are detectable due to differences in the wood produced early and late in the growing season. The wood produced early, called earlywood or springwood, is considerably lighter in color than the wood termed latewood or summerwood which is produced late in the growing season. The color difference is due primarily to the different kinds of cells produced either early or late in the growing season. Figure 3 depicts three growth rings and part of a fourth. Note that at this low magnification you can detect the individual cells which constitute the springwood area; however, the summerwood zone appears merely as a dark zone. At a higher magnification, as shown in Figure 4, individual cells of both springwood and summerwood are evident. The springwood cells have a large cross-sectional area with thin walls and a large, open center. The large, open center, called the lumen, permits efficient conduction of water. The summerwood cells have a smaller cross-sectional area, thicker walls and a smaller lumen than springwood cells. Obviously, this type of cell provides substantial support for the tree stem but is not as efficient as springwood cells in conduction. In some softwood species the gradation between springwood and summerwood is gradual and fewer summerwood cells are produced. When this occurs, it is often difficult to distinguish growth increments as this type of wood possesses a more

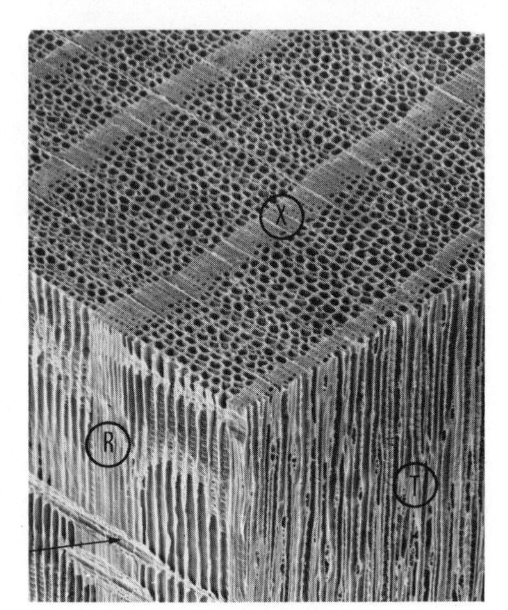

Figure 3. Softwood block showing three complete and part of two other growth rings in the cross-sectional plane (X). Individual springwood cells can be detected, whereas the smaller summerwood cells cannot be seen as individual cells. Also note the absence of vessels and the uniformity of the wood. Two longitudinal surfaces (R—radial; T—tangential) are illustrated. Food-storing cells can be easily detected on the radial surface (arrow). 47× (Courtesy of N. C. Brown Center for Ultrastructural Studies, S.U.N.Y. College of Environmental Science and Forestry)

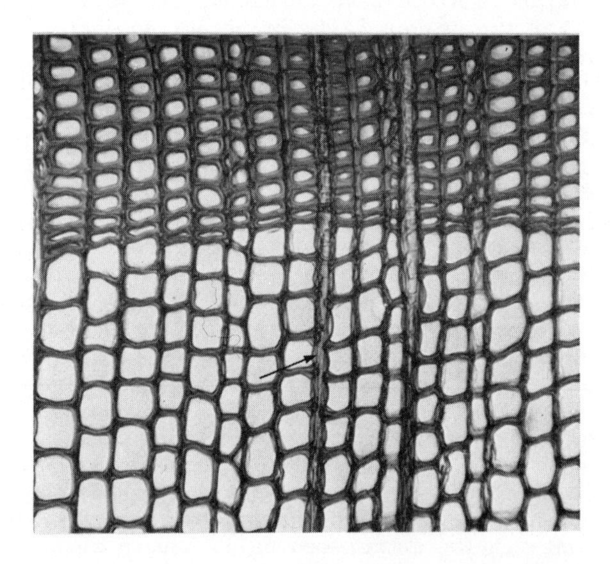

Figure 4. Cross-sectional view of springwood and summerwood cells. The springwood cells are larger and characterized by thin cell walls and a large central opening. The smaller cells with thick cell walls and a small central opening are summerwood cells. The narrow ribbon-like rows of cells traversing both the springwood and summerwood zones (arrow) are called rays. They consist of transversely oriented food-storage cells. 135×

uniform structure.

As indicated earlier, hardwoods are characterized by the presence of vessels or pores. Vessels are cells which occupy a large cross-sectional area and can in most species be detected with the unaided eye. The wood of hardwood trees is classified as either ring-porous or diffuse-porous depending upon the arrangement of the vessels. In a typical ring-porous wood, as illustrated in Figure 5, the vessels formed in the springwood are considerably larger than those formed in the summerwood. Figure 6 reveals a diffuse-porous wood in which the vessels are essentially the same size throughout the growth ring. As a result, in many diffuse-porous woods it is difficult to distinguish growth rings. A comparison of Figures 3, 5 and 6 will reveal the major difference between hardwood and softwood anatomy. Note the lack of vessels and the relatively uniform appearance of the softwood in Figure 3, compared to the hardwoods depicted in Figures 5 and 6.

Wood rays are found in all species of wood and consist of ribbon-like aggregations of food-storing cells extending in the transverse direction from the bark toward the center of the tree. In the cross-sectional view, rays take the form of lines of varying width running at right angles to the growth rings (Figures 3, 4, 5 and 6).

Softwood Anatomy

The anatomy of softwoods will be described first because it is less complex than hardwoods. The two main cell types which constitute softwoods are tracheids, which conduct and support, and parenchyma which store food. These two cell types can be further classified as to their orientation, that is longitudinal or transverse. Cells oriented in the longitudinal direction have the long axis of the cell oriented parallel to the vertical axis of the tree trunk whereas transversely oriented cells have their long axis at right angles to the vertical axis of the tree stem. A greatly simplified model of softwood anatomy can be made by gluing together a group of soda straws along their length and dispersing throughout this group matches, laid end to end with the long axis of the matches oriented at right angles to the long axis of the soda straws. The soda straws represent the longitudinally oriented tracheids which occupy about 90% of the volume whereas the matches represent transversely oriented parenchyma cells occupying about 10% of the volume. The transversely oriented parenchyma are the cells which constitute wood rays. As the matches in the model are considerably shorter than the soda straws, so parenchyma cells are considerably shorter than longitudinal tracheids.

Since longitudinal tracheids constitute about 90% of the volume and are therefore largely responsible for the resulting physical and chemical properties of softwoods, a detailed

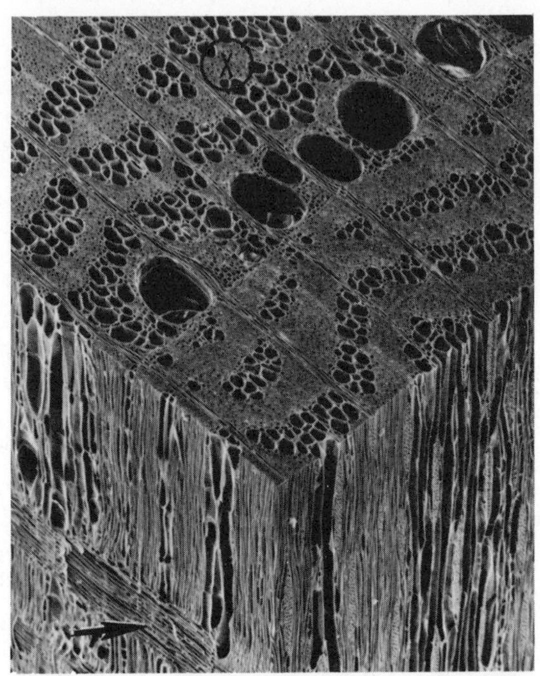

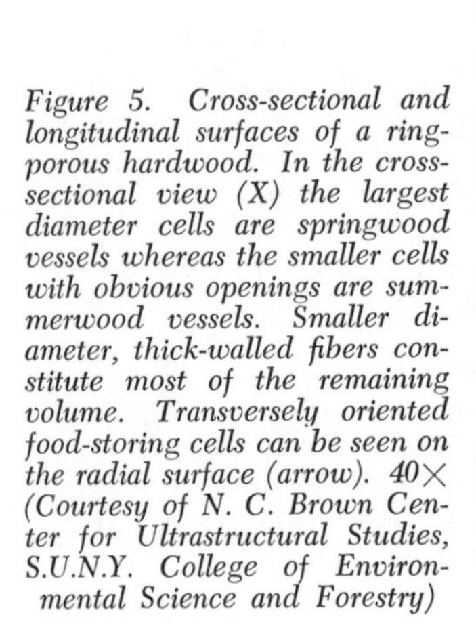

Figure 5. Cross-sectional and longitudinal surfaces of a ring-porous hardwood. In the cross-sectional view (X) the largest diameter cells are springwood vessels whereas the smaller cells with obvious openings are summerwood vessels. Smaller diameter, thick-walled fibers constitute most of the remaining volume. Transversely oriented food-storing cells can be seen on the radial surface (arrow). 40× (Courtesy of N. C. Brown Center for Ultrastructural Studies, S.U.N.Y. College of Environmental Science and Forestry)

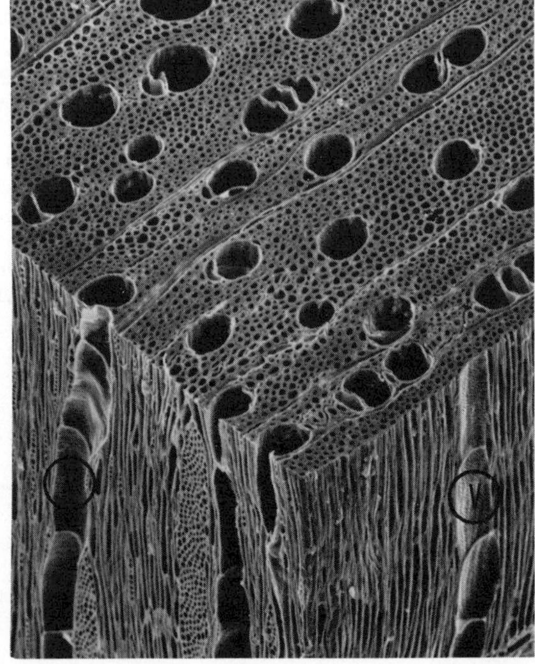

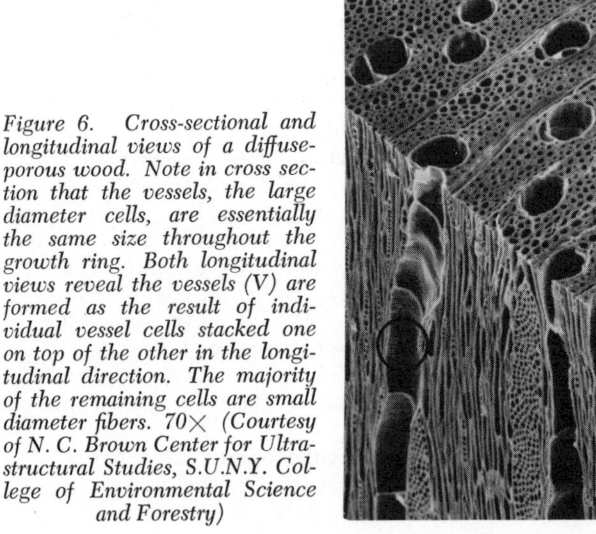

Figure 6. Cross-sectional and longitudinal views of a diffuse-porous wood. Note in cross section that the vessels, the large diameter cells, are essentially the same size throughout the growth ring. Both longitudinal views reveal the vessels (V) are formed as the result of individual vessel cells stacked one on top of the other in the longitudinal direction. The majority of the remaining cells are small diameter fibers. 70× (Courtesy of N. C. Brown Center for Ultrastructural Studies, S.U.N.Y. College of Environmental Science and Forestry)

description of their anatomy is merited. Figure 3, showing a small block of pine magnified 50 times, clearly illustrates that springwood and summerwood longitudinal tracheids make up the largest volume of the wood. Note that in addition to the cross-sectional view, two longitudinal views are exposed. Transversely oriented parenchyma cells, which make up the rays, can also be seen in the transverse and longitudinal views. In the cross-sectional view, three complete and two partial growth rings are depicted.

The longitudinal tracheid is about 100 times longer than wide. Depending on species, most domestic softwoods have longitudinal tracheids ranging from 3 to 5 mm in length. Redwood has the longest tracheids, up to 7.3 mm, and cedar the shortest, about 1.18 mm in length. The width of longitudinal tracheids for domestic species ranges from 20 μm for cedar up to 80 μm for redwood.

Since the longitudinal tracheid is a long, thin, cylindrical, tube-like cell, its appearance depends upon how it is viewed. For example, if the cell is cut at right angles to its long axis, the cross-sectional view is exposed on the cut surface. The cross-sectional view exposed in Figure 3 reveals in the springwood a square or polygonal shape predominating while in the summerwood a more rectangular shape is apparent. In the longitudinal, radial plane view, the ends of springwood tracheids are rounded, while in tangential view they are pointed (Figure 7). The ends of summerwood tracheids are pointed in both radial and tangential views (Figure 7).

It should be obvious that the structure of the longitudinal tracheid is well suited to perform the dual roles of conduction and support. Since water is translocated up the tree via the tracheids, the orientation of the long axis of the tracheid parallel to the vertical stem permits a longer passageway prior to interruption by a cell wall. The rigid cell walls, of varying thickness, provide adequate support.

Of the several types of markings found on the longitudinal tracheid wall (Figures 7 and 8), the most obvious are the circular dome-like structures called bordered pits. Also depicted are clusters of egg shaped pits which interconnect longitudinal tracheids to transversely oriented parenchyma cells. The structure of the bordered pit facilitates liquid flow between cells. An opening termed the pit aperture is located in the center of the dome-like structure which is called the pit border (Figure 9). Removal of the border reveals the pit membrane which resembles a wheel with a hub and spokes (Figure 10). The portion of the membrane similar to the part of the wheel with spokes is called the margo. Note the many openings in the margo through which liquid can flow. The central portion, resembling the hub of a wheel, has no detectable openings and is termed the torus. Since each bordered pit within a cell usually has a complimentary pit in the contiguous cell, liquid can flow from the lumen through the pit

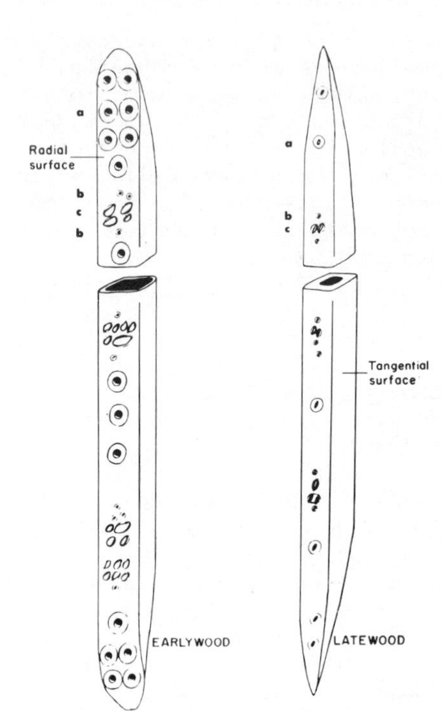

Wood Science

Figure 7. Isolated springwood (earlywood) and summerwood (latewood) longitudinal tracheids. Note resemblance of tracheids to long cylindrical tubes. Tracheid lengths in this figure are considerably reduced as tracheids are normally about 100 times longer than wide. Note the rounded end of the springwood tracheid in the radial view and the pointed end in the tangential view. Summerwood tracheid ends tend to be pointed in both views. a. Bordered pits to adjacent longitudinal tracheids; b. and c. pits to adjacent ray cells. (Drawing from: Howard, E. T., Manwiller, E. G., Wood Science *(1969)* 2, *77–86)*

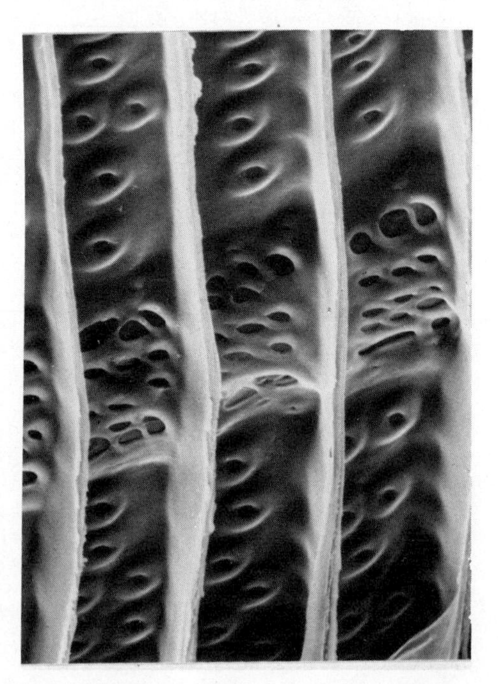

Figure 8. View of internal cell walls of springwood longitudinal tracheids. The circular dome-like structures are bordered pits which permit liquid flow between contiguous longitudinal tracheids. The smaller egg-shaped pits in clusters lead to adjacent transversely oriented ray cells. 400× (Courtesy of N. C. Brown Center for Ultrastructural Studies, S.U.N.Y. College of Environmental Science and Forestry)

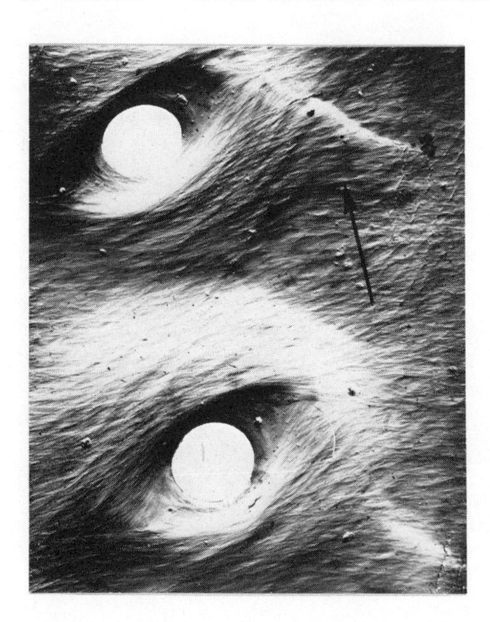

Figure 9. Innermost wall layer as seen from the inside of a springwood longitudinal tracheid showing two bordered pits. Note the circular pit apertures and the stringlike microfibrils which are oriented at approximately 90° to the long axis of the cell. Arrow indicates longitudinal axis of the cell. 3,000×

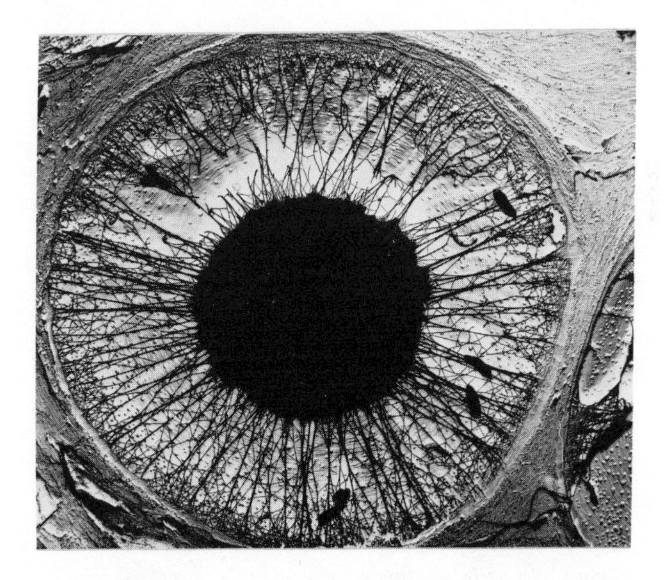

Figure 10. View of a bordered pit membrane with the dome-shaped pit border removed. The dark central portion is the torus. The stringlike microfibrils radiating from the torus constitute the margo portion of the pit membrane. Water flows freely from cell to cell through the openings between the margo microfibrils. 3,000×

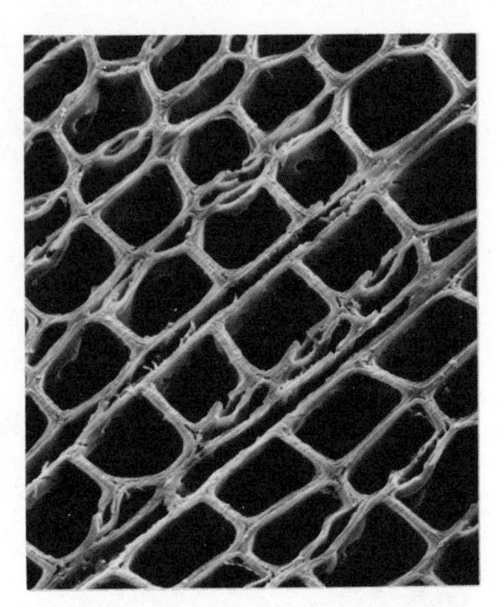

Figure 11. Springwood longitudinal tracheids showing bordered pits in cross-sectional views. Note also the longitudinal surface views of bordered pits inside of the tracheids. In addition, narrow, elongated, transversely oriented food-storage cells which constitute rays are visible. 400×
(Courtesy of N. C. Brown Center for Ultrastructural Studies, S.U.N.Y. College of Environmental Science and Forestry)

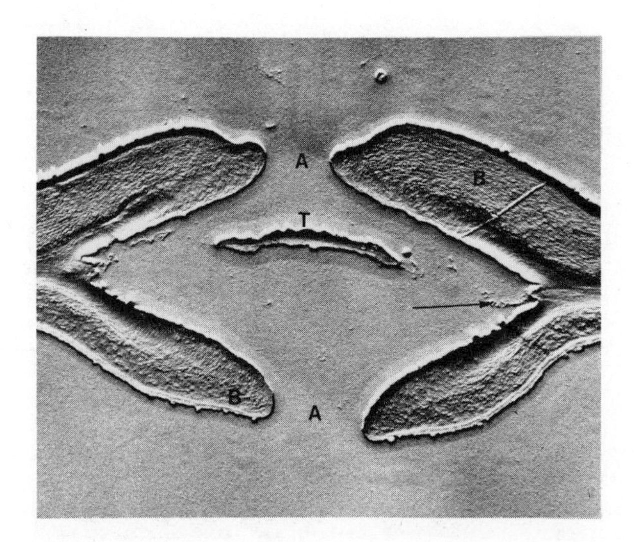

Figure 12. Cross-sectional view of a bordered pit-pair. A: pit aperture; B: pit border; T: torus. Note the thick, nonperforated torus. Most of the thin margo was destroyed during specimen preparation, and only portions of it remain (arrow). Liquid flow occurs through the pit aperture, around the torus through the margo, and out the other pit aperture into the adjacent cell. 5,000×

aperture of one cell, through the margo portion of the membrane and out the pit aperture into the lumen of the adjacent cell. Cross-sectional views of bordered pit-pairs interconnecting tracheids are illustrated in Figure 11. A view of a single bordered pit-pair in Figure 12 reveals an extremely thin and disrupted margo and a relatively thick torus. Because of the thin and porous nature of the margo region, it is often disrupted when preparing specimens for microscopic examination.

Notice in Figure 12 that the torus is in the central position while in Figure 13 it has moved to one side and effectively sealed one of the apertures. In this condition the pit is in the aspirated state. Comparison of an aspirated pit-pair (Figures 13 and 14) with a non-aspirated pit-pair (Figures 11 and 12) clearly illustrates the displacement of the pit membrane against the pit border and the extremely tight seal which exists between the torus and pit border. Aspiration of the pit membrane occurs as water is removed from the cell. In the living tree, aspiration prevents, in the event a tree is wounded, air embolisms from occurring throughout the tree and interrupting all water conduction. Pit membrane aspiration, which occurs during the drying of wood, reduces wood permeability as liquid flow between cells is prevented. Thus any processing of wood involving the penetration of liquids after drying is more difficult.

The other major cell wall structure found on longitudinal tracheids is termed a ray crossing and is illustrated in Figures 7 and 8. Ray crossings consist of pits which interconnect longitudinal tracheids to ray parenchyma. Due to the diverse structure of ray crossing pits they are extremely useful in the identification of wood and wood fibers. However, since identification is beyond the scope of this review, a description of the different types of pits found in ray crossings is not included.

Figures 8 and 15 reveal ray crossing pits as seen from the inside of longitudinal tracheids. The considerably higher magnification in Figure 15 shows a solid pit membrane. Openings in the pit membrane would expose the cytoplasm to the hostile environment of the longitudinal tracheid lumen and result in the death of the parenchyma cell. Thus, the membranes are solid and do not provide a passageway for free liquid flow.

Hardwood Anatomy

Obviously, softwood anatomy is relatively simple as only two types of cells, longitudinal tracheids and ray parenchyma, constitute the bulk of the wood. Hardwoods have a more complex anatomy as more kinds of cells are present. The roles of conduction and support are carried out by different cells and in addition to the transverse ray parenchyma, food-storage cells oriented in the longitudinal direction are present. Parenchyma oriented longitudinally are called longitudinal or axial parenchyma. Vessel segments perform the conduction role, and fibers the support role.

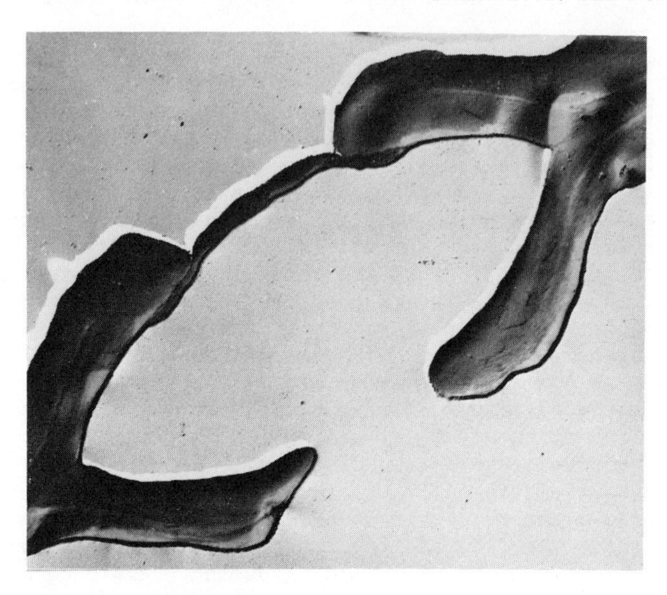

Figure 13. Cross-sectional view of an aspirated bordered pit-pair. The pit membrane has moved to the border and sealed a pit aperture with the torus. In this condition, liquid flow no longer occurs between contiguous cells. 5,000×

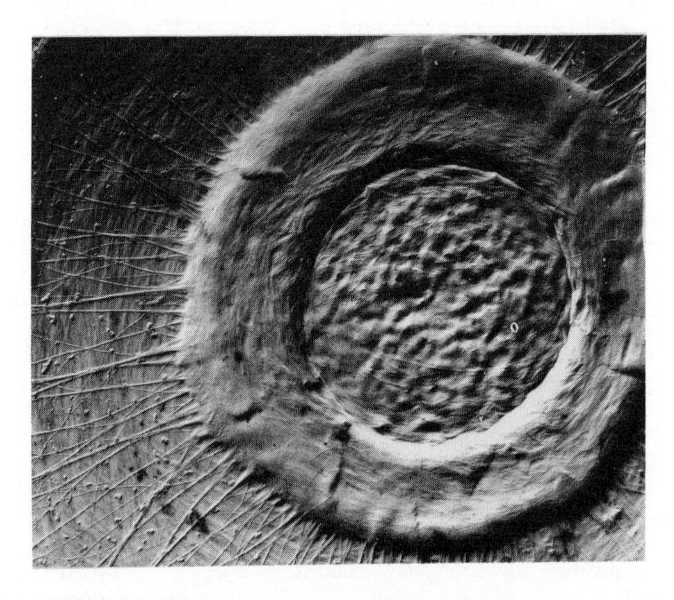

Figure 14. Surface view of an aspirated bordered pit membrane. The imprint of the pit aperture through the torus is the result of an extremely tight seal. 6,200×

Thus, most hardwood species contain four types of cells, vessel segments, fibers, transverse and axial parenchyma, whereas most softwood species possess two types; longitudinal tracheids and transverse parenchyma.

Vessels are structures uniquely designed to carry out the conduction role. A vessel consists of individual vessel cells stacked one on top of the other in the vertical direction of the stem (Figure 6). Vessel lengths up to three meters have been noted. The vessel cells or segments, which constitute a vessel, differ widely in their size and shape. The length of vessel segments varies from 0.18 mm to 1.3 mm and from 20 μm to 330 μm in width. The shortest and widest vessel segments are found in the springwood of ring-porous woods where often the width of the vessel segment is greater than the length. However, in diffuse-porous woods the vessel segments are usually 8 to 10 times longer than they are wide. Note in Figures 5 and 6 that vessel segments in cross-sectional view have a more or less rounded shape. In this view they are often called pores which is a term of convenience to describe the cross section of a vessel. As seen longitudinally in Figure 16, they range from drum-shaped to barrel-shaped to oblong, linear-shaped cells.

Two obvious structural features of vessel segments are perforation plates and pits. Perforation plates are distinct openings found at both ends of the vessel segment which lead to adjacent vessel segments. In some species a number of openings arranged parallel to each other form the perforation plate. These are termed scalariform perforation plates. A single opening is called a simple perforation plate. Both types of perforation plates are illustrated in Figure 16. The creation of openings between vessel segments provides an elongated tubelike structure of considerable length highly suited for the longitudinal translocation of water.

When vessels end, they rarely do so in isolation but rather among a group of vessels. Translocation continues into the adjacent vessels via the intervessel pits. These pits differ from softwood bordered pits in that they lack a torus and openings large enough to be readily detected with an electron microscope. Figure 17 depicts a typical intervessel pit membrane. Different arrangements of intervessel pits can be detected and are useful in the identification of hardwood species.

The longitudinal cell types responsible for the support role in hardwoods are fibers. Fibers are thick-walled, elongated cells with closed pointed ends. It should be noted that frequently the term "fibers" is used loosely for all types of wood cells. However, specifically it refers only to those cell types found in hardwoods which meet the above definition. Fibers range in length from 0.7 mm to 3 mm with an average slightly less than 2 mm for domestic species. In diameter, an average of less than 20 μm can be expected. The percentage of the volume of wood occupied by fibers varies considerably. In sweetgum, fibers constitute only

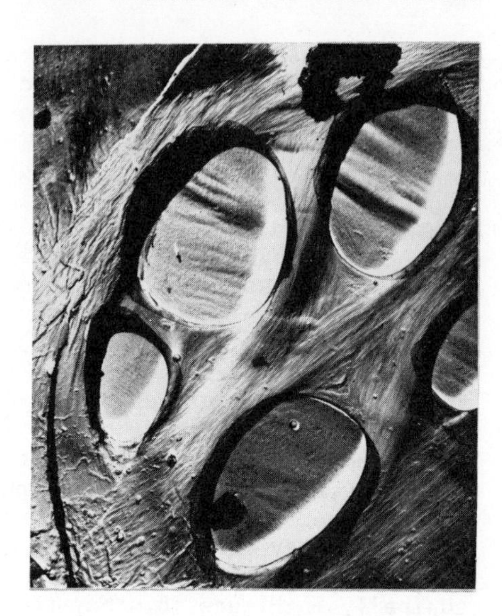

Figure 15. View from the inside of a longitudinal tracheid showing pits connecting a longitudinal tracheid to a ray cell. Note the lack of openings within the pit membrane. 2,500×

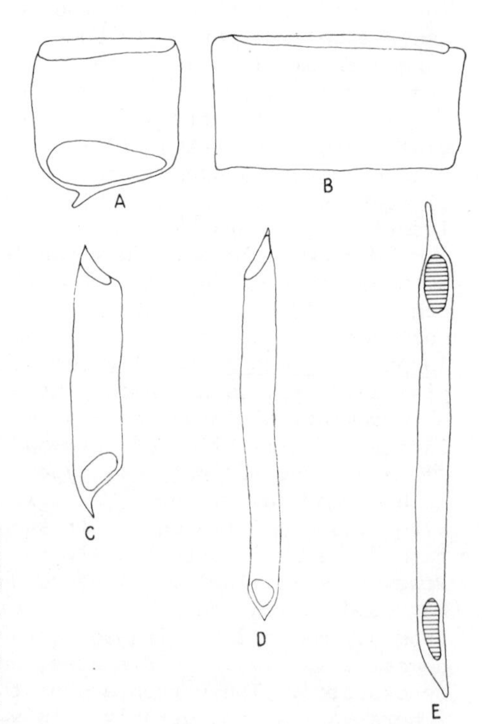

Figure 16. Types of vessel segments found in hardwoods. A and B: springwood vessels from a ring-porous wood. Note the short length compared with the diameter. C and D: typical vessel elements from diffuse-porous woods with simple perforation plates at each end. E: typical diffuse-porous vessel element with scalariform perforation plates at each end. 140×

26% whereas in hickory up to 67% of the volume is made up of fibers. In most cases the higher the percentage volume of fibers present the higher the specific gravity. Exceptions, when noted, are usually due to the fibers being relatively thin-walled. Figures 6 and 18 depict a wood in which fibers occupy more than 50% of the volume. Also illustrated are the obvious differences in the relative size, shape and wall thickness between fibers and vessels.

Considerable variation in the amount of transverse and longitudinal parenchyma exists among hardwood species. For example, basswood has approximately the same as softwoods, that is about 10%, while some oak species approach 40% parenchyma. As in softwoods, the parenchyma are usually brick-shaped cells although some variations of this shape occurs. The rays, composed of transverse parenchyma, range from one to thirty-plus cells wide. The ray illustrated in Figure 18 is seven cells wide. Thus the higher parenchyma volume is due to wider rays and the additional presence of axial parenchyma which is rather rare in softwood species.

Based on the wood anatomical descriptions presented, it is obvious that hardwoods and softwoods differ considerably from each other. For example, vessels are present in hardwoods and absent in softwoods. In hardwoods more cell types, shorter cells, more parenchyma and a more variable arrangement of cell types occur. The relative uniformity of softwood anatomy is the result of the preponderance of a single type cell, the longitudinal tracheid.

Cell Wall Structure

Although some variability exists, the internal cell wall structure described below represents the typical structure found in most woody plant cells. At the time the cell is produced by cell division, it consists of a primary wall which is capable of enlarging both longitudinally and transversely. After the cell reaches full size, a secondary wall is deposited internal to the primary wall adding thickness and rigidity to the cell wall. Figure 19 illustrates the cell walls of two mature, contiguous cells from a softwood species. Note the three distinct layers which make up the cell wall. Adjacent to and on each side of the middle lamella, a primary wall from both cells is present. However, this wall is too thin to be easily observed. Therefore, the term compound middle lamella, which refers to the middle lamella and the two primary walls, is often utilized. Adjacent to the compound middle lamella and easily detected is the first layer of the secondary wall termed the S_1 layer. The next and central layer, which is the largest, is called S_2. The innermost layer adjacent to the lumen is termed S_3.

The total cell wall thickness is largely controlled by the thickness of the S_2 layer. That is, thick-walled cells result

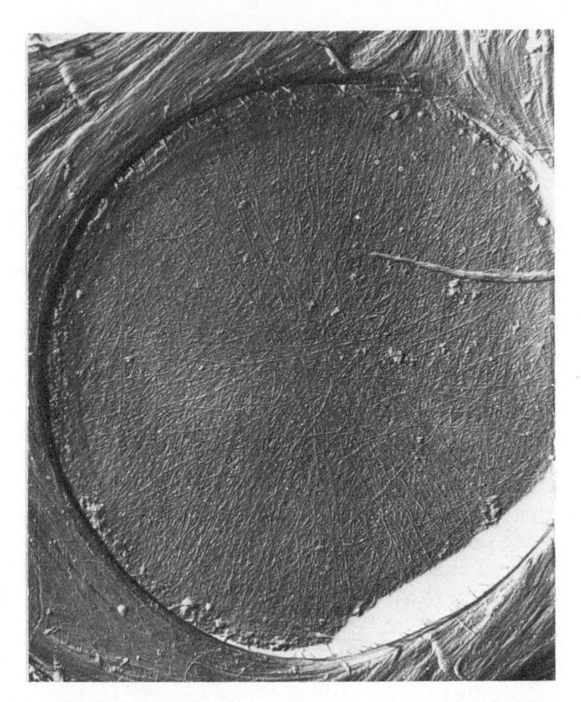

Figure 17. Pit membrane from an intervessel bordered pit. Note the absence of a torus and detectable openings in the membrane. 2,400×

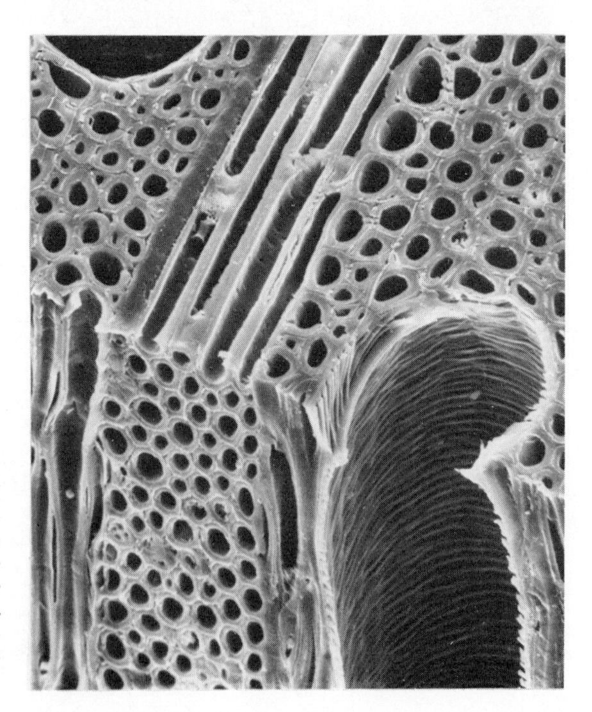

Figure 18. Hardwood specimen showing a vessel cell, fibers, and ray cells. Note the relative differences in size and shape of the cells. The tangential view shows a ray which is up to seven cells wide. (Courtesy of N. C. Brown Center for Ultrastructural Studies, S.U.N.Y. College of Environmental Science and Forestry)

from a substantial increase in the thickness of the S_2 layer and little or no increase in the S_1 and S_3 layers. For example, note in Figure 20, which depicts adjacent vessel and fiber cell walls, the thin S_2 layer in the thin vessel wall and the thick S_2 in the thick fiber wall. Notice also that the S_1 and S_3 of both cells are essentially the same size. The average relative size of the various cell wall layers is indicated in Table I. The thinnest is of course the primary wall with the S_3, S_1 and S_2 in order of increasing thickness.

Table I

Thickness of Various Cell Wall Layers and Microfibril Angle Within the Layers

Wall Layer	Relative Thickness (%)	Average Angle of Microfibrils
P.W.	>1	Random
S_1	10-22	50-70°
S_2	70-90	10-20°
S_3	2-8	60-90°

Detection of the various cell wall layers is due to the presence of microfibrils which are oriented at different angles within each layer. A microfibril is the basic naturally occurring unit which can be easily seen with an electron microscope. Note in Figure 21 the string-like appearance of the microfibrils. In size, microfibrils range from 100 to 300 Å in diameter, whereas their length has not been determined. Within each of the cell wall layers, the microfibrils are oriented at different angles to the long axis of the cell. Table I indicates the average microfibril orientation within the various cell wall layers. Figure 22 presents an idealized drawing of the microfibril orientation within the various cell wall layers.

Chemical Composition of Cell Wall

Chemically the cell wall is rather heterogeneous, consisting primarily of three polymeric materials: cellulose, hemicellulose and lignin. These materials are composed of large molecules and constitute from 95 to 98% of the cell wall. The remaining 2-5% are lower molecular weight compounds called extractives. The amount of each component, especially the hemicelluloses, lignin and extractives, varies considerably between hardwoods and softwoods (Table II). Other factors such as species, location of cells within the tree and growth environment also influence the final chemical composition.

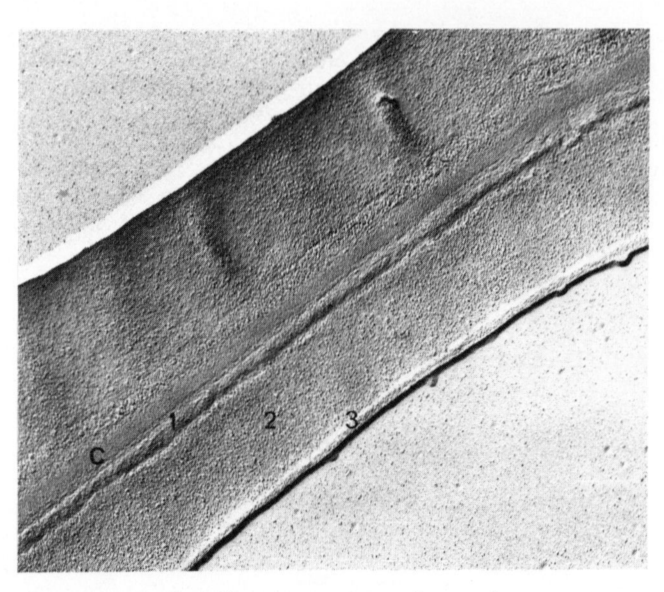

Figure 19. Cell walls in cross-sectional view from contiguous springwood longitudinal tracheids depicting wall layering. C: compound middle lamella. 1: S_1 layer; 2: S_2 layer; and 3: S_3 layer. Note the S_2 layers are the largest. 16,000×

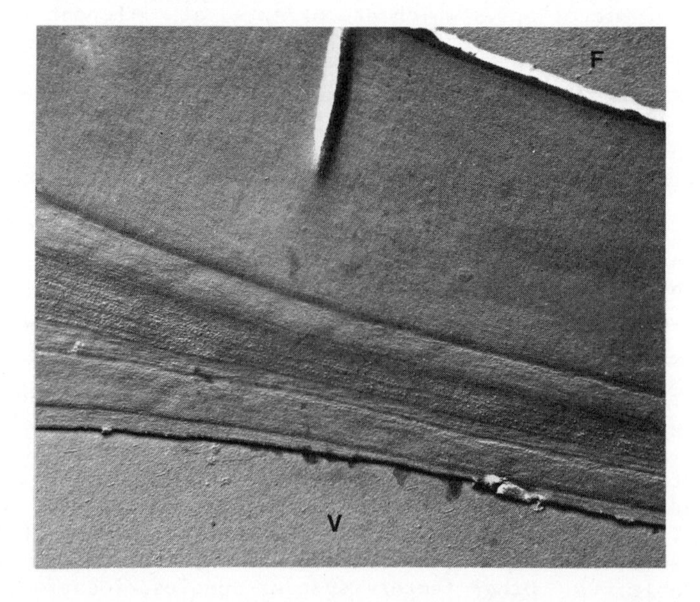

Figure 20. Cell walls of a vessel (V) and adjacent fiber (F) in cross-sectional view. Note the very large S_2 layer in the thick-walled fiber and the small S_2 layer in the thin-walled vessel. 13,000×

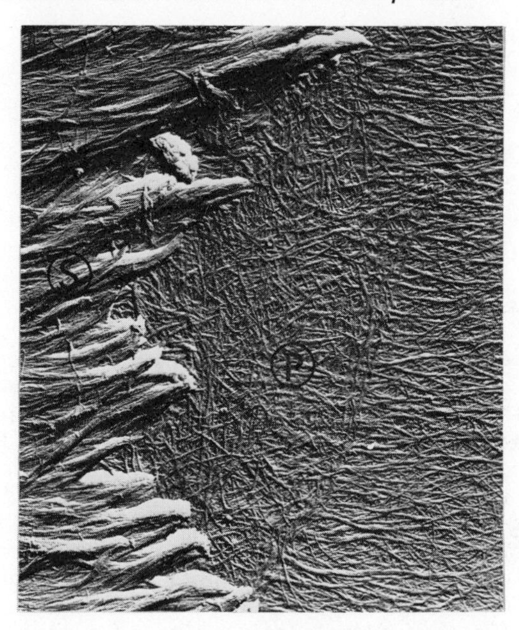

Figure 21. Cellulose microfibrils in the primary wall and S_1 portion of the secondary wall from a longitudinal tracheid. Note the loosely packed and randomly arranged microfibrils in the primary wall (P). The S layer (S) consists of tightly packed, parallel microfibrils. 12,000×

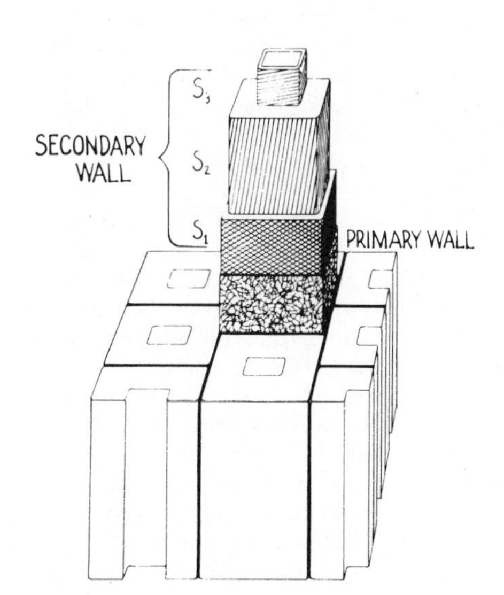

Figure 22. Idealized drawing of cell wall layering showing microfibril orientation and the relative size of each layer

Table II

Average % Chemical Composition of Softwoods and Hardwoods

	Softwoods	Hardwoods
Cellulose	42 ± 2	45 ± 2
Hemicellulose	27 ± 2	30 ± 5
Lignin	28 ± 3	20 ± 4
Extractives	3 ± 2	5 ± 3

Cellulose is a linear polymer of anhydro-D-glucopyranose units linked by β-(1→4) glycosidic bonds. The number of glucose residues varies from 7,000 to 10,000. The cellulose molecules are linked laterally by hydrogen bonds into linear bundles. The extremely large number of hydrogen bonds results in a strong lateral association of the linear cellulose molecules. This strong association and almost perfect alignment of the cellulose molecules gives rise to crystallinity. X-ray measurements show that the crystalline regions are interrupted every 600 angstroms with non-crystalline (amorphous) regions. Whether this is due to minor imperfections in the crystalline lattice or a real structural entity is not completely clear. The most widely held concept is the latter in which the cellulose molecules are highly oriented (crystalline) for a distance of about 600 Å, then pass through an area of poor orientation (amorphous) and re-enter a crystalline region. The pattern repeats throughout the length of the cellulose molecule.

Electron microscopy studies revealed threadlike structures called microfibrils (Figure 21). Microfibril widths, depending upon material source and methods used, vary from 100 to 300 angstroms and are about half as thick as they are wide. Microfibril length has not been determined nor has the internal structure of the microfibril been clearly established. Some investigators believe that the microfibril consists of a crystalline core of cellulose surrounded by an amorphous region containing hemicelluloses. Depending upon the investigator, estimates of the crystalline core size ranges from 30 by 50 Å to 40 by 100 Å. Further variation in the size of the microfibril is due to the amount of hemicelluloses surrounding the crystalline core as well as aggregations of single microfibrils to form microfibrils with larger diameters. Studies with very high resolution electron microscopes coupled with negative staining techniques has led to the view that a 35 Å wide cellulose fibril called the elementary fibril is the basic structure. One proposal from this work views the microfibril as consisting of four elementary fibrils with hemicelluloses, lignin and water around the outside as well as between the elementary fibrils. More recent work has indicated so-called sub-elementary fibrils in the 10 Å range. Obviously the physical structure of cellulose in the woody plant cell wall is far from certain.

Hemicelluloses, like cellulose, are polymers of anhydrosugar units. They differ from cellulose in that a given hemicellulose molecule may contain several different sugar units. In addition, hemicelluloses are usually branched molecules containing only 150 to 200 sugar units. Hemicelluloses found in wood are polymers of D-glucose, D-galactose, D-xylose, D-mannose, L-arabinose and 4-O-methyl-D-glucuronic acid. The hemicelluloses along with lignin surround the crystalline cellulose.

Lignins are the major non-carbohydrate component of wood. They are very complex, crosslinked, three dimensional polymers formed from phenolic units. The number of building units varies considerably from a few up to several hundreds. The aromatic nature of the phenolic units renders lignin hydrophobic and the three dimensional network provides rigidity and optimum resistance to compressive forces. The structural make-up of lignin from hardwoods differs from softwood lignin in that the basic building unit is syringyl whereas guaiacyl is the primary building unit for softwoods. Because the guaiacyl unit has a greater number of potential reactive sites, a higher degree of crosslinking exists. Furthermore lignin formed primarily from guaiacyl has, on the average, a higher molecular weight. Typically, softwood lignin has about ten times more guaiacyl than syringyl units whereas in hardwoods the ratio is usually one to one. Thus hardwood lignins are more easily degraded than softwood lignins.

The distribution of the chemical constituents across the cell wall is not uniform. With regard to cellulose, the primary wall contains about 10% increasing to more than 50% in the S_2 layer and decreasing slightly in the S_3 layer. The lignin content of the middle lamella and primary wall is on the order of 70%. The S_2 contains about 22% and the S_3 approximately 15%. It should be noted that due to the very large volume of the S_2 layer, over half of the total lignin is found in this layer although the concentration is lower than in the compound middle lamella. Also, because the compound middle lamella is so thin only 10% of the total lignin is present despite the very high concentration. The hemicellulose fraction tends to vary about the same as the cellulose across the cell wall.

In addition to the major cell wall components of cellulose, hemicellulose and lignin, wood contains varying amounts of substances termed extractives. The term extractives includes a wide range of chemical types and a very large number of individual compounds. Some of the major chemical types are 1) Terpenes and related compounds, 2) Fatty acids, 3) Aromatic compounds and 4) Volatile oils. Species differ widely in the type and amount of extractives present. Also there is considerable variation in the distribution of extractives throughout the wood of individual trees. Although some extractives are found in sapwood, the heartwood usually contains the largest amount.

Although the exact cause of heartwood formation is not known, the resulting changes have been well documented. The parenchyma cells change their metabolic activity and tend to produce extractives from stored carbohydrates. The parenchyma cells die and the extractives diffuse into adjacent cells. Thus, the heartwood contains no living cells and in many species is discolored as a result of the extractive content. In certain softwoods resin canals are another source of extractives. These structures contain considerable quantities of resin in the living tree.

Despite the relatively low percentage content of extractives (Table II), they very often influence wood properties and thus play a role in utilization. Advantages accrue from the presence of colored and volatile extractives which provide esthetic values. Some of the phenolic compounds provide resistance to insect and fungal attack. Other extractives provide useful products. From tall oil, products such as turpentine, rosin and fatty acids are produced. In addition, tannins, camphor, gum arabic, natural rubber and flavonoids are some of the many products from extractives.

Extractives sometimes prevent or inhibit the utilization of wood. For example, woods containing phenolic type extractives cannot be pulped via the sulfite process. The so-called "pitch troubles" in the pulp and paper industry result from the tendency of the resin type extractives to coagulate and adhere to metal and fibrous surfaces. Also the presence of extractives result in a higher consumption of pulping chemicals and in lower pulp yields.

Despite the numerous products derived from extractives, much of the basic chemistry of the numerous compounds found in many species is still virtually unknown. Future work in this area should provide many new products.

Literature Cited

Wood Anatomy

1. Isenberg, I. H. "The Structure of Wood in The Chemistry of Wood" edited by B. L. Browning. Interscience Publishers, John Wiley and Sons. New York, New York. 1963.
2. Jane, F. W. "The Structure of Wood" 2nd Edition. Adam and Charles Black. London, England. 1970.
3. Panshin, A. G. and Carl deZeeuw. "Textbook of Wood Technology" 3rd Edition. McGraw-Hill. New York, New York. 1970.

Wood Chemistry

1. Browning, B. L. Editor. "The Chemistry of Wood" Interscience Publishers, John Wiley and Sons. New York, New York. 1963.
2. Hanna, R. B. and W. A. Côté, Jr. The Sub-elementary Fibril of Plant Cell Wall Cellulose. Cytobiologie (1974) 10(1):102.

3. Hillis, E. W. "Wood Extractives and Their Significance to the Pulp and Paper Industry" Academic Press. New York, New York. 1962.
4. Preston, R. D. "The Physical Biology of Plant Cell Walls" Chapman and Hall. London. 1974.
5. Wise, L. E. and E. C. Jahn. "Wood Chemistry Volume I" ACS Monogram Series. Reinhold Publishing Corporation. New York, New York. 1952.

2

Prevention of Stain and Mould in Lumber and Board Products

A. J. CSERJESI

Western Forest Products Laboratory, Vancouver, British Columbia, V6T 1X2, Canada

Wood, like any other natural organic material is perishable, and may serve as a nutrient source for different microorganisms. Some wood-inhabiting microorganisms (decay fungi) can utilize as nutrients those wood components which form the structural elements of wood (cellulose, lignin), while others (sapstain fungi and moulds) consume those wood components which represent the cell contents of the tree (starch, other carbohydrates, etc.) and thus no strength loss of wood occurs.

To prevent growth of wood-inhabiting fungi it is customary to treat wood with chemicals which are toxic to fungi. To protect wood in service from deterioration by decay fungi, the chemicals, called wood preservatives, are applied in most cases by pressure treatment because deep penetration into the wood is essential. Wood products intended for use where the decay hazard is low are unlikely to require pressure treatment but in many instances may require superficial fungicidal treatment to prevent fungal discoloration.

History of Sapstain and Mould Prevention

Wood-inhabiting fungi need water, air, nutrients and a suitable temperature for growth. Sapwood in freshly cut lumber has the proper balance of air and water, contains nutrients, and therefore in warm weather it is readily attacked by fungi. Absence of any one of these four requirements stops the growth of these fungi, however the fungi are not necessarily killed. Among these four requirements the moisture content (M.C.) of the wood is the easiest to control, and if it is below about 20% fungal growth is stopped.

In the early days air-drying was the method used to reduce the M.C. of wood. However, air-drying in humid and rainy weather was relatively slow, often taking as long as two or three months to dry lumber to less than 20% M.C. During this period fungi were able to grow on (moulds) and in (staining fungi) the lumber. This period however, is too short for decay fungi to cause an appreciable amount of decay.

Planing air-dried lumber largely removes surface discoloration caused by moulds, but discoloration in the wood caused by staining fungi cannot be removed.

Kiln-drying does not depend on the weather and, in addition, due to the high temperature used, the wood is essentially sterilized. However, if kiln drying is delayed mould and sapstain fungi may discolor the wood.

For the prevention of sapstain development during air-drying, sawmills started to use fungicides as early as the end of the last century. The first attempt to prevent sapstain in lumber was in 1888, according to Bryant (4). At the beginning of this century, the use of sodium carbonate and bicarbonate became widespread in sawmills, and remained the sapstain and mould preventive until about 1930 (4). The studies carried out by the Southern Forest Experiment Station and the Forest Products Laboratory in the U.S.A. from 1928 to early 1930's (31) resulted in a change over from the use of sodium carbonate to the use of the sodium salts of chlorinated phenols and organo-mercuric compounds. Since sapstain and mould preventives based on the toxicity of individual compounds were reported to fail occasionally, the use of mixtures of mercurials and the sodium salts of chlorinated phenols was suggested (39,40).

Since then the sodium salts of chlorinated phenols, sodium pentachlorophenate and sodium tetrachlorophenate, buffered in most formulations with borax, with or without organo-mercurials, has dominated the sapstain and mould preventive market. In British Columbia the lumber industry voluntarily stopped using formulations containing mercury in the late 1960's. However sapstain and mould preventives containing mercurials are still in use in the U.S.A.

The original requirement for a sapstain and mould preventive treatment was to prevent fungal growth on lumber for the period of air-drying, or for the storage period between sawing and kiln-drying. Thus, the chemicals were required to be effective for about a two month period. Traditionally lumber from the Pacific Northwest is shipped unseasoned, by ocean transport. Shipping charges are based on volume not on weight, and neither air nor kiln-drying would be justified economically. After the Second World War, in order to reduce handling costs, packaging of lumber was introduced. Although it was recommended that the importer break up the packages on arrival and pile the lumber for drying, in practice it became customary to break up the packages only when the lumber was actually used. This period is generally less than one year, but is sometimes as much as two years (26).

During storage of packaged green lumber, conditions remain favourable for fungal growth and since the pieces are in close contact in the packages, fungi, including decay fungi, may spread from one piece to another if fungicidal treatment was inadequate.

This extended time requirement for effective control of fungal growth by sapstain and mould preventives put more stringent requirements on the fungicides as well as on the application methods. Chlorinated phenols are effective fungicides for this purpose, as proven by use (17,26), however they may cause skin irritation (4), and an increase of the concentration in the treating solution without the proper precautions may cause other problems in handling freshly-treated lumber (17,35). Searching for safer and more effective fungicides to replace chlorinated phenols for the prevention of sapstain and mould, therefore is important for both the chemical companies manufacturing sapstain and mould preventives, as well as for the lumber industry using them.

An extensive study in Germany in the early 1960's (28,29) was carried out by testing about 300 chemicals for their toxicity to fungi and studying their physical and chemical properties. None of the tested chemicals were found to be more effective than chlorinated phenols.

In New Zealand, Butcher found that captafol is a suitable replacement for chlorinated phenols (5,6,7) and it has been widely accepted for use in that country.

In North America, Chapman Chemical Co. is promoting different metal complexes of 8-hydroxyquinoline (9), but the acceptance of these formulations is slow.

The Eastern Forest Products Laboratory, Ottawa, found that ammoniacal zinc oxide is an inhibitor to fungi in pine lumber (16,33). Sanford Products Corp. is offering 2-(thiocyanomethyl)(thio) benzothiazole for prevention of sapstain and mould in Canada and U.S.A., and tributyl tinoxide emulsion is marketed in Japan.

Experiments were carried out in Europe to use benomyl, di(guanidino-octyl) amine and p-chlorophenyl-3-iodo-propargyl formal to be used in sapstain and mould preventive formulations. Laboratories, including the Western Forest Products Laboratory, are also testing new chemicals.

However sapstain and mould preventive formulations containing chlorinated phenols as the main fungicidal components, still dominate the market.

Sapstain and Mould Preventive Treatment in Practice

An effective sapstain and mould preventive, does not necessarily make a successful treatment. For a successful treatment the following requirements are also necessary:

1. Lumber free from sapstain, mould and active decay;
2. Not more than 24 hours delay of the treatment after sawing;
3. The surface of lumber should be uniformly covered with the treating solution (application method);
4. Proper concentration of the fungicide in the treating solution;

5. At least a temporary protection of lumber from the weather after treatment.

Points 1 and 4 need no further comment. The requirement of 2 is also obvious, considering that this treatment is restricted only to a thin surface layer of the wood and a delay of the treatment allows staining fungi to penetrate beyond the treated layer, where their growth is uninhibited.

The third requirement refers to the application method which is used to apply the sapstain and mould preventives. According to the author's experience, the use of a good application technique is as important as, or perhaps more important than, the effectiveness of the fungicide in the formulation. The type of application methods used are:

(1) Dipping. (a) hand dipping; (b) dipping on the sorting chain; and (c) bulk dipping (of packaged lumber).

(2) Spraying. The author's preference for sapstain and mould preventive treatment is dipping. Experience, in analyzing commercially treated lumber, indicates that the quality of treatment by dipping is superior to spraying (12).

Hand dipping is used today only for experimental purposes. Dipping on sorting chain is one of the earliest automated application methods (31). The advantage of this latter is, that it requires a relatively small volume of treating solution, which is easy to handle. Although it is a good treating method, because of some problems in handling wet lumber on the sorting chain, it had not been used in B.C. for many years. Recently however it was reintroduced and with the development of automatic sorting equipment, this treating method probably will become more popular in the future.

Bulk dipping of packaged lumber was introduced in B.C. in the 1960's, primarily for the treatment of rough sawn lumber. Analyzing commercially and experimentally treated lumber for chlorinated phenols after bulk dipping showed that this treatment resulted in retention comparable to that of hand-dipped rough-sawn lumber. It was also observed that with surfaced lumber, although each piece in the package received some treatment, the concentrations of the fungicide were much lower than in hand dipped lumber, possibly due to the restricted space between pieces (25).

Bulk dipping is done either by driving a carrier with the lumber into the "dip tank", or dipping on a platform. The latter method usually utilizing automatic equipment, which lowers the lumber placed on the platform into the solution, keeps it down for a short time, and brings it up. In some small sawmills, the package of lumber is held under the forks of a forklift, which push them under the solution in an above ground "dip tank".

A major advantage of bulk dipping is that the lumber is not handled manually after treatment. A disadvantage is the large volume of treating solution needed to fill the "dip tank".

Spraying is used mainly for the treatment of surfaced lumber and thus the equipment is usually installed behind a high speed planer. The main problem with this method is that the nozzles of the spray system often become temporarily plugged. Even a very short stoppage of spray, considering the high speed of the lumber (180-350 m/min), may result in large surface areas of the pieces remaining untreated. These untreated areas are responsible for failure in the protection of lumber treated by this method.

Retention of Sapstain and Mould Preventives on Lumber

According to Verrell (39) the effectiveness of the sapstain and mould preventive treatment is correlated to the retention of fungicides on the lumber. This retention is affected by some physical and possibly chemical properties of wood in addition to the application of fungicide, as well as by the method of its application.

In laboratory experiments by the author it was found that the moisture content of the wood has little effect on the retention of the treating solution on the lumber following a 15 second dip (14). In another laboratory study (11) it was found that although alternate wetting and drying caused complete loss of pentachlorophenol from wood, neither leaching nor drying alone caused significant loss of pentachlorophenol after a 1 to 2 day fixation period in the wet condition. This suggests that lumber needs only temporary protection from weather, although complete loss may occur from the surfaces of lumber if exposed to weather due to repeated wetting and drying (12).

Using the results of a large scale field test (26), a number of conclusions about sapstain and mould preventive treatments can be made. In this experiment more than 10,000 pieces of 2" x 4" and 8 ft long lumber were treated by hand dipping into two sapstain and mould preventives, each used at four concentrations. Both contained sodium tetrachlorophenate as the main fungicide. All pieces were inspected for growth of sapstain and mould after different storage periods, and more than 3000 pieces were analyzed for tetrachlorophenol. The concentrations of tetrachlorophenol were calculated on weight per surface area basis (13) and were reported as mg/cm^2.

Considering the large number of observations the following conclusions were well documented:

1. The same protection was observed against sapstain and mould fungi with the same retention (mg/cm^2) regardless of whether the lumber was rough sawn or surfaced (15).
2. Rough-sawn lumber retained an average of 2.5 to 3 times as much tetrachlorophenol as surfaced lumber did, when they were treated in the same solution. This means that surfaced lumber should be treated in a solution which is

three times more concentrated than that used for rough-sawn lumber for equivalent treatment (15).

3. Large variations in the retentions (the maximum being 10 x higher than the minimum) were found in the lumber treated in the same solution (26). This variation was about the same on both, surfaced and rough-sawn lumber (hand-dipped!).
4. On the premise that the protection depended only on the retention of tetrachlorophenol, we calculated that for close to 100% protection, a minimum retention of 0.05 mg/cm^2 is necessary for a 2 year period and a minimum retention of 0.04 mg/cm^2 tetrachlorophenol for a one year period (26).

There is no standard for sapstain and mould preventive treatment, but for export lumber most contracts specify that the lumber should be "effectively treated". Savory and Cockroft in 1961 (30) came to the conclusion: " .... it is doubtful whether a certificate of treatment employing the words 'effectively treated' has any real meaning. It is suggested that certificates of treatment would be more valuable if they stated the treating chemical and the concentration at which it was applied." This suggestion is not yet accepted. But following up their suggestion based on the results of our experiment we calculated that 13 randomly taken samples would be required to estimate with reasonable certainty whether the retention was within the desired target amount (15).

Another factor which may reduce the effectiveness of a sapstain and mould preventive treatment is the tolerance and the adaptation of fungi to fungicides (10,39). However the tolerance of fungi is considered when the effective concentration of a fungicide is determined.

Experimental Methods

The experiment which made it possible to draw several conclusions on sapstain and mould preventive treatment was a large scale test that needed about three years to complete. The quantitative conclusions justified such extensive experimentation. However, for screening new fungicides, such a large experiment would be too expensive and time consuming. Field tests using commercially produced lumber with much shorter storage periods (4 months) give a very good indication of the usefulness of a fungicide for sapstain and mould prevention (7,25,31,40). Experiments were also carried out with smaller samples, up to 3 ft long and shorter storage periods (2-3 months). The results of these tests also gave a good indication of the effectiveness of fungicides (5,31,40). As an alternative to these tests, where untreated pieces were used as controls, pieces half dipped into the treating solution were tested against the untreated other halves, which served as controls (Chapman Chem. Co.).

For preliminary tests of fungicides, different laboratory methods were developed. Several methods were described using artificial media to test the toxicity of chemicals to microorganisms. One of the first tests of this kind was to use malt agar mixed with the chemicals (19). The reproducibility of these tests is much better than those which use wood as the medium (21). However, the results obtained with artificial media rarely correlate with those which are obtained using wood, and therefore artificial media are best used for special experiments (e.g. adaptation of fungi to fungicides (10). Most of the laboratory research with fungicides intended to be used for wood protection has been carried out with a wood medium. Since no standard exists to test sapstain and mould preventives, almost as many methods exist as experiments carried out (3,12,13,19,20,23,24,30,38).

Discoloration of Non-Fungal Origin on Lumber

Discoloration, other than sapstain and mould (or decay) also occurs on lumber. Some of this discoloration originates from the trees and is restricted mainly to the heartwood. It may be caused by chemical reactions which do not continue after the trees are harvested (1,31,34) or it may be caused by fungi in the living tree, which may or may not continue to develop after harvesting (31,34). The only protection against discoloration which is caused by fungi that continue after harvesting, is kiln-drying of the lumber (1).

A non-fungal discoloration developing in sapwood of lumber of several species is due to the oxidation of certain wood extractives (2,22,31). Colourless carbohydrates or phenolic wood extractives, either present in sound wood or produced by microorganisms e.g. bacteria in ponded logs (18,37), migrate to the surface, mainly onto the cross-cut ends of lumber. On the surface these extractives are converted to tannin-like, coloured materials by oxidation, in most cases catalyzed by enzymes (2,22). This discoloration is mostly restricted to the surfaces of sapwood, and more pronounced on the ends of pines (22,36) western hemlock (2,18) and a few other wood species. Similar discoloration occurs during kiln-drying in pine lumber, and is caused by a similar process. But instead of enzymatic oxidation, the high temperature accelerates the formation of the coloured compounds. This discoloration is usually not restricted to the surfaces of the lumber (39).

To control this discoloration in pines, buffered sodium azide (36,37), and more recently the less hazardous sodium fluoride have been used (8). Recently ammoniacal zinc oxide (32) and sodium carbonate or bicarbonate (41) have been suggested to prevent this discoloration.

The brown stain on western hemlock sapwood is not controlled by sodium azide (18). A number of chemicals (reducing agents and acids) prevent the colour formation in laboratory experiments but in practice they only delay it. The ineffectiveness of these chemicals in practice is probably due to the lowering of the concentration of the chemicals below their effective concentrations by rain and by the continuous movement of the extractives to the surface during drying (2). Since this discoloration is restricted to the surface, no preventive measures are taken against its formation.

This type of discoloration is less important in British Columbia than the discoloration caused by fungi. Surface discoloration can be readily removed by surfacing of lumber, and usually objection is raised only when it's observed on an unfamiliar wood species as was the case with western hemlock.

Literature Cited

1. Ananthamarayana, S., IPIRI J. (1975) 5(1) 20-25.
2. Barton, G.M. and Gardner, J.A.F., Can. Dep. For. Pub. 1147 (1966).
3. Boocock, D., B.W.P.A. News Sheet No. 30 (1963) 4 pp.
4. Bryant, R.C., "Lumber: its manufacture and distribution", 2nd ed. Wiley, New York. (1938) pp. 213-216.
5. Butcher, J.A., Mat. U. Org. (1973) 8(1) 51-70.
6. Butcher, J.A., N.Z. Wood Industries (1974) 20(10) 9-11.
7. Butcher, J.A. and Drysdale, J., For. Prod. J. (1974) 24(11) 28-30.
8. Cech, M.Y., For. Prod. J. (1966) 16(11) 22-27.
9. Anon., B.C. Lumberman (1974) 58(10) 60-61.
10. Cserjesi, A.J., Can. J. Microbiol (1967) 13 1243-1249.
11. Cserjesi, A.J. and Roff, J.W., For. Prod. J. (1964) 14(8) 373-376.
12. Cserjesi, A.J. and Roff, J.W., B.C. Lumberman (1966) 50(6) 64-66.
13. Cserjesi, A.J. and Roff, J.W., Mater. Res. Stand. (1970) 10(3) 18-20.
14. Cserjesi, A.J., Quon. K.K. and Kozak, A., J. Inst. Wood Sci. (1967) 4(1) 27-33.
15. Cserjesi, A.J., Roff, J.W. and Warren, W.G., Mater. Res. Stand. (1971) 11(8) 29-32.
16. Desai, R.L. and Shields, J.K., Can. For. Ind. (1975) 94(4) 43.
17. Eades, H.W., "Sap stain and mould prevention on British Columbia softwoods." Can. Dep. Northern Aff. Nat. Res., For. Branch, Bull. 116 (1956) 39 pp.
18. Evans, R.S. and Halvorson, H.N., For. Prod. J. (1962) 12(8) 367-373.
19. Hatfield, I., A.W.P.A. Proc. (1931) pp. 305-315.
20. Hatfield, I., Phytopathology (1950) 40(7) 653-663.
21. Horsfall, J.G., Bot. Rev. (1945) 11(7) 357-397.
22. Millett, M.A., J. For. Prod. Res. Soc. (1952) 6(12) 232-236.

23. Momoh, Z.O. and Oluyide, A.O., Dep. For. Res., Nigeria, Tech. Note No. 38 (1967) 10 pp.
24. Richardson, B.A., Papper och Tra (1972) 10 613-624.
25. Roff, J.W. and Cserjesi, A.J., B.C. Lumberman (1965) 49(5) 90-98.
26. Roff, J.W., Cserjesi, A.J. and Swann, G.W., "Prevention of sapstain and mould in packaged lumber." Can. Dep. Environment, Can. For. Serv. Publ. 1325 (1974) 43 pp.
27. Sandermann, W. and Casten, R., Holz als Roh-u. Werkstoff (1961) 19(1) 20-21.
28. Sandermann, W., Casten, R. and Thode, H., Holzforchung (1963) 17(4) 97-105.
29. Sandermann, W., Thode, H. and Casten, R., Holzforchung (1965) 19(2) 43-47.
30. Savory, J.G. and Cockcroft, R., Timber Trade J. (1961) 239(4440) 67-68.
31. Scheffer, T.C. and Lindgren, R.M., U.S.D.A. Tech. Bull. No. 714, (1940) 124 pp.
32. Shields, J.K., Desai, R.L. and Clarke, M.R., For. Prod. J. (1973) 23(10) 28-30.
33. Shields, J.K., Desai, R.L. and Clarke, M.R., For. Prod. J. (1974) 24(2) 54-57.
34. Siegle, H., Can. J. Bot. (1967) 45(2) 147-154.
35. Smith, R.S., "Responsibilities and risks involved in the use of wood-protecting chemicals." Can. Dep. Nat. Health Welfare, Occup. Health Rev. (1970) 21(3-4) 1-6.
36. Stutz, R.E., For. Prod. J. (1959) 9(12) 459-463
37. Stutz, R.E., Koch, P. and Oldham, M.L., For. Prod. J. (1961) 11(6) 258-260.
38. Unligil, H.H., For. Prod . J. (1976) 26(1) 32-33.
39. Verrall, A.F., Bot. Review (1945) 11(7) 398-415.
40. Verrall, A.F. and Mook, P.V., "Research on chemical control of fungi in green lumber, 1940-51." U.S.D.A. Tech. Bull No. 1046 (1951) 60 pp.
41. Hulme, M.A., For. Prod. J. (1975) 25(8) 38-41.

3

Chemical Methods of Improving the Permeability of Wood

DARREL D. NICHOLAS

Institute of Wood Research, Michigan Technological University, Houghton, Mich. 49931

In the past, it has been well documented that the relative permeability of wood has a significant influence on the effectiveness of preservative treatments. Without adequate penetration, even the best preservatives will not provide sufficient protection and the wood will fail prematurely. Consequently, any developments in methods which increase the treatability of wood which are difficult to treat could have a significant impact on the wood preserving industry.

The objective of this paper is to review methods of improving the permeability of wood. Both mechanical and chemical methods are possibilities, but this discussion will be limited to the latter.

Structural and Chemical Factors in Wood Which Affect Flow

A complete review of the structure of wood is presented in another paper in this symposium. Hence, this discussion will be limited to a review of the principal structural and chemical factors which could possibly influence the flow of liquids in wood.

Structural Factors. Wood is essentially a closed cellular system, and the cell walls are characterized by the presence of numerous pit pairs which serve as flow paths for liquids in living cells. After the cells die and they are transformed into heartwood, the pits undergo aspiration and become occluded with wood extractives. This results in a reduction in the effective pore size which in turn restricts flow of materials. Nevertheless, evidence suggests that the major flow path from cell to cell (in either rays or tracheids) is through the pit membrane since this is the path of least resistance. Consequently, an understanding of the structure and chemical composition of the pit membrane is paramount in our attempt to improve the permeability of wood.

Pit Structure. Because the pit membrane appears to be the controlling factor in flow through wood, an examination of its structure would be pertinent. Simple pit pairs having continuous pit membranes (1) are found between parenchyma cells. Therefore, the flow through this type of cell must pass through the pit membrane. Conversely, most softwood tracheids have bordered pit pairs which have a differentiated membrane composed of a network-like open margo and a thickened central portion called the torus (1). As long as the bordered pit pair is in the unaspirated state, flow can occur relatively easily through the porous margo. However, the pit pairs of heartwood are frequently closed to flow due to a combination of aspiration and occlusion by extractives or lignin-like substances (2, 3). In sapwood the pits may or may not be aspirated, but do not contain occlusions so flow is generally less restricted than in the heartwood. In an aspirated pit, the flow could occur either through the thickened torus, or possibly between the torus and the overhanging border. Based on evidence accumulated to date, it is not possible to determine which of these paths is the major pathway of flow through aspirated pit pairs, but both are probably functional and vary in importance within and among species.

Electron microscopic studies (4, 5, 6, 7) reveal that the bordered pit membrane is fibrillar in nature. Until recently, the margo was considered to be an open network of microfibrils in the green state. However, the work by Sachs and Kinney (8) indicates that the margo is essentially a continuous membrane in the green state but becomes quite porous as a result of drying stresses during seasoning. As long as wood is dried prior to treatment, which is normally the case, this point is superfluous with respect to permeability.

Although there are no visible openings in the torus, evidence suggests that openings do exist (9). It is envisioned that flow through the torus would occur through tortuous paths between randomly oriented microfibrils similar to the openings between the fibers in filter paper. Extending this analogy, the structure and composition of the filter paper determines the rate of flow. The finer and more compact the elements of the paper are, the slower the flow and the more susceptible it is to plugging by particulate matter. This situation would also apply to the structure of the membrane torus. Hence, the relative packing density of the microfibrils and the amount and type of encrustance present would be expected to have a significant influence on the flow of fluids.

Chemical Factors. Based on the structure and flow paths in wood, the pit membrane appears to be the component which should be chemically altered in order to increase the permeability. In order to accomplish this, knowledge of the chemical composition of the pit membrane is desirable.

A number of investigators have studied the chemical composition of the pit membrane (10, 11, 12, 13, 14) using combinations of microscopy, specific enzymes, histochemical methods and UV microspectrophotometry. From this work, it appears that the pit membrane undergoes chemical modification during heartwood formation. It has been postulated that the mechanism for this transformation is that the parenchyma cells produce compounds which migrate to the pit membrane and serve as precursors for the formation of polyphenolic compounds (10). Peroxidase, which is frequently present in sapwood pits, probably serves as a catalyst for these reactions.

In sapwood, the pit membranes appear to be principally composed of cellulose and pectin (13). However, in a number of genera, the sapwood pit membranes also contain polyphenols in some cases, but the distribution is not uniform even within species.

In heartwood, all pit membranes of the genera studied appear to contain polyphenols in addition to the other major chemical components. The presence of lignin in the pit membrane has not been positively established. However, the fact that a monomeric C6-C3 compound has been found in the capillary liquid of sapwood tracheids, strongly suggests that lignification may occur in the pit membranes. In any event, it appears that the polyphenols present in the heartwood undergo polymerization reactions to higher molecular weight compounds which resist extraction by neutral solvents. Hence, they exhibit properties similar to that of lignin.

One of the significant findings by Bauch, et. al. (13) is the fact that the presence of lignin-like compounds is not uniform within samples from a given species. There also appears to be a considerable variation in the chemical composition of the pits within a single tracheid. This suggests that treatment with chemicals that degrade lignin may not be necessary in all cases, since modification of a few pit membranes per tracheid should be sufficient to significantly improve the permeability of wood.

It should be pointed out that most of the research on chemical composition of the pit membrane has been limited to the bordered pits. The chemical composition of the simple pit membranes in the ray parenchyma cells may or may not be the same. However, since the parenchyma cells produce the precursors for the formation of polyphenolic compounds, it is anticipated that the membrane occlusions would be similar.

Methods of Increasing the Permeability of Wood

There are a number of possible chemical methods which could be employed to increase the permeability of wood. These are: a) chemical treatments, b) modification of treating solution properties, and c) biological treatments. The potential of each of these methods will now be examined in detail.

Chemical Treatments. Over the years, a considerable amount of research has been conducted on the possibility of using various chemical pretreatments to improve the permeability of wood. The basic principle behind such treatments is either to extract extraneous material from the pit membrane or degrade the pit membrane in order to enlarge the openings.

Pre-extraction of Wood. Since the major factor causing a reduction in the permeability of wood during heartwood formation is occlusion of the pit membranes with extraneous material, one would anticipate that pre-extraction of the wood with a suitable solvent would be a method of increasing the permeability of wood. This contention has been verified by a number of studies (15, 16, 17, 18, 19, 20, 3, 21). However, it appears doubtful that such treatments would be commercially feasible since the solvents are expensive and excessive time is required for the additional step in the treating process.

Chemical Degradation of the Pit Membranes. Chemical modification of the pit membranes is a possible method of increasing the permeability of wood as long as it can be accomplished without incurring excessive strength loss. In this regard, sodium chlorite, pulping liquors, acids and bases have been used to increase the permeability of wood (19, 22, 3, 23). Unfortunately, excessive strength loss of the wood resulted from these treatments. Nevertheless, it may be possible to achieve the desired results by selecting the proper chemicals so this approach should not be abandoned. For example, the work by Emery and Schroeder (24) indicates that wood can be chemically oxidized with an iron catalyzed reaction under acidic conditions. It is conceivable that this type of treatment could degrade the pit membrane and increase the permeability. Furthermore, Tschernitz (25) has shown that treatment of Rocky Mountain Douglas fir sapwood with hot ammonium oxalate improved the treatability of this material. In this latter case, the ammonium oxalate probably solubilized the pectins in the pit membrane.

Another chemical method of increasing the permeability of sapwood is to steam the green wood. This technique is frequently used in processing southern pine where it aids in the drying step as well as increasing the permeability. It has been shown that the probable mechanism involved in the change in permeability is acid hydrolysis which reduces the effectiveness of pit aspiration significantly (26).

Attempts have been made to increase the permeability of heartwood but without success. Furthermore, steaming has not been effective on species other than the southern pines (27).

Modification of Liquid Properties. The relative permeability of wood varies with the type and condition of the impregnating liquid. Consequently, by selecting the appropriate parameters, it is possible to alter the apparent permeability of wood. These factors will now be discussed in detail.

Type of Liquid. Both petroleum hydrocarbons and water are used as carriers for wood preservatives. A number of studies have clearly shown that petroleum hydrocarbons penetrate wood much more rapidly than water (28, 29, 20, 27). The reason for this difference is not entirely clear, but it has been proposed that hydrogen bonding ability of the liquid is the major factor that influences its ability to penetrate wood (28, 29, 27). Water has the ability to form strong hydrogen bonds between molecules which results in a structured medium. In addition, water has the ability to form hydrogen bonds with the hydroxyl groups in wood. Hence, these two factors could produce a frictional drag as water moves through wood and effectively reduce the flow rate.

Since it has never been proven conclusively that hydrogen bonding ability is the reason for differences in penetrability of liquids, other possibilities should not be prematurely ruled out. For example, the ability of a liquid to form bubbles may be significant since it is known that the presence of bubbles reduces the flow rate because capillary pressure must be overcome during impregnation (30). In this regard, there is a fundamental difference between water and organic liquids since it has been shown that stabilized gas micronuclei exists in the former but not in the latter (31). Hence, water would have a tendency to form bubbles whereas organic liquids would not. Another possible explanation of the difference between the two liquids has been advanced by Bailey and Preston (32). They contend that the difference is attributable to the deposition of hydrophobic material in the pores which effectively increases the contact angle for water. As a result, more pressure is required to impregnate wood in accordance with Jurins equation (30).

The above are only 3 possible factors which could influence the penetrability of liquids into wood. Other possibilities are: a) surface tension, b) molecular size, c) chemical activity, d) solvency, and e) ability to swell wood. Some of these factors may be operative in treatments with propylene oxide which have been carried out by Rowell (33). In this study, he used a mixture of 95% propylene oxide and 5% triethylamine (v/v). He was able to completely treat southern pine and red pine heartwood, both of which were classified as being refractory. This work clearly shows that with the proper treating medium, heartwood can be fully penetrated.

In summary, it can be concluded that the treating liquid characteristics have a significant influence on the treatment results. Consequently, a better understanding of the factors which affect penetrability could lead to improved treatment methods.

Liquid Contamination. As treating solutions are continuously reused, they become contaminated with particulate matter from chemical reactions and extraneous sources. It has been shown that this particulate matter can significantly reduce the penetration of preservative solutions into wood. Consequently, it appears that methods for continuously removing particulate matter could result in improved penetration of preservative solutions.

Chemical Additives. Previous research has shown that the addition of chemicals to water has an effect on its ability to penetrate wood (7). In this regard, standard preservative chemicals and wood extractives tend to reduce the flow of liquids in wood. Hence, proper selection of chemicals and the removal of extractives could insure that the treatment is maximized.

The fact that there is a significant difference in the penetrability of polar and non-polar liquids suggests that alterations of treating solutions with additives may be a method of improving treatment. In this regard, Buckman, et. al. (28) and Hartmann-Dick (20) have shown that the flow of water in wood can be significantly increased by the addition of the zinc chloride. The reason for this improvement is not known, but it suggests that there may be some potential in this area and other additives should be investigated.

Biological Treatments. The possibility of using biological treatments to increase the permeability of wood has received considerable attention by researchers. In general, this type of treatment can be separated into 3 categories; namely, a) treatment with fungi, b) treatment with bacteria, and c) enzyme treatments.

Treatment with Fungi. The use of fungi to increase the permeability of wood was initiated by Lindgren and Harvey (34) and Blew (35). In these studies, *Trichoderma* mold was used to increase the permeability of southern pine sapwood. Following this work, other studies were conducted with Douglas fir (36, 37), jack pine (38), and spruce and aspen (39). All of these studies involved the use of *Trichoderma* mold. However, in a later study by Erickson and Depreitas (40), *Gliocladium fusarium*, and *Chaetomium* were used along with *Trichoderma*.

Improved penetration, higher preservative retention, and more uniform treatments were the general rule in the above studies. Furthermore, the results were similar among the species of fungi tested.

The mechanism for increasing permeability is illustrated in Figure 1. This shows a fungal hyphae growing inside a cell lumen and a branch has formed at the pit aperature so that it can penetrate into the next cell. If this occurs frequently enough, it would effectively increase the permeability.

There are two major limitations in the processes described above. First of all, the fact that these fungi are effective only in the sapwood, which in most instances is readily treatable, reduces its usefulness. Secondly, it has been shown by Johnson and Gjovik (41) that extraneous bacteria, rather than fungi, may actually be responsible for the pit membrane degradation.

Consequently, it does not appear that this particular approach will prove to be practical in commercial applications.

Since it has been shown that the degradation of the pit membranes in wood is the most logical method of increasing permeability, fungi must be capable of attacking this structure in order to be successful. As is shown in Figure 1, the fungal hyphae are able to locate and penetrate the pit membrane so this is a viable method. Because the heartwood pits contain polyphenolic substances, the fungi must degrade these compounds. In this regard, the work by Kirk (42) is significant. Kirk has found that the fungus *Phanerochaete chrysosporium* readily degrades lignin. Hence, it might be possible to use this or a similar fungus to improve the permeability of wood. Since it may be desirable to degrade the cellulose as well as polyphenolic compounds, it might be necessary to use a mixed culture (43). That such a fungus exists was verified by Greaves and Barnicle (44) when they discovered that a fungus was responsible for the increased permeability of karri heartwood by selective attack on the pit membrane and wood rays.

If the right fungi could be found, then it may be possible to pre-treat wood with a spore suspension which would be allowed to react sufficiently to open up the structure before treatment. This, of course, assumes that the fungi will sufficiently degrade the pit membranes before significantly damaging the cell wall structure and cause excessive strength loss. This may be possible because of the accessibility of the pit membranes. Furthermore, the incubation period must be relatively short in order for such a process to be feasible. Kirk (42) indicates that these reactions proceed rapidly under the proper conditions so this may not be a serious problem.

In order to make it feasible to treat heartwood to satisfactory depths, it may be necessary to incise with an open pattern prior to applying this spore suspension. Since this could be done in a single operation, it would be economically attractive. Such a system might provide a means of achieving a uniform preservative distribution, which is currently not possible with incising alone without incurring excessive damage to the wood.

Treatment with Bacteria. Several years ago, it was discovered that the permeability of wood was increased by bacterial attack when logs were soaked in water for a period of time (45). Following this, additional studies were conducted to determine which bacteria were involved, their mode of action and optimum operating conditions (11, 12, 46, 47, 48, 45, 49, 50, 51, 52, 53, 54, 55, 56, 57, 58, 59, 60, 61).

In pine, it was found that *Bacillus polymyxa* was the major species involved (45, 55), whereas, in spruce the major species were *Bacillus subtilis* and *Flavobacterium pectinovorum* (49). In another study (53), *Clostridium omelianskii* was identified as the species attacking softwoods. In all studies, it was found that the bacterial attack on the pit membranes was the reason for increased permeability of the wood. Furthermore, it was shown by Fogarty and Ward (49) that bacteria degraded the pit membranes by excreting the enzymes amylase xylanase, and pectinase. A typical sapwood pit membrane that has been attacked by bacteria is shown in Figure 2. As can be seen, the torus is severely degraded and has well defined openings.

The important thing to remember in most of these experiments is that the bacteria have been used to improve the permeability of sapwood. For species like Sitka spruce which has sapwood that is impermeable to creosote, this type of treatment may prove to be commercially feasible. However, in order to fully exploit this method for increasing permeability, it will be necessary to find bacteria which have the ability to degrade heartwood pit membranes. Indeed, this may be possible since Greaves (50) has shown that some bacteria can increase the permeability of heartwood. In addition to this, it would be desirable to accelerate the reaction as much as possible in order to make it compatible with a commercial operation.

Possible strength loss is important whenever an attempt is made to increase the permeability of wood. Variable effects have been reported by researchers. For example, Bauch, et. al. (11, 12) and Unligil (58) observed a significant reduction in the impact strength of wood that had been ponded to improve permeability. On the other hand, Dunleavy and Fogarty (48) reported no significant strength loss in spruce poles that were exposed to bacterial attack. The reason for this difference in results may be due to the bacterial species involved. Greaves (51) studied this in detail and found that some bacteria were able to increase the permeability of wood without affecting the strength. Conversely, other bacteria were able to increase the permeability but at the same time they decreased the strength significantly. This difference may also be due to the presence of other microorganisms since one

Figure 1. A fungal hyphae growing inside the lumen of a tracheid. Note how the hyphae is branched to penetrate the pit membrane. (Courtesy of I. B. Sachs)

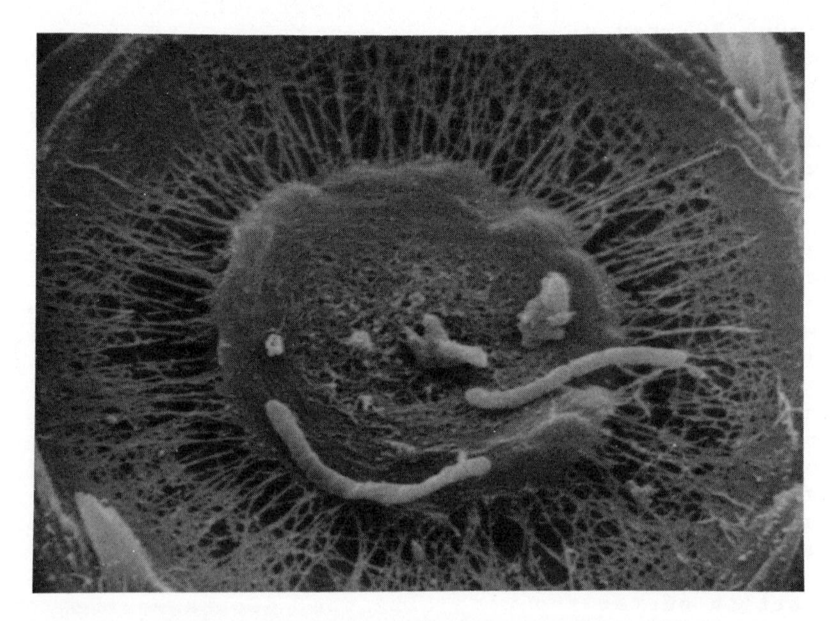

Figure 2. A sapwood pit membrane that has been degraded by bacteria. (Courtesy of I. B. Sachs)

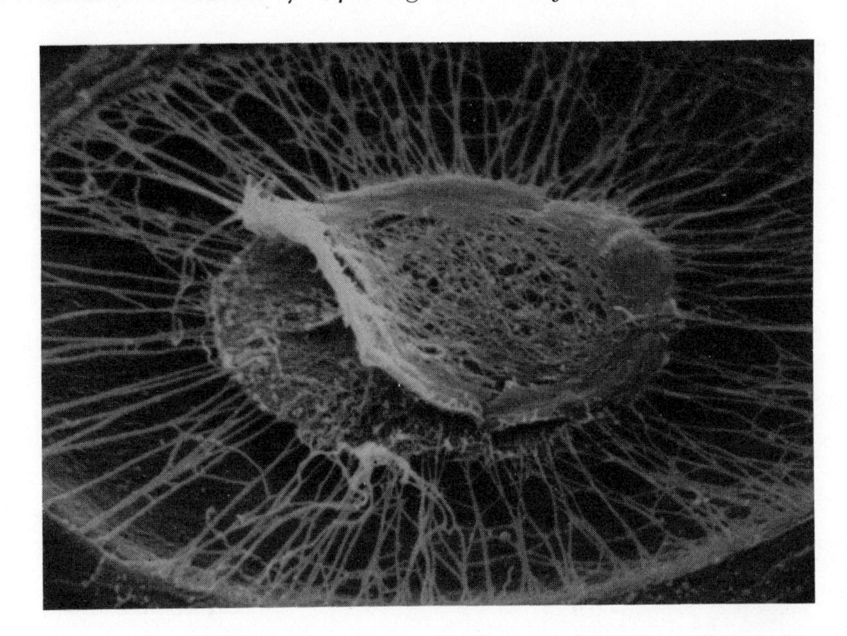

Figure 3. A Douglas-fir sapwood pit membrane which has been degraded by pectinase. (Courtesy of I. B. Sachs)

study (59) revealed that the increased permeability was due to fungi rather than bacteria.

Enzymes. Since both bacteria and fungi utilize enzymes to degrade the pit membrane in wood, it is not surprising that treatment with isolated enzymes produces similar effects. This was shown to be the case by Nicholas and Thomas (26) using cellulase, hemicellulase and pectinase. Similar results were subsequently obtained by other researchers (62, 63, 64, 65).

In a recent study by Tshernitz (25), it was verified that enzymes could be used to increase the permeability of Rocky Mountain Douglas fir. By pretreating the wood with pectinase, a completely uniform treatment of the sapwood zone was possible with creosote. This is in contrast to erratic treatment normally obtained when this material is treated. A typical pit membrane which has been degraded by pectinase is shown in Figure 3.

With regard to the effect of enzymes on the strength of treated wood, the work by Meyer (65) indicated that it is possible to significantly increase the permeability of wood without any appreciable strength loss. Hence, it does not appear that this would be a problem with the use of enzymes.

To date, it has not been possible to degrade heartwood pit membranes with isolated enzymes. Since co-factors are probably required (42), it appears that use of a whole organism rather than isolated enzymes may be required if the permeability of heartwood is to be increased.

LITERATURE CITED

1. Panshin, A. J., DeZeeuw, C., Brown, H. P., "Textbook of Wood Technology," Vol. 1, McGraw-Hill Book Company, Inc., New York, N. Y., 1964.
2. Cote, W. A., J. Polymer Sci. C., (1963), (2):231-42.
3. Krahmer, R. L., Cote, W. A., Tappi, (1963), 46 (1):42-49.
4. Erickson, H. D., Wood Science, (1970), 2(3):149-158.
5. Cote, W. A., Forest Prod. J., (1958), 8(10):296-301.
6. Cote, W. A., Krahmer, R. L., Tappi, (1962), 45(2): 119-22.
7. Nicholas, D. D., Forest Prod. J., (1972), 22(5):31.
8. Sachs, I. B., Kinney, R. E., Wood Science, (1974), 6(3):200-205.
9. Sachs, I. B., Personal communication, (1975).
10. Bauch, J., Berndt, H., Wood Science and Technology, (1973), 7(1):6-19.
11. Bauch, J. Adolf, P., Liese W., Holz als Roh-und Werkstoff, (1973), 31(3):115-120.
12. Bauch, J., Liese, W., Berndt, H., Holzforschung, (1970), 24(6):199-205.
13. Bauch, J., Schweers, W., Berndt, H., Holzforschung, (1974), 28(3):86-91.
14. Nicholas, D. D., Thomas, R. J., Forest Prod. J., (1968), 18(1):57-59).
15. Benvenuti, R. R., Unpublished Masters thesis, North Carolina State Univ., Raleigh, N. C. (1963).
16. Buro, A., Buro, E. A., Holz als Roh-und Werkstoff, (1959), 17(12):461-74.
17. Charuk, E. V., Razumova, A. F., Holztechnologie, (1974), 15(1):3-6.
18. Comstock, G. L., Forest Prod. J., (1965), 15(10): 441-49.
19. Erdtman, H., Svensk Papperstidning, (1958), 61625-32.
20. Hartmann-Dick, V., Forstwiss. CBI., (1955), 74(5/6):163-83.
21. Scarth, G. W., Spier, J. D., Royal Soc. of Can. Proc. and Trans., (1929), 23:281-88.

22. Johnston, H. W., Maass, O., Can. J. Res., (1930), 3:140-73.
23. Yoshimoto, T., Hayashi, S., Kishima, T., Wood Research, (1970), (52):90-105.
24. Emery, J. A., Schroeder, H. A., Wood Sci. and Technology, (1974), 8(2):123-137.
25. Tschernitz, J. L., Forest Prod. J., (1973), 23(3):30-38.
26. Nicholas, D. D., Thomas, R. J., Proc. Am. Wood-Preservers' Assoc., (1968), 64:1-7.
27. Nicholas, D. D., Siou, J. F., "Wood Deterioration and Its Prevention by Preservative Treatments," D. D. Nicholas, Ed., Vol. II, Syracuse University Press, Syracuse, N. Y., 1973.
28. Buckman, S. J., Schmitz, H., Gortner, R. A., Chem., (1935), 39:103-19.
29. Erickson, H. D., Schmitz, H., Gortner, R. A., Minnesota Ag. Exp. Sta. Tech. Bull. No. 22, (1937).
30. Siau, J. F., "Flow in Wood," Syracuse Univ. Press, Syracuse, N. Y., 1971.
31. Hayward, A. T. J., Am. Scientist, (1971), 59:434-443.
32. Bailey, P. J., Preston, R. D., Holzforschung, (1969), 23(4):113-120.
33. Rowell, R. M., Personal communication, (1975).
34. Lindgren, R. A., Harvey, G. M., Forest Prod. J., (1952), 2(5):250-56.
35. Blew, J. O., Jr., Forest Prod. J., (1952), 2(3):85-86.
36. Graham, R. D., Forest Prod. J., (1954), 4(4):164-166.
37. Lindgren, R. M., Wright, E., Forest Prod. J., (1954), 4(4):162-164.
38. Panek, E., Forest Prod. J., (1957), 7(4):124-127.
39. Schulz, G., Forest Prod. J., (1956), 6(2):77-80.
40. Erickson, H. D., Defreitas, A. R., Forest Prod. J., (1971), 21(4):53-58.
41. Johnson, B. R., Gjovik, L. R., Proc. Amer. Wood-Preservers' Assoc., (1970), 66:234-240.
42. Kirk, T. K., Personal communication, (1975).
43. Skinner, K. J., Chem. and Engr. News, (1975), 53(Aug. 18):22-41.
44. Greaves, H., Barnacle, J. E., Forest Prod. J., (1970), 20(8):47-51.
45. Ellwood, E. L., Ecklund, B. A., Forest Prod. J., (1959), 9(9):283-92.
46. DeGroot, R. C., Scheld, H. W., Forest Prod. J., (1973), 23(4):43-46.
47. Dunleavy, J. A., McQuire, A. J., J. Inst. Wood Sci., (1970), 5(2):20-28.

48. Dunleavy, J. A., Fogarty, W. M., Proc. British Wood Preserving Assoc., (1971), 1-28.
49. Fogarty, W. M., Ward, O. P., Wood Sci. and Tech., (1973), 7(4):261-270.
50. Greaves, H., Holzforschung, (1970), 24(1):6-14.
51. Greaves, H., Wood Sci. and Tech., (1971), 5(1): 6-16.
52. Greaves, H., Levy, J. F., "Proc. of the First International Biodeterioration Symposium," p. 429-441, Elsevier Publishing Company, Ltd., Amsterdam, 1968.
53. Karnop, G., Material and Organism, (1972), 7(3):189-203.
54. Knuth, D. T., Forest Prod. J., (1964), 12(9):437-442.
55. Knuth, D. T., Unpublished Ph.D. dissertation, Univ. of Wisconsin, Madison . Univ. Microfilm, Ann Arbor, Michigan, (1964).
56. Suolahti, O., Wallen, A., Holz als Roh-und Werkstoff, (1958), 16:8-20.
57. Unligil, H. H., J. Inst. Wood Sci., (1971), 5(6):30-35.
58. Unligil, H. H., Forest Prod. J., (1972), 22(9): 92-100.
59. Unligil, H. H., Krzyzewski, J., Newfoundland bimonthly Research Notes, (1972), 28(2/3):11-12.
60. Ward, O. P., Fogarty, W. M., J. of the Inst. of Wood Sci., (1973), 6(2):8-12.
61. Banks, W. B., Dearling, T. B., Mater. Organismen, (1973), 8(1):39-49.
62. Adolf, F. P., Holzforschung, (1975), 29(5):181-186.
63. Imamura, Y., Harada, H., Saiki, H., Wood Sci. and Tech., (1974), 8(4):243-254.
64. Jutte, S. M., Levy, J. F., Acta Bot. Neerl., (1971), 20(5):453-466.
65. Meyer, R. W., Wood Sci., (1974), 6(3):220-230.

4

Nonconventional Wood Preservation Methods

ROGER M. ROWELL

Forest Products Laboratory, P.O. Box 5130, Madison, Wis. 53705

A most effective way to extend the Nation's timber supply is to use wood so that its service life is increased. The service life of wood in hazardous use conditions can be increased severalfold by the proper use of wood preservatives. It is estimated that the preservative treatment of railway ties results in an annual savings of 2.4 billion board feet of lumber and that, if utility poles were not treated, an additional 20 million mature trees of pole-stock quality softwoods would be needed each year simply as replacements for those destroyed by decay and termites.

Some 275 million cubic feet of wood are treated annually for protection against decay fungi, insects, or marine borers. Nevertheless, losses to these agents are still large and may amount to between $1 and $2 billion in the United States annually. These losses may be attributed to either inadequate or no preservative treatment. Although conventional preservatives are generally effective, they are coming under increasing attack because of their toxicity, so information is urgently needed on newer, safer, more environmentally acceptable, effective preservatives.

All of the commercial wood preservatives presently used in the United States are effective in preventing attack by microorganisms because of their toxic nature. Most of these preservatives are classified as broad spectrum preservatives, that is, effective against several different types of living systems. Because of the toxic hazards and environmental concerns and because prevention of wood decay is needed if we are to extend our timber resources by increasing its service life, we have investigated alternative methods of wood preservation not based on toxicity for their effectiveness.

This paper is not meant to present finished data as much as it is to present ideas, concepts and research approaches in the area of new methods for wood preservation. For this discussion, a nonconventional preservation method will be defined simply as a concept or process not presently in use but with future potential.

Controlling micro-organisms by means other than poisoning them can be investigated by considering the basic needs of the organism. In order for wood-destroying micro-organisms to thrive, they require: (1) oxygen, (2) water, (3) food, including all essential trace compounds, and (4) a favorable environment. To eliminate any one of these will effectively control the growth of the organism.

It is very difficult to restrict the oxygen from micro-organisms so control measures based on this approach would probably be fruitless. Wood below the fiber saturation point does not decay. Therefore, by restricting the amount of water in the wood cell wall below the fiber saturation point, the micro-organisms will not thrive.

Restricting or eliminating an essential component in the micro-organism food chain such as metals, vitamins, etc., would cause the organisms to look elsewhere for nourishment. Modifying the wood so the organism did not recognize it as food would also protect it from attack. Inhibiting the enzyme systems such as the cellulases unique to those organisms capable of breaking wood down would also protect the wood without the treatment being harmful to humans.

The final consideration, a favorable environment, would capitalize on creating a hostile environment for the organism. For higher organisms that destroy wood, such as rodents, deer, birds, etc., repellents, which are not toxic to the organisms, but because of their smell, taste, or texture, cause the prospective diner to leave the treated material alone. Changing the pH of the wood or maintaining a temperature above or below that required for organisms to thrive will effectively control their growth. The problem with this approach is that some of the conditions which are not favorable for the organisms are also not favorable to maintain the desirable properties of the wood. For example, at a low pH, the wood components undergo hydrolysis causing severe strength losses.

In discussing concepts for preventing attack by organisms not based on toxicity, the mechanism of effectiveness quite likely is not based on any single mechanism, but a combination of several factors. For example, in the discussion to follow on chemical modification of wood, the mechanism for the protective action may be due to: (a) blocking conformational sites required for the highly specific enzyme-substrate reactions to take place, (b) plugging holes in the lignin-hemicellulose shield protecting the cellulose, (c) stabilizing labile polymer units which may be the point of the fungus' first attack, (d) removal of soluble chemicals in the wood which are required by the micro-organisms to start or sustain the attack, (e) changing the wood-water relationship as to be inimicable to microbial life, and (f) combinations of these or other possibilities.

Research Approaches

A. Irradiation. Ponderosa pine, red and white oak, sweetgum and Douglas-fir have been treated with highly penetrating gamma radiation emitted by Cobalt^{60} in an attempt to alter the polymer structure of the wood. After irradiation of the wood, the decay resistance was determined using: (a) the soil-block test employing the fungus *Poria monticola* (*Poria placenta*) (Madison 698) (1) or agar-block tests using *Lenzites trabea* (*Gloeophyllum trabeum*) (Madison 617) (2). Radiation levels from 10^2-10^7 reps showed no change in decay resistance in the irradiated wood over nonirradiated control blocks. It has been found by several workers (3-5) that the primary effects of high-energy radiation on wood polymers are depolymerization, decrystallization, and degradation. It would be expected that the effects of irradiation would cause a decrease in the decay resistance of wood rather than an increase.

B. Thiamine Destruction. Farrer (6) showed that one of the essential metabolites for fungal growth, thiamine, was destroyed in 2 hours at 100°C at pH 7. At the same temperature but at pH 8, destruction was complete in 1 hour and at pH 9 in 15 minutes. These results encouraged Baechler, *et al*. (7,8) to treat wood with either ammonia or sodium hydroxide to presumably destroy the thiamine, thus protecting wood by removing an essential trace compound so long as outside sources of thiamine were excluded.

Douglas-fir, birch, southern pine, and sweetgum blocks were treated with 1% aqueous ammonia or sodium hydroxide for various times, temperatures, and pressures (9). These samples were submitted to soil-block tests with two brown-rot fungi *Poria monticola* (Madison 698) and *Lentinus lepideus* (Madison 534) and two white-rot fungi *Polyporus versicolor* (*Coriolus versicolor*) (Madison 697) and *P. anceps* (F 784-5) as well as outside exposure tests (10). In the soil-block tests, the treated wood was resistant to the two brown rotters, but was not resistant to the two white rotters. In the outdoor stake tests, the average lifetime was 3.5 years while untreated controls had an average lifetime of 3.6 years. The outdoor tests show that there is no increase in rot resistance by this treatment.

C. Heat Treatments. Several woods have been heated under wet and dry heating conditions to determine the effect heat has on the decay resistance of these woods. Alaska-cedar, Atlantic white-cedar, bald cypress, Douglas-fir, mahogany, redwood, white oak, Sitka spruce, and western redcedar were heated under dry conditions or wet conditions at temperatures of 80-180°C for varying lengths of time. Boyce (11) found that dry heat at 100°C or steam heat at 120°C for 20 minutes had no effect on the decay resistance. Similar results were observed by Scheffer and Eslyn (12) in soil-block tests with *Lenzites trabea* for the heated softwoods and *Polyporus versicolor* for the heated hardwoods. Thus, heat treatments do not increase the decay resistance of the heat-treated wood. In some cases, a slight loss in decay resistance was observed.

D. Plastic Composites. Many different woods have been treated with organic monomers and the monomers catalytically polymerized within the wood structure. The subject of treating wood to form plastic composites is covered by Dr. John Meyer in another section of this publication. For the most part, these composites have been prepared and studied for their use in dimensional-stabilized products (for example, see 13-17).

Southern pine, Douglas-fir, and yellow poplar stakes were impregnated with phenolic resin and cured (impreg) or impregnated with phenolic resin, compressed, and cured (compreg). Separate samples were treated with urea-formaldehyde and cured. These samples were placed in the ground and their average lifetime determined. The results are shown in Table I (18).

Table I. Average Lifetime of Resin-Impregnated Wood in Ground Contact

Treatment	Retention	Average Life
	Lb/ft^3	Yr.
Control	--	1.8-2.7
Impreg-phenolic resin	5	6.8-11.7
Impreg-phenolic resin	10	12.4-19.5
Compreg-phenol resin	10	19.5
Urea-formaldehyde	6	9.1

E. Repellents. It is questionable whether a repellent would have any effect on micro-organisms, but they have been studied for application for protecting wood against higher life forms. The amount of damage to wooden structures each year by animals is considerable.

The basic approach here is to repel the prospective diner, not to kill it. Various repellents have been tested for different animals (19). Solutions or slurries of these compounds have been painted on wooden structures. Since no bonding to the wood components takes place, these repellents are leached, volatilized, broken down, and weathered out of the wood.

A new approach in this area is to encapsulate the repellent in a resistant or slow release shell. This way there is very little, if any, repellency until the animal comes into contact with the wood. This contact causes the shell to be broken and releases the repellent. The slow release type would slowly break down and release the repellent over a period of time depending on the stability or weathering characteristics of the capsule. The encapsulated chemicals could be added to a dispersant or paint and applied to the wood surface. If the capsules could be made small enough, deep penetration by pressure impregnation might be possible.

F. Bound Toxins. Another approach to more environmentally acceptable preservatives is to chemically bond a toxic compound onto a wood component so that it cannot be leached out. The compound, once reacted, would have to retain its toxic properties. Compounds now used as wood preservatives are toxic to the organism because they are ingested by the organism. If the toxic compound were bound to the wood, they may be toxic to the organism only when ingested. Because of this, the approach of permanently bound toxins may not be a fruitful research area.

It is also possible to react acid chlorides (20) or anhydride-containing compounds so as to form ester bonds with hydroxyl groups on one of the wood components. Ester bonds could slowly hydrolyze and release the bound toxin. In this case, the release of the preservatives would be a function of the rate of hydrolysis and not directly related to weathering effects (for example, water solubility, vapor pressure, UV degradation, etc.). Controlled release fungicides based either on slow hydrolysis or capsule erosion could greatly decrease the quantity of preservative needed to adequately protect a wooden structure, since leaching could be controlled.

G. Metabolic Difference. Micro-organisms attack wood by secreting enzymes into the immediate structure which in turn break down the wood components into small, soluble units that become nutrients for the organism. The main destructive enzyme system the wood-rotters contain is a class of proteins known as cellulases. These enzymes break down the polymeric cellulose, the strong backbone of wood, into digestible units. Humans do not possess this enzyme system; consequently, we cannot degrade cellulose-containing materials.

Capitalizing on this metabolic difference between higher forms of life and micro-organisms is the basis for this research approach to wood protection. Compounds are available which inhibit the cellulase enzyme systems; however, their specificity has not been determined. Mandels and Reese (21,22) found that the extracts from the immature fruit of persimmon or the extract from leaves of bayberry were very effective inhibitors of the cellulase system. At concentration levels of .00005 and .00018%, respectively, these two extracts inhibited the cellulase enzymes isolated from *Trichoderma viride*. It is not known what the active component(s) are in these two extracts.

New materials need to be investigated as possible specific inhibitors to the cellulase enzymes. This research approach will require the screening of chemicals against pure enzyme solutions of known activity. Specificity must be determined using a variety of human-type enzymes such as transferases, phosphorylases, dehydrogenases, etc. Ultimate success will depend on finding compounds which are only inhibitory to the cellulase enzymes.

A more basic approach to this area would be to study the cellulase enzymes themselves. If the active sites and true nature of this protein were known, selective inhibition could be determined.

H. Chemical Modification. The chemical modification of wood involves a chemical reaction between some reactive part of a wood component and a simple single chemical reagent, with or without catalyst, to form a covalent bond between the two. The wood component may be cellulose, hemicellulose, or lignin. The objective of the reaction is to render the wood decay resistant. The mechanism of the effectiveness is not known, but some possible explanations were given earlier.

By far the most abundant reactive chemical sites in wood are the hydroxyl groups on cellulose, hemicellulose, and lignin. The types of covalent chemical bonds of the carbon-oxygen-carbon type that are of major importance are ethers, acetals, and esters. The ether bond is stable to bases, but labile to acids, and the ester bond is labile to both acids and bases.

The treated wood must still possess the desirable properties of untreated wood; the strength must remain high, little or no color change (unless a color change is desirable), good electrical insulator, not dangerous to handle, gluable, paintable, etc. For this reason, the chemicals to be considered for the modification of wood must be capable of reacting with wood hydroxyl groups under neutral or mildly alkaline conditions at temperatures below 120°C. The chemical system should be simple and capable of swelling the wood structure to facilitate penetration. The complete reagent molecules should react quickly with the wood components yielding stable chemical bonds that will resist weathering (23,24).

These chemicals, once reacted, are effective in preventing attack by micro-organisms, but they are not toxic to the decay organisms. The important factor in preventing attack is to attain a treatment level which inhibits the growth of the organisms. A recent review on this subject (23) shows that reaction with acetic anhydride, dimethyl sulfate, acrylonitrile, butylene oxide, phenyl isocyanate, and β-propiolactone all give good rot resistance at 17-25 weight percent gain (WPG). The exception to this is formaldehyde where a 2-5 WPG gives decay resistance. In this case, there may be crosslinking of larger wood units which gives it different properties (25).

Figure 1 shows that the decay resistance of acetylated wood is directly proportional to the WPG (26). The degree of dimensional stability is also proportional to the WPG so the exclusion of cell wall or biological water may be a very important factor in the decay resistance mechanism.

The average service life of acetylated yellow birch and cyanoethylated southern pine stakes in ground contact is shown in Table II (27).

Table II. Average Lifetime of Chemically Modified Wood in Ground Contact

Treatment	Level (WPG)	Average Life (Years)
Control	--	2.7
Acetylation	19.2	17.5
Control	--	3.6
Cyanoethylation	11	3.9
Cyanoethylation	15	5.3

In preliminary tests, alkylene oxide-treated southern pine (28) was found to be resistant to termite attack and attack from the marine borers, *Teredo* (shipworm) and *Limnoria*.

I. Basic Mechanism of Attack. The ultimate solution for preventing attack by micro-organisms will come once we know how an organism breaks wood down. How does the organism know wood is something to eat? What does it recognize first to start the attack? What enzymes are vital in the initial and sustained attack? Is there a specific weak link in those important enzymes that can be used to develop selective inhibitors?

It is also possible that enzymatic reactions are not the only degrading reactions in the deterioration of wood by organisms. Figure 2 shows in the left hand graph that only a 10-15% weight loss occurs in the first 2 weeks of attack by brown-rot fungi. The graph on the right, however, shows that with only a 10% weight loss, there is a drop in the average degree of polymerization of the holocellulose from 1,600 to 400 (29). This represents a four-fold decrease in DP, which affects the strength of the wood with very little total weight loss.

These results would indicate that, at least in the initial attack, hydrolytic chemical reactions play an important part. It has been suggested that hydrogen peroxide and iron are the cause of this rapid depolymerization (30). It is possible that the initial attack by micro-organisms is not enzymatic but hydrolytic and oxidative in nature. If this is true, then a preservative system could be based on anti-oxidant properties of the chemical. If the initial attack can be stopped, then the total attack has been stopped. It is also possible that the initial and sustained attacks are caused by a combination of chemical

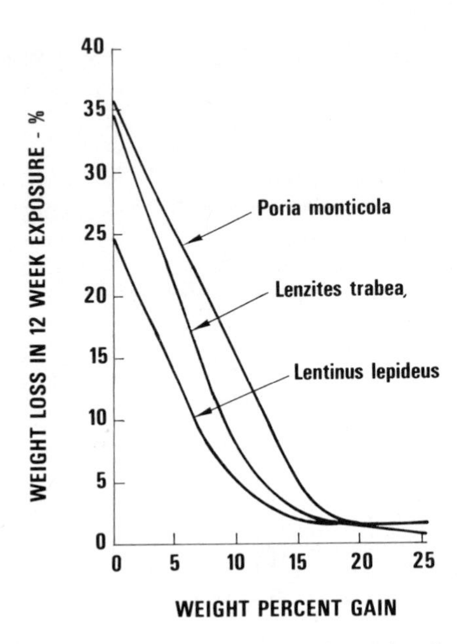

Figure 1. Decay resistance of acetylated wood

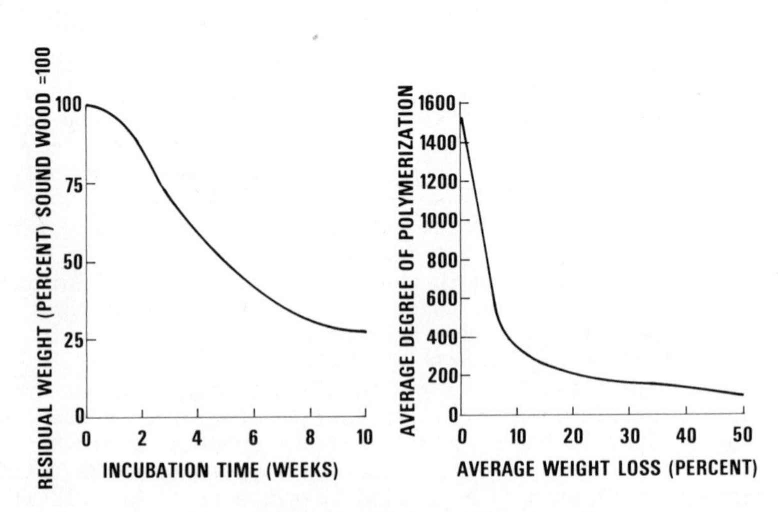

Figure 2. Action of brown-rot fungi on pine

(hydrolytic, oxidative, etc.) and enzymatic reactions.

Conclusions

The purpose of this paper was to plant seeds for thought in the area of new methods of wood preservation, which are not based on broad spectrum toxicity for their effectiveness. The present concern for our environment has created the opportunity for research in this area to find alternative wood preservatives which are effective in preventing attack by organisms and, at the same time, not harmful to the environment or man.

Chemical modification of wood does result in a treatment which is nontoxic, effective, and nonleachable. The high chemical treatment level required for effectiveness, however, results in a rather expensive treatment. Dimensional stability is also obtained at these high (17-25 WPG) substitution levels so, for those products where both rot resistance and dimensional stability are important, the present state of the technology is close to a viable industrial process.

Toxic chemicals which are permanently bound to the wood components may be an environmentally acceptable preservation method. The actual effectiveness of a bound toxicant, however, still needs to be investigated.

The encapsulation of preservatives is another interesting area for research. Procedures for encapsulation, capsule properties, and capsule size are important factors to be determined. Slow release fungicides by means of hydrolyzable linkages is also an interesting possibility.

Basic knowledge of the nature of the attack of microorganisms on wood, the enzymes involved which are unique to micro-organisms, the chemical reaction which takes place in the initial and sustained attack, and an investigation of specific inhibitors for these reactions is the most promising long-range approach.

Literature Cited

1. Scheffer, T. C., Forest Prod. J. (1963), 13(5): 208.
2. Kenaga, D. L., and Cowling, E. B., Forest Prod. J. (1959), 9(3): 112-116.
3. Lawton, E. J., Bellamy, W. D., Hungate, R. E., Bryant, M.P., and Hall, E., Science (1951), 113: 380-382.
4. Saeman, J. F., Millett, M. A., and Lawton, E. J., Ind. Eng. Chem. (1952), 44: 2848-2852.
5. Mater, J., Forest Prod. J. (1957), 7(6): 208-209.
6. Farrer, K.T.H., J. Proc. Austral. Chem. Inst. (1941), 8: 113.
7. Baechler, R., Forest Prod. J. (1959), 9(5): 166-171.
8. Gjovik, L. R., and Baechler, R. H., Forest Prod. J. (1968), 18(1): 25-27.
9. Highley, T. L., Phytopathology (1970), 60(11): 1660-1661.

10. Comparison of wood preservatives in stake tests, USDA Forest Service Research Note FPL-02 (1975), 52.
11. Boyce, J. S., Jr., J. of Forestry (1950), 48(1): 10.
12. Scheffer, T. C., and Eslyn, W. E., Forest Prod. J. (1961), 11(10): 485-490.
13. Stamm, A. J., Forest Prod. J. (1959), 9(3): 107-110.
14. Stamm, A. J., and Vallier, A. E., Forest Prod. J. (1954), 4(5): 305-312.
15. Meyer, J. A., and Loos, W. E., Forest Prod. J. (1969), 19(12): 32-38.
16. Loos, W. E., Wood Sci. & Tech. (1968), 2(4): 308-312.
17. Choong, E. T., and Barnes, H. M., Forest Prod. J. (1969), 19(6): 55-60.
18. Ref. 10, 22: 13.
19. Hampel, C. A., and Hawley, G. G., "The Encyclopedia of Chemistry," Van Nostrand Reinhold Co., 3rd ed., p. 968, 1973.
20. Allan, G. G., Chopra, C. S., Neogi, A. N., and Wilkins, R. M. Tappi (1971), 54(8): 1293-1294.
21. Mandels, M., and Reese, E. T., Ann. Rev. of Phytopathology (1965), 3: 85-102.
22. Mandels, M., and Reese, E. T., "Enzymic Hydrolysis of Cellulose and Related Materials," The MacMillan Co., New York, ed., E. T. Reese, pp. 115-157, 1963.
23. Rowell, R. M., Proc. Amer. Wood-Preservers Assoc. (1975), 71: 41-51.
24. Rowell, R. M., Amer. Chem. Soc. Symp. Ser. No. 10 (1975), pp. 116-124.
25. Stamm, A. J., and Baechler, R. H., Forest Prod. J. (1960), 10(1): 22-26.
26. Goldstein, I. S., Jeroski, E. G., Lund, A. E., Nielson, J.F., and Weaver, J. W., Forest Prod. J. (1961), 11(8): 363-370.
27. Ref. 10, 52: 26.
28. Rowell, R. M., and Gutzmer, D. I., Wood Sci. (1975), 7(3): 240-246.
29. Cowling, E. B., "Comparative Biochemistry of the Decay of Sweetgum Sapwood by White-Rot and Brown-Rot Fungi," USDA Forest Service Tech. Bull. No. 1258, p. 50, 1961.
30. Koenings, J. W., Wood & Fiber (1974), 6(1).

5

Thermal Deterioration of Wood

FRED SHAFIZADEH and PETER P. S. CHIN

Wood Chemistry Laboratory, University of Montana, Missoula, Mont. 59801

When wood is heated at elevated temperatures, it will show a permanent loss of strength resulting from chemical changes in its components. The thermal decomposition can start at temperatures below 100°C if wood is heated for an extended period of time. Figure 1 shows that wood heated at 120° loses 10% of its strength in about one month, but it takes only one week to obtain the same loss of strength if it is heated at 140° (1). Heating at higher temperatures gives volatile decomposition products and a charred residue. The pyrolytic reactions and products control the combustion process and relate to the problems of cellulosic fires, chemical conversion of cellulosic wastes and utilization of wood residues as an alternative energy source. In our laboratory, the pyrolytic reactions of wood and its major components have been investigated by a variety of analytical methods.

Thermal analysis of cottonwood and its major components (2), as shown in Figures 2 and 3, indicates that the thermal behavior of wood reflects the sum of the thermal responses of its three major components, cellulose, hemicellulose (xylan) and lignin. All these substrates are initially dried on heating at 50-100°. The hemicellulose component is the least stable and decomposes at 225-325°. Cellulose decomposes at higher temperatures within the narrower range of 325-375°. Lignin, however, decomposes gradually within the temperature range of 250-500°. The cell wall polysaccharides provide most of the volatile pyrolysis products, while lignin predominantly forms a charred residue. Since the thermal reactions of the wood components are highly complex and additive, they have been studied individually.

Among the three major components, the pyrolysis reactions of cellulose have been most extensively studied. At temperatures above 300°, rapid cleavage of the glycosidic bond takes place,

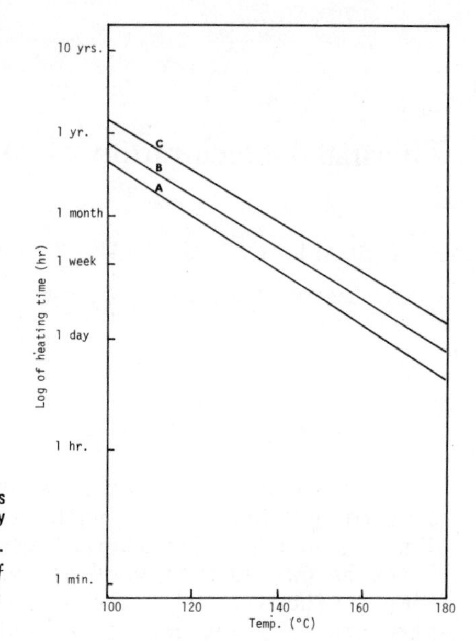

Journal of the Forest Products Research Society

Figure 1. Effect of time and temperature on the modulus of elasticity of wood: (A) 10%, (B) 20%, and (C) 40% loss of modulus of elasticity (1).

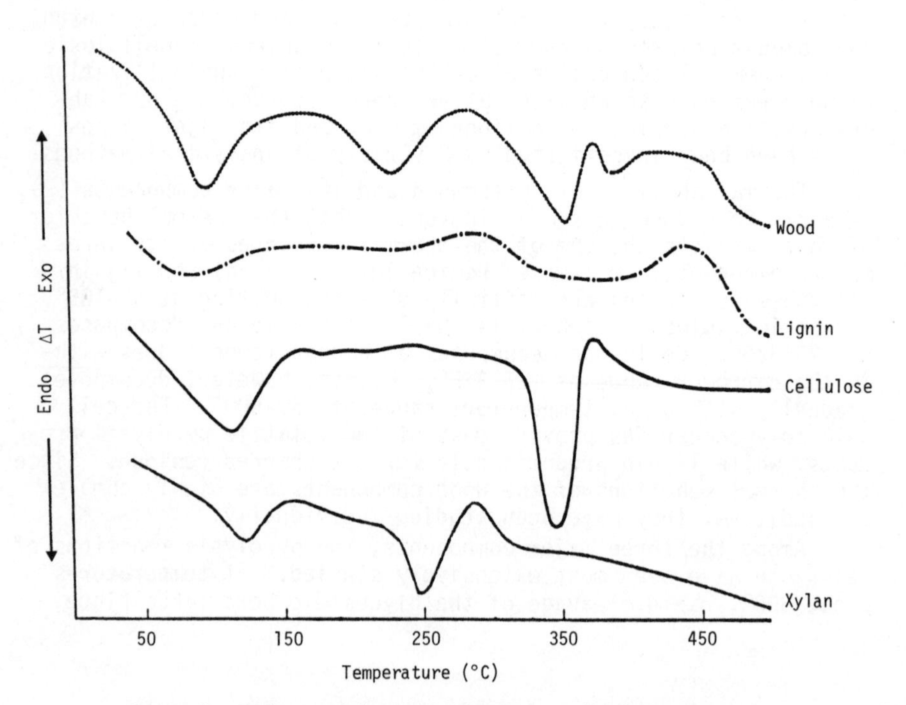

Figure 2. Differential thermal analysis of wood and its components

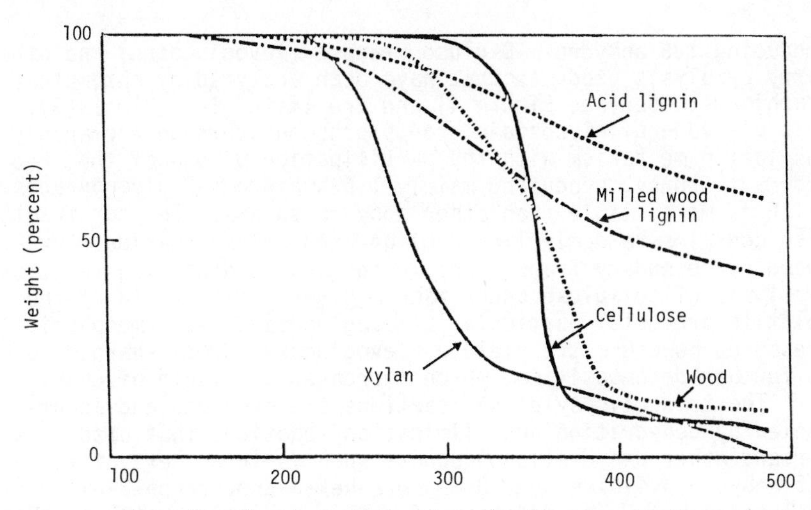

Figure 3. Thermogravimetry of wood and its components

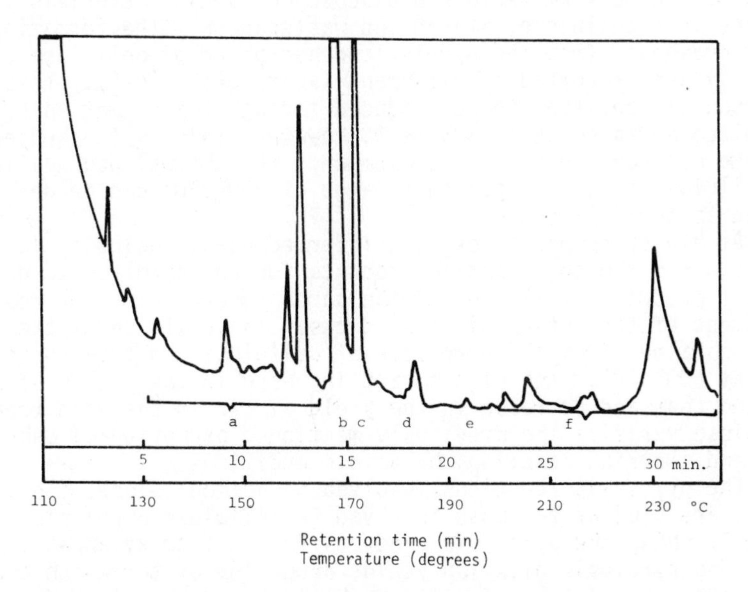

Figure 4. Chromatogram of cellulose pyrolysis tar after reduction with $NaBH_4$: a, unknown; b, 1,6-anhydro-β-D-glucopyranose; c, 1,6-anhydro-β-D-glucofuranose; d, 3-deoxyhexitols; e, D-glucitol; f, oligosaccharide derivatives

producing 1,6-anhydro-β-D-glucopyranose (levoglucosan) and other tarry pyrolysis products that have been analyzed by chromatographic methods (see Figure 4) and are listed in Table I (3). This cleavage of glycosidic groups proceeds through a transglycosylation mechanism with the participation of one of the free hydroxyl groups, producing mainly 1,6-anhydro-β-D-glucopyranose, which is more stable than other anhydro sugars. The tar fraction also contains randomly linked oligo- and polysaccharides, produced by secondary transglycosylation and condensation reactions. Pyrolysis of cellulose under vacuum gives a high yield of the volatile products, particularly levoglucosan. At atmospheric pressure, however, the yield of levoglucosan drops sharply, due to further decomposition, which increases the yield of char.

The transglycosylation reactions are preceded and accompanied by dehydration and elimination reactions that produce water and other dehydration products such as furan derivatives and 1,6-anhydro-3,4-dideoxy-β-D-*glycero*-hex-3-enopyranos-2-ulose (levoglucosenone). The addition of acidic additives, such as phosphoric acid, diammonium phosphate, diphenyl phosphate and zinc chloride, can significantly catalyze the latter reactions. This is illustrated in Figure 5 and Table II. Figure 5 shows the enhanced yields of levoglucosenone and 2-furaldehyde due to the addition of diphenyl phosphate, a strong Arrhenius acid, and Table II shows the effect of H_3PO_4 on promoting the production of levoglucosenone from various D-glucose-containing materials including pure cellulose, starch and wastepapers. The formation of levoglucosenone from the pyrolytic dehydration of cellulose has recently been reported by different laboratories (4-6). This compound was believed to be produced through the formation of levoglucosan as shown in Scheme 1, however, this is the subject of some controversy (6,7). In summary, the thermal degradation of cellulose in the temperature range of 300-350° can be depicted as shown in Scheme 2.

At higher temperatures, the intermediates, including levoglucosan and the condensation products further pyrolyze to give various products by fission of the carbohydrate units and rearrangement of the intermediate products. Table III shows the products obtained from the pyrolysis of cellulose and treated cellulose at 600° (8). The significant increase in the yields of water and char and decrease in the yield of tar in the acid treated cellulose verifies the previously mentioned promotion of dehydration and charring reactions by acidic additives.

The pyrolysis reactions involved in hemicellulose, *i.e.*, xylan, are similar to those involved in cellulose pyrolysis. Table IV shows the pyrolysis products formed from xylan at 300° (9). The pyrolysis of xylan yields about 16% of tar which contains 17% of a mixture of oligosaccharides. Upon acid hydrolysis, they give an approximately 54% yield of D-xylose. Structural analysis of the polymers shows that they are branched-chain

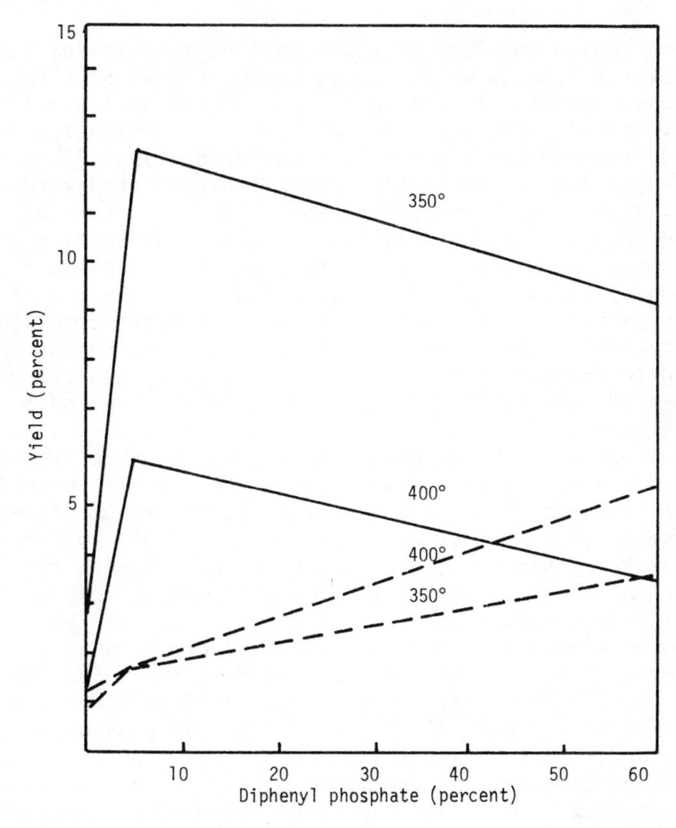

Figure 5. Production of levoglucosenone (——) and 2-furaldehyde (— —) from pyrolysis of cellulose containing diphenyl phosphate

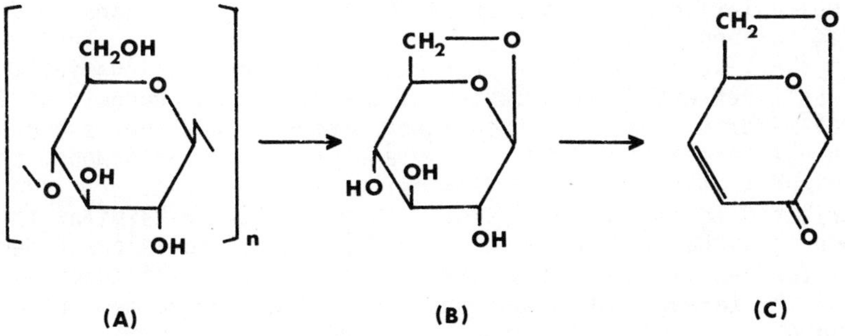

CH_2OH OH   OH   n   (A)

CH_2—O   HO   OH   OH   (B)

CH_2—O   O   (C)

Scheme 1. Pyrolysis of cellulose (A) to levoglucosan (B) and levoglucosenone (C) polymers, indicating that they are derived from random condensation of xylosyl units which are formed by cleavage of the glycosidic groups similar to that occuring in cellulose pyrolysis. The addition of a Lewis acid, i.e., $ZnCl_2$ significantly decreases the production of tar and enhances the production of char due to the enhanced dehydration reactions. At higher temperatures the glycosyl units and the random condensation products are further degraded to a variety of volatile products, as shown in Table V (9). Comparison of this table with the high temperature pyrolysis products listed for cellulose in Table III shows that the products of both fractions are basically similar. The significant increase in the yields of 2-furaldehyde, water and char and decrease in the yield of tar by the addition of $ZnCl_2$ verifies the enhanced dehydration and is similar to observed effects in cellulose pyrolysis.

Compared with the pyrolysis of polysaccharides, the pyrolysis of lignin is relatively unexplored. While the thermal reactions of lignin occur over a wide temperature range of 250-500°, the decomposition is most rapid between 310-420°, as indicated by the yield of gas and distillate produced. Due to the complex structure of lignin, the mechanism of its thermal degradation is not well understood. Table VI lists the major pyrolytic products from lignin (10). The most abundant product is char, a highly condensed carbonaceous residue, obtained in about 55% yield. The second fraction of the pyrolytic products is an aqueous distillate, produced in about 20% yield. It contains mainly water and some methanol, acetone and acetic acid. The yield of methanol for hardwood lignin is about 2%, twice as much as for softwood lignin because it contains a syringyl rather than guaiacyl structure. The yield of acetic acid from hardwood lignin is also significantly higher than from softwood lignin; presumably it originates from the propanoid side chains. The third fraction of the pyrolytic products is tar, which is produced in about 15% yield. It is a mixture of phenolic compounds closely related to phenol-guaiacol and 2,6-dimethoxy-phenol, with substituents at the position para to the hydroxyl group. The last pyrolytic fraction includes volatile products such as CO, CH_4, CO_2 and ethane, produced in about 12% yield.

The list of pyrolysis products of cottonwood shown in Table VII (11) reflects the summation of the pyrolysis products of its three major components. The higher yields of acetone, propenal, methanol, acetic acid, CO_2, water and char from cottonwood, as compared to those obtained from cellulose and xylan, are likely attributed to lignin pyrolysis. Other results are similar to those obtained from the pyrolysis of cell-wall polysaccharides. This further verifies that there is no significant interaction among the three major components during the thermal degradation of wood.

The pyrolysis products of wood can be broadly grouped into three categories as shown in Scheme 3, i.e., the combustible

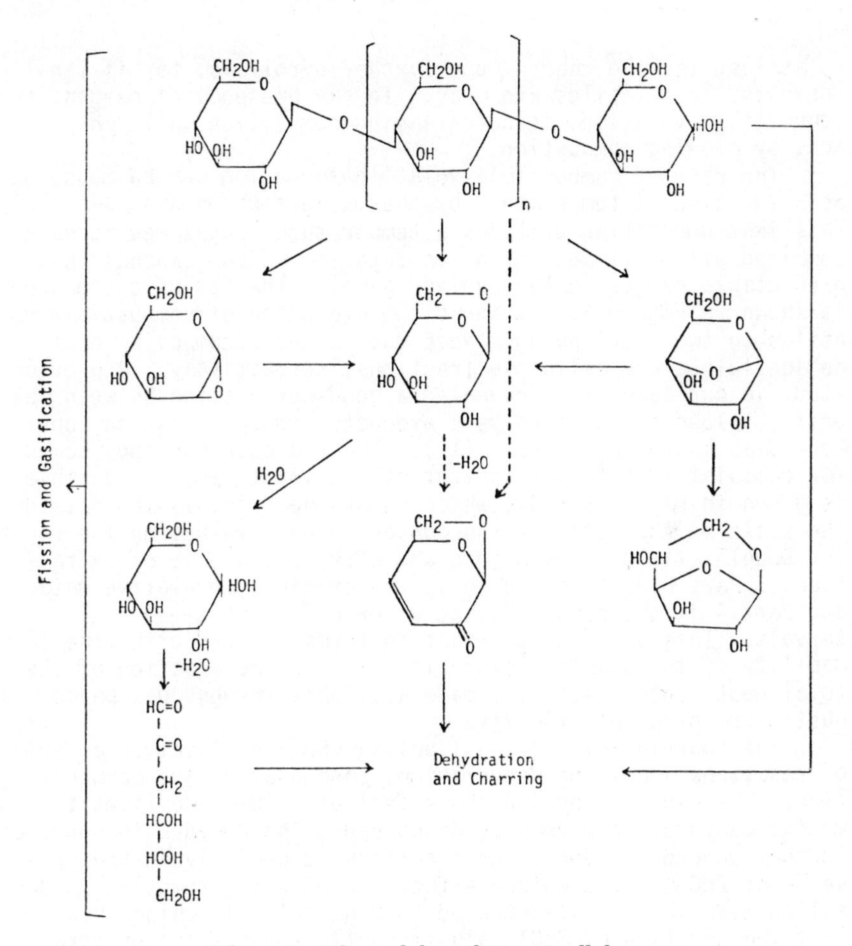

Scheme 2. Thermal degradation of cellulose

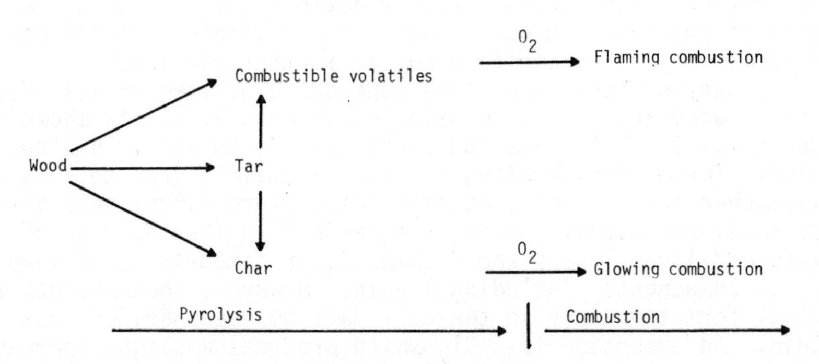

Scheme 3. The pyrolysis and combustion of wood

volatiles, tar and char. Upon further pyrolysis, tar is finally converted to volatiles and char. In the presence of oxygen, the combustible volatiles lead to flaming combustion while char reacts by glowing combustion.

The rate of combustible volatile formation can be measured as a function of temperature by thermal evolution analysis (TEA). This instrumentation utilizes a temperature programmed furnace combined with a flame ionization detector which responds in a predictable manner to the evolved gases. The TEA of cottonwood is shown in Figure 6. It shows the evolution of combustible volatiles in two overlapping stages due to the decomposition of hemicellulose and cellulose fractions, respectively. A previous study in our laboratory revealed a good correlation between calorific values of the pyrolysis products and their carbon contents, as shown in Figure 7 (12). The TEA data can thus be used for calculating the heat content of the volatiles. An example is given in Figure 8 (13), which shows the original TEA data on the scale on the left and the converted heat values on the right for Douglas fir needles before and after a sequence of extractions. Part B of this figure is the original cumulative data, and Part A is the data calculated for different temperature intervals. This data is important in terms of predicting the flammability of the samples, since it reflects the fraction of the total heat content actually made available through gas phase combustion to propagate the fire.

The charring of cell wall polysaccharides involves a series of reactions including dehydration, condensation and carbonization. The dehydration and the effect of acidic additives in promoting dehydration have been discussed. The dehydration products further undergo condensation reactions, especially in the presence of $ZnCl_2$. The unique effect of $ZnCl_2$ on promoting condensation reactions is illustrated in Figure 9 (5), which shows that the addition of $ZnCl_2$ significantly reduces the evaporation of levoglucosenone by promoting the condensation of this dehydration product to nonvolatile materials that are charred on further heating. The carbonization reactions involve further elimination of the substituents, production of stable free radicals and formation of new carbon bonds on further heating.

A study of the temperature dependence of free radical formation of wood and its three major components by ESR is shown in Figures 10-13. This data indicates that up to 350°, the free radicals formed from heating of wood are mainly from cell wall polysaccharides. Lignin, at this temperature range, generates very small concentrations of free radicals. The addition of acidic additives lowers the decomposition temperature of wood and its components, including lignin. However, they promote free radical formation only in the cell wall polysaccharides, not in lignin. An exception is $ZnCl_2$ which produces a slight increase in free radical formation in lignin at temperatures above 300°.

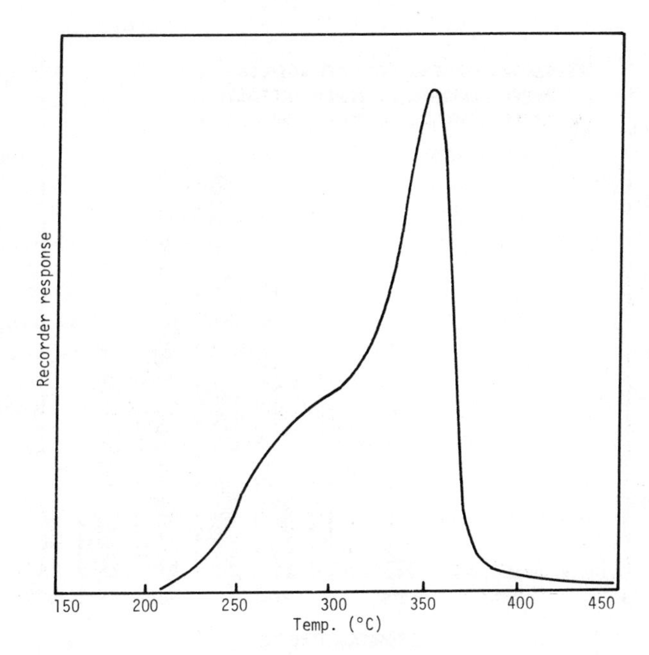

Figure 6. Thermal evolution analysis (TEA) of cottonwood

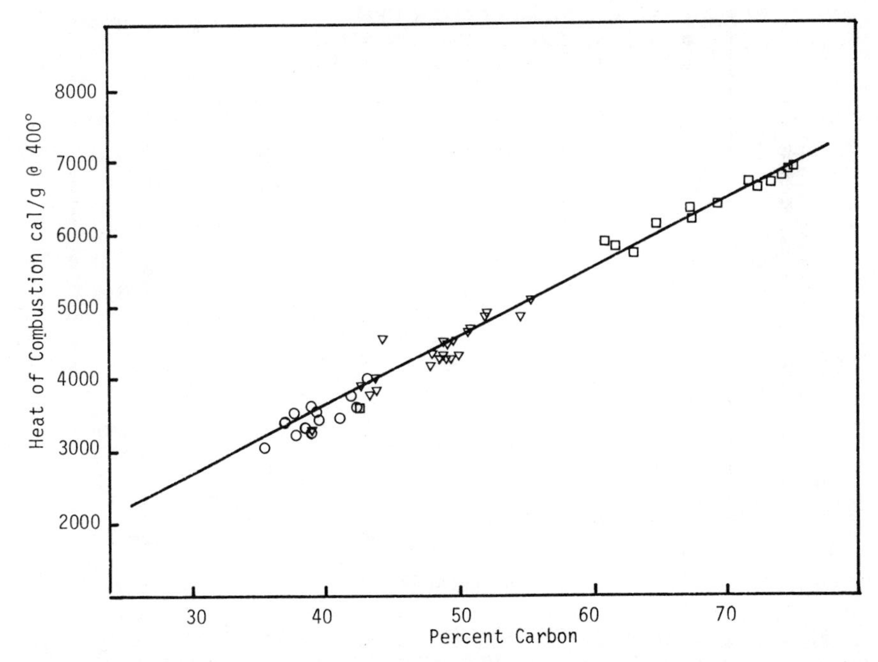

Figure 7. Heat of combustion at 400°C vs. percent carbon: ▽ fuels, □ char, ○ volatiles

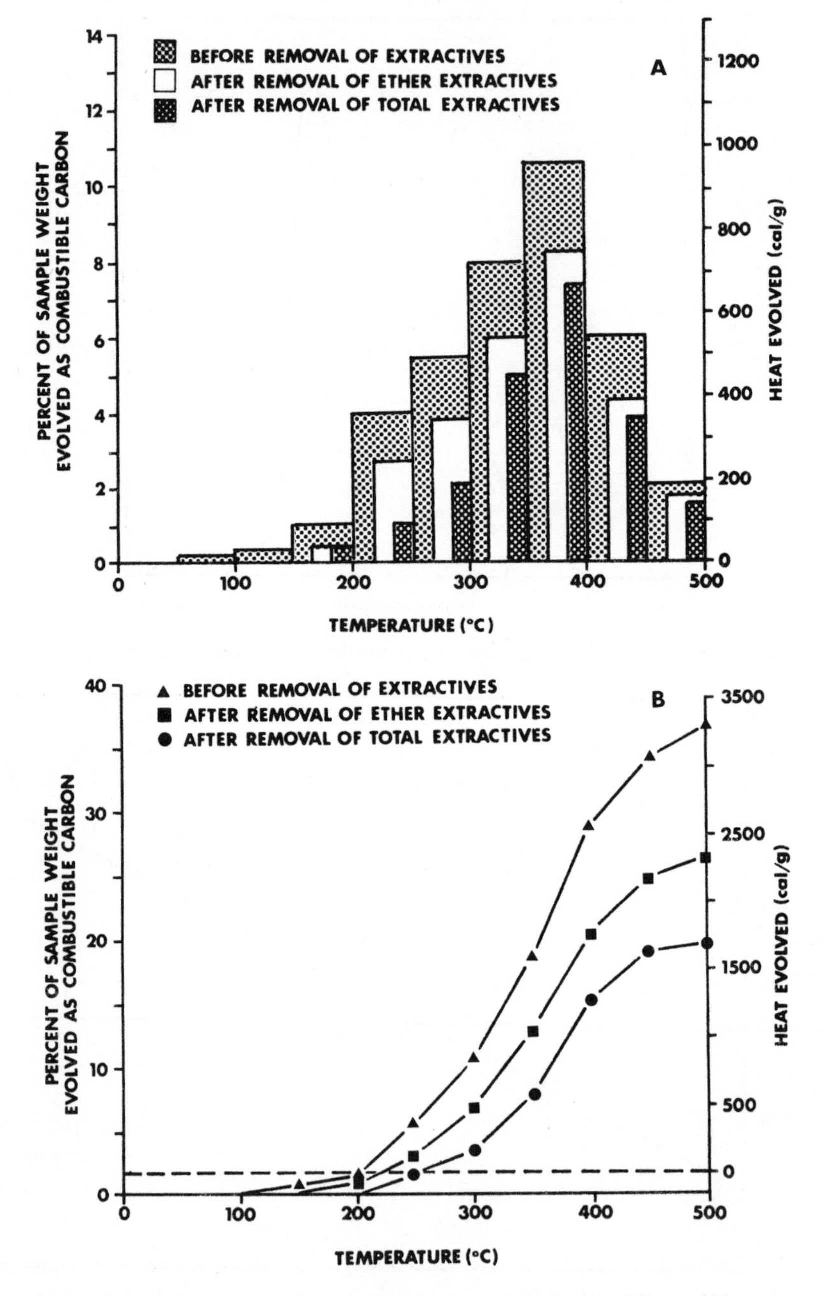

Figure 8. Evolution of carbon and heat from Douglas-fir foliage (A) in temperature intervals and (B) cumulative, based on dry weight of the unextracted sample

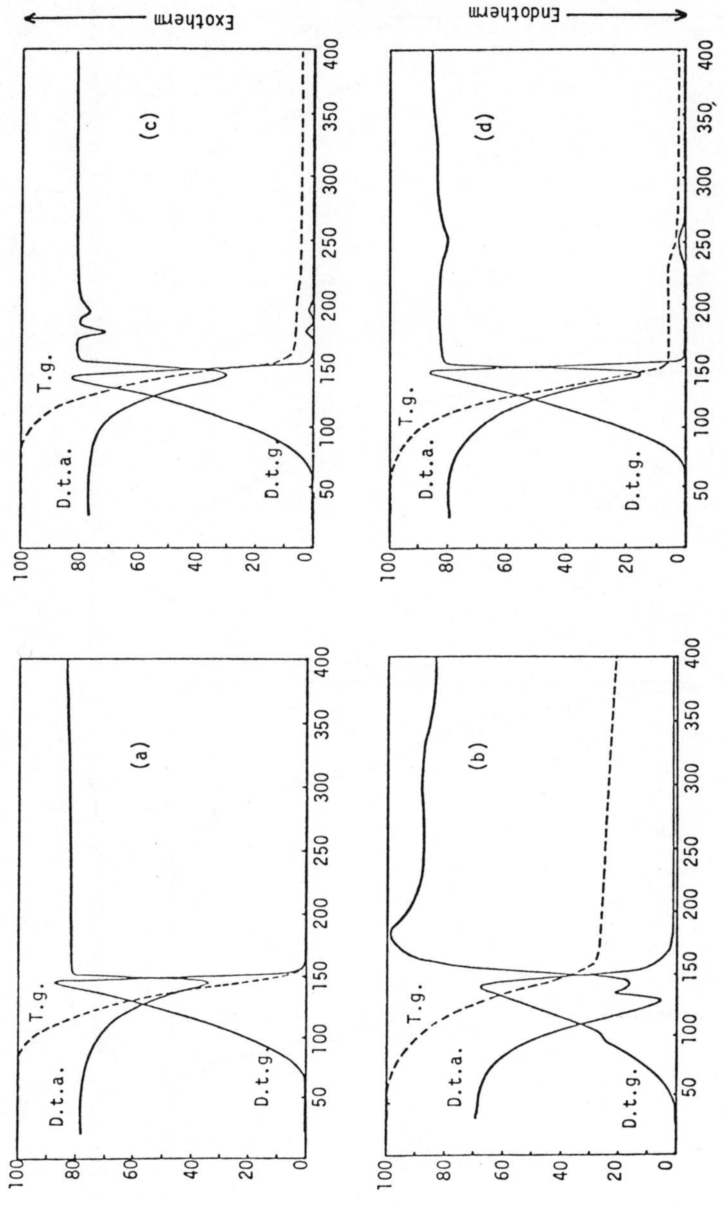

Figure 9. Thermal analysis of levoglucosenone samples: (a) neat, (b) +5% zinc chloride, (c) +5% diammonium phosphate, and (d) +5% diphenyl phosphate (D.t.a., differential thermal analysis; T.g., thermogravimetry; D.t.g., derivative of thermogravimetry)

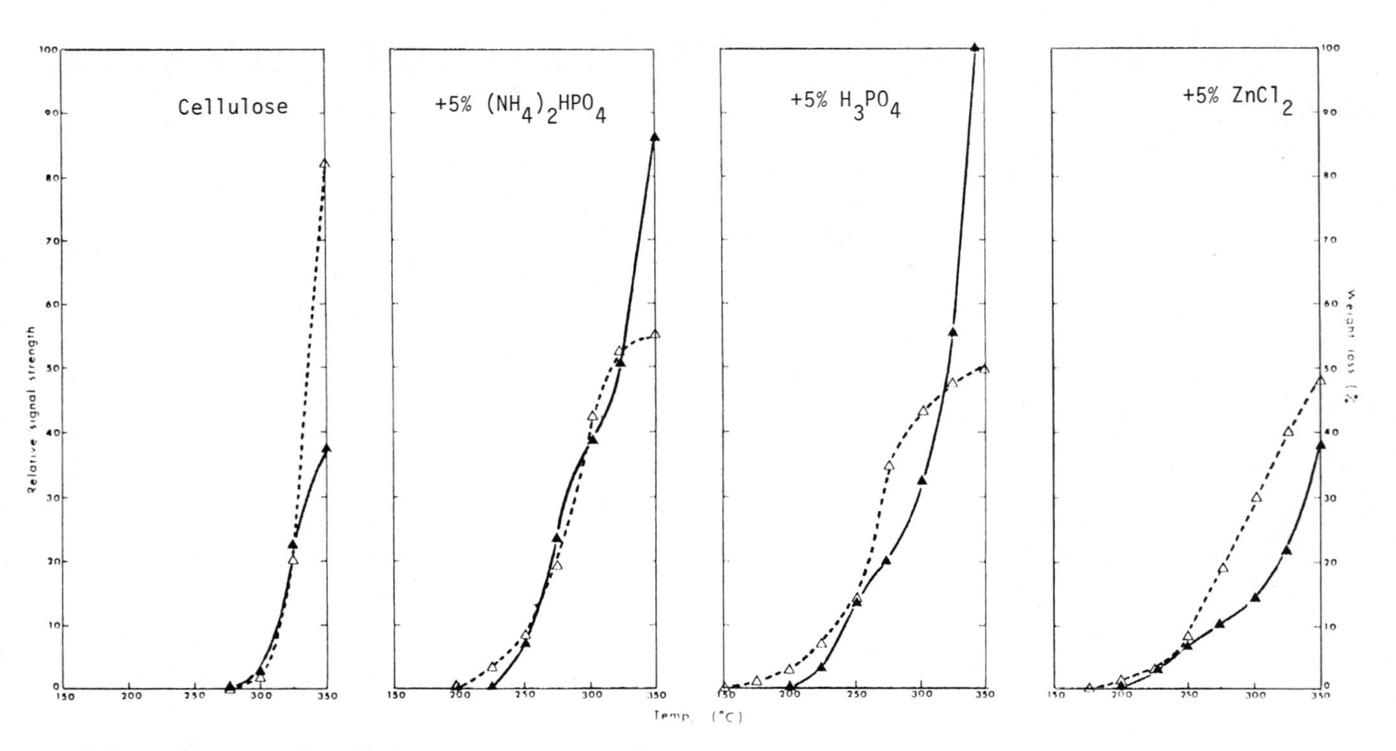

Figure 10. The rate of free radical formation (—▲—) and weight loss (– –△– –) on heating of cellulose and treated cellulose

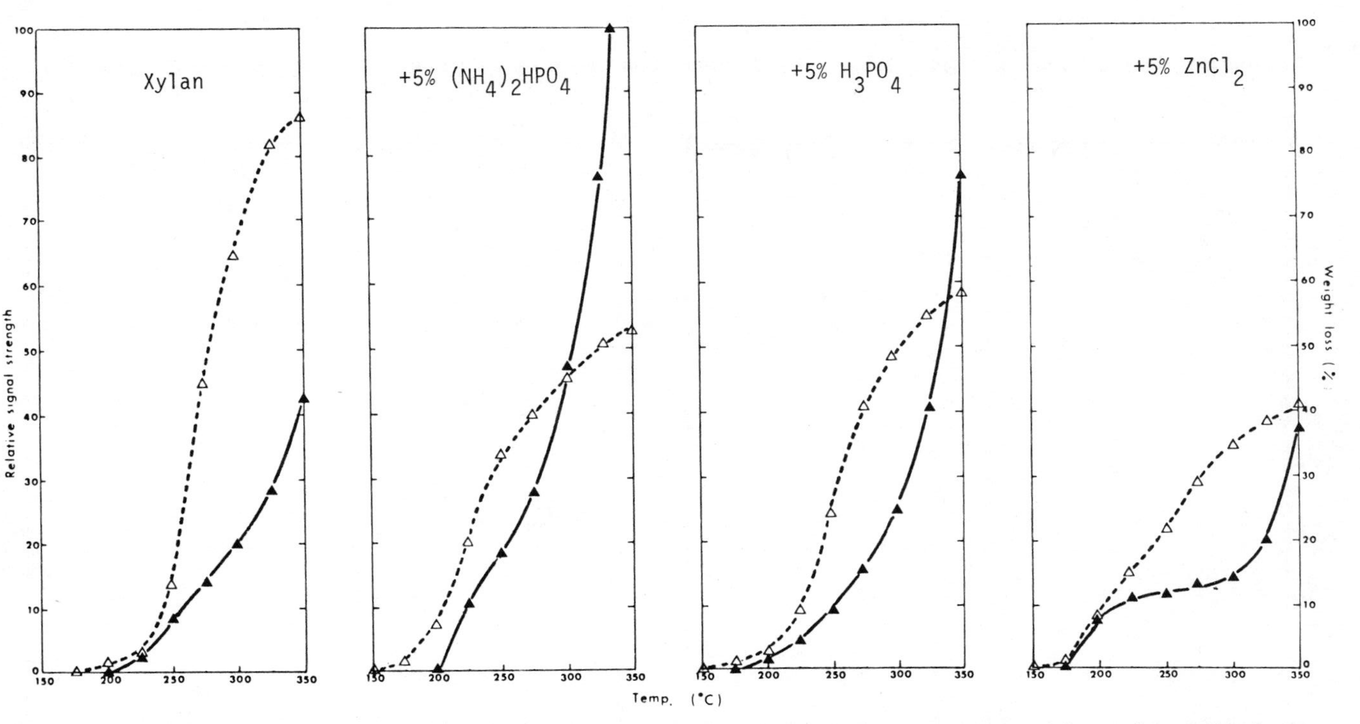

Figure 11. The rate of free radical formation (—▲—) and weight loss (– –△– –) on heating xylan and treated xylan

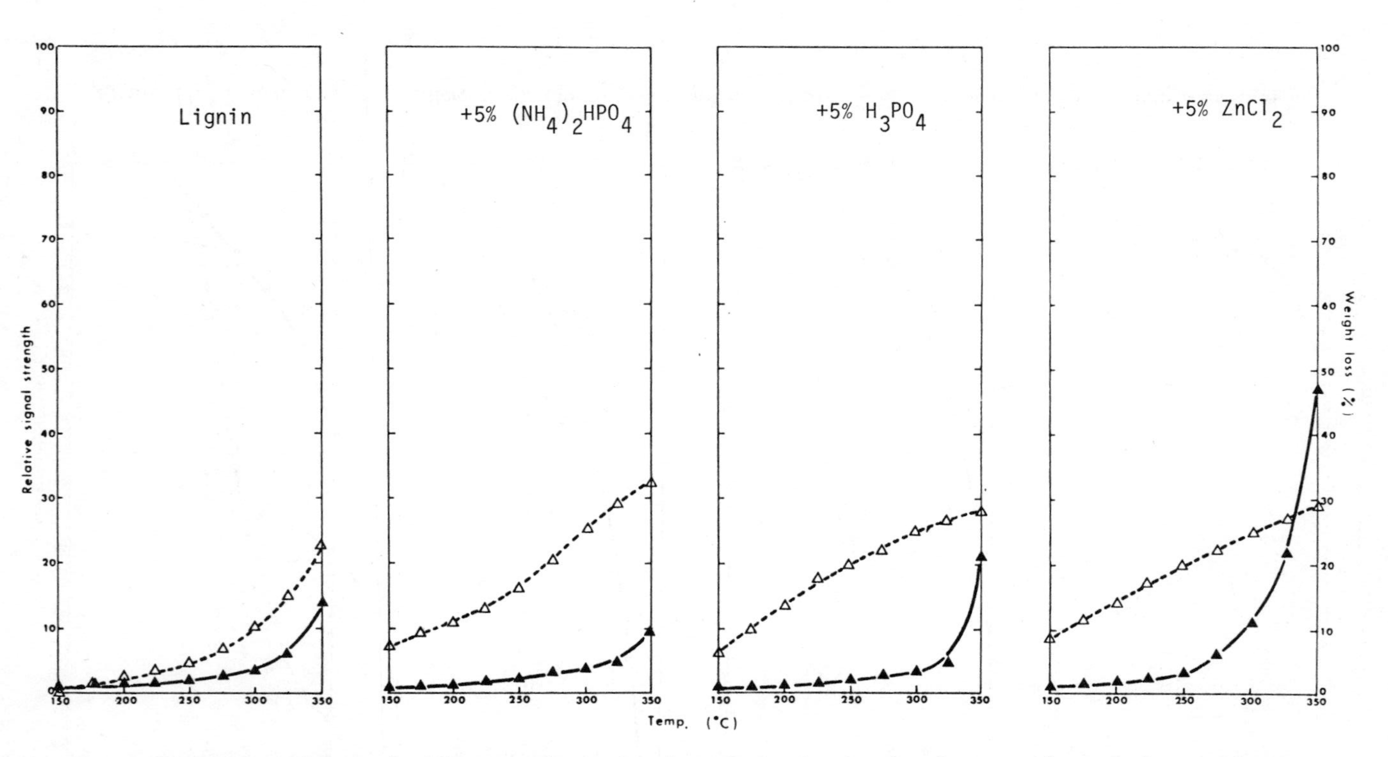

Figure 12. The rate of free radical formation (—▲—) and weight loss (- -△- -) on heating of lignin and treated lignin

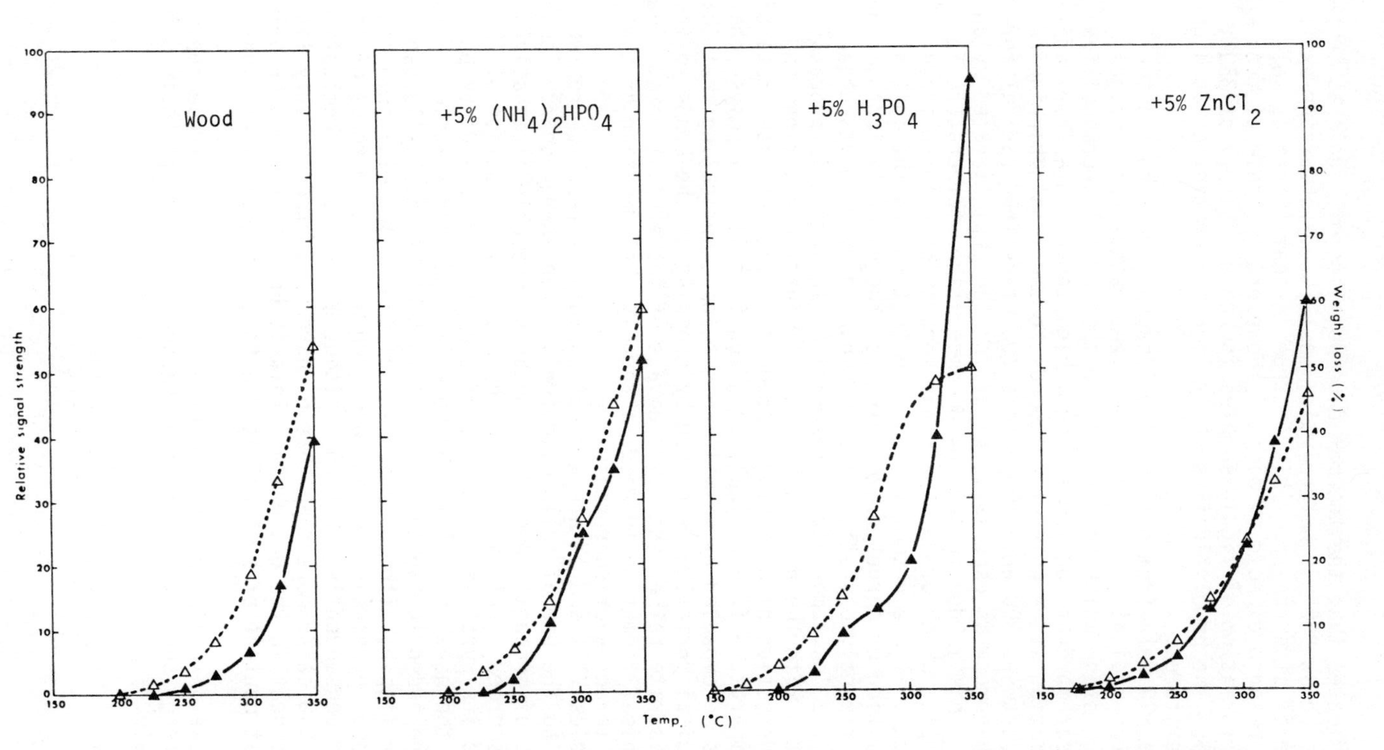

Figure 13. The rate of free radical formation (—▲—) *and weight loss* (--△--) *on heating of wood and treated wood*

This indicates that the thermal decomposition of lignin at temperatures below 350° is heterolytic, most likely occurring by the cleavage of side chains. This is true in neat lignin as well as in the presence of acid catalysts. However, the rate of free radical formation of lignin samples increases sharply at 350°, indicating the homolytic fission of lignin predominates at higher temperatures.

The ESR data from cell wall polysaccharides shows a two-stage free radical formation; a low temperature free radical formation stage which corresponds to the initial decomposition of the carbohydrate polymers, and a high temperature free radical formation stage which corresponds to the final charring reactions between 300 and 350°. This two-stage free radical formation phenomena is especially clear for the $ZnCl_2$ treated xylan sample. This is due to the low decomposition temperature for this sample, which produces a clear separation between these two stages. These phenomena indicate that during the initial decomposition of carbohydrate polymers, the heterolytic reactions, such as transglycosylation, are accompanied by homolytic reactions. The low temperature free radical formation is probably associated with the dehydration and elimination reactions and the condensation of unsaturated products. The increased rate of high temperature free radical formation in cell wall polysaccharides is accompanied by a small weight loss, indicating that this free radical formation is caused by cracking of the bonds in the char substrate rather than by cleavage of the substituents. The free radical formation in wood is roughly the summation of that for its three major components.

Tables VIII and IX are a brief summary of results from work related to the use of cellulosic fuels as an energy source, both in terms of propagation of fire and as a renewable alternative energy source. Table VIII shows the heats of combustion of the fuel and its pyrolysis products. Table IX shows the distribution of the heat content in the volatile and char fractions. The energy released in the gas phase is much higher for cellulose than for lignin, although the heats of combustion of lignin and its gaseous pyrolysis products are much higher than those of cellulose. Consequently, softwood, although its heat of combustion is over 500 cal/g more than that of hardwood, produces very little more heat in the gas phase. This is because the higher original heat content of softwood is due to its higher lignin content.

These data also clearly point out the value of these fuels as an energy source, either after carbonization or in their original state.

Appendix

TABLE I. ANALYSIS OF THE PYROLYSIS PRODUCTS OF CELLULOSE AT 300° UNDER NITROGEN

Condition	Atm. pressure	1.5 Mm Hg	1.5 Mm Hg, 5% $SbCl_3$
Char	34.2%	17.8%	25.8%
Tar	19.1	55.8	32.5
levoglucosan	3.57	28.1	6.68
1,6-anhydro-β-D-glucofuranose	0.38	5.6	0.91
D-glucose	trace	trace	2.68
hydrolyzable materials	6.08	20.9	11.8

TABLE II. YIELDS OF LEVOGLUCOSENONE FROM THE PYROLYSIS OF DIFFERENT MATERIALS AT 350°[a]

Material	Neat (%)	5% H_3PO_4-treated (%)
Cellulose	1.2	11.1
Starch	0.3	9.0
Newsprint (with ink)	T[b]	9.1
Kraft shopping bags	T	10.2

a. Determined by pyrolyzing 5 mg samples and directly analyzing the volatiles by GLC.

b. T = trace amount.

TABLE III. PYROLYSIS PRODUCTS OF CELLULOSE AND TREATED CELLULOSE AT 600°

Product	Neat	+5% H_3PO_4	+5% $(NH_4)_2HPO_4$	+5% $ZnCl_2$
Acetaldehyde	1.5[a]	0.9	0.4	1.0
Furan	0.7	0.7	0.5	3.2
Propenal	0.8	0.4	0.2	T
Methanol	1.1	0.7	0.9	0.5
2-Methylfuran	T	0.5	0.5	2.1
2,3-Butanedione	2.0	2.0	1.6	1.2
1-Hydroxy-2-propanone, Glyoxal	2.8	0.2	T	0.4
Acetic acid	1.0	1.0	0.9	0.8
2-Furaldehyde	1.3	1.3	1.3	2.1
5-Methyl-2-furaldehyde	0.5	1.1	1.0	0.3
Carbon dioxide	6	5	6	3
Water	11	21	26	23
Char	5	24	35	31
Balance (tar)	66	41	26	31

[a]Percentage, yield based on the weight of the sample; T = trace amounts.

TABLE IV. PYROLYSIS PRODUCTS OF XYLAN AND TREATED XYLAN AT 300°

Product	Neat	+10% $ZnCl_2$
Liquid condensate	30.6[a]	45.3
Carbon dioxide	7.9	7.5
Char	31.1	42.2
Tar	15.7	3.2
High mol. wt. component	(17)[b]	
D-xylose from hydrolysis	(54)[c]	

[a]Percentage, yield based on the weight of the sample.

[b]Based on the weight of the tar

[c]Based on the weight of oligosaccharides.

TABLE V. PYROLYSIS PRODUCTS OF XYLAN AND TREATED XYLAN AT 500°

Product	Xylan		O-Acetylxylan	
	Neat	+10% $ZnCl_2$	Neat	+10% $ZnCl_2$
Acetaldehyde	2.4[a]	0.1	1.0	1.9
Furan	T	2.0	2.2	3.5
Acetone / Propionaldehyde	0.3	T	1.4	T
Methanol	1.3	1.0	1.0	1.0
2,3-Butanedione	T	T	T	T
1-Hydroxy-2-propanone	0.4	T	0.5	T
3-Hydroxy-2-butanone	0.6	T	0.6	T
Acetic acid	1.5	T	10.3	9.3
2-Furaldehyde	4.5	10.4	2.2	5.0
Carbon dioxide	8	7	8	6
Water	7	21	14	15
Char	10	26	10	23
Balance (tar)	64	32	49	35

[a]Percentage, yield based on the weight of the sample; T = trace amounts.

TABLE VI. PYROLYSIS PRODUCTS OF LIGNIN AT 450-550°

Fraction	Products	Yield (%)
Volatile	carbon monoxide, methane, carbon dioxide, ethane	12
Liquid	water, methanol, acetone, acetic acid	20
Tar	phenolic compounds	15
Char	carbonaceous residue	55

TABLE VII. PYROLYSIS PRODUCTS OF WOOD AND TREATED WOOD AT 600°

Product	Neat	+5% $ZnCl_2$
Acetaldehyde	2.3[a]	4.4
Furan	1.6	7.9
Acetone / Propionaldehyde	1.5	0.9
Propenal	3.2	0.9
Methanol	2.1	2.7
2-Methylfuran	*b*	*b*
2,3-Butanedione	2.0	1.0
1-Hydroxy-2-propanone	2.1	T
Glyoxal	2.2	T
Acetic acid	6.7	5.4
2-Furaldehyde	1.1	5.2
Formic acid	0.9	0.5
5-Methyl-2-furaldehyde	0.7	0.9
2-Furfuryl alcohol	0.5	T
Carbon dioxide	12	6
Water	18	18
Char	15	24
Balance (tar)	28	22

[a]Percentage, yield based on the weight of the sample; T = trace amounts.

[b]Not clearly identifiable for wood.

TABLE VIII. THE HEAT OF COMBUSTION OF WOOD AND ITS PYROLYSIS PRODUCTS

Fuel			Char		Combustible volatiles	
Source	Type	$\Delta H^{25°}$ comb (cal/g)	Yield[a] (%)	$\Delta H^{25°}$ comb (cal/g)	Yield[a] (%)	$\Delta H^{25°}$ comb (cal/g)
Cellulose	Filter paper	-4143	14.9	-7052	85.1	-3634
Lignin	Klason	-6371	59.0	-7416	41.0	-4867
Softwood	Douglas Fir	-5156	27.4	-7259	72.6	-4362
Hardwood	Populus ssp.	-4618	21.7	-7124	78.3	-3923

[a]heated at 400° for 10 min.

TABLE IX. DISTRIBUTION OF THE HEAT OF COMBUSTION OF WOOD AND ITS COMPONENTS

Fuel Source	Fuel Type	Char[a] (cal/g fuel)	Gas[a] (cal/g fuel)	Total (cal/g)
Cellulose	Filter paper	-1050	-3093	-4143
Lignin	Klason	-4375	-1995	-6370
Softwood	Douglas Fir	-1987	-3169	-5156
Hardwood	Populus ssp.	-1546	-3072	-4618

[a]heated at 400° for 10 min.

Literature Cited

1. Seborg, R.M., Tarkow, H., and Stamm, A.J., *J. Forest Prod. Res. Soc.*, (1953), 3, 59.
2. Shafizadeh, F. and McGinnis, G.D., *Carbohyd. Res.*, (1971), 16, 273-277.
3. Shafizadeh, F. and Fu. Y.L., *Carbohyd. Res.*, (1973), 29, 113-122.
4. Halpern, Y., Riffer, R., and Broido, A., *J. Org. Chem.*, (1973), 38, 204-209.
5. Shafizadeh, F. and Chin, P.P.S., *Carbohyd. Res.*, (1976), 46, 149-154.
6. Fung, D.P.C., *Wood Science*, (1976), 9, 55-57.
7. Broido, A., Evett, M., and Hodges, C.C., *Carbohyd. Res.*, (1975), 44, 267-274.
8. Chin, P.P.S., Ph.D. Dissertation, University of Montana (1973).
9. Shafizadeh, F., McGinnis, G.D., and Philpot, C.W., *Carbohyd. Res.*, (1972), 25, 23-33.
10. Allan, G.G. and Mattila, T., Lignins, Sarkanen, K.V. and Ludwig, C.H., Eds., Wiley-Interscience Publishers, New York, 1971, p. 575.
11. Philpot, C.W., Ph.D. Dissertation, University of Montana (1970).
12. Susott, R.A., DeGroot, W.F., and Shafizadeh, F. *J. Fire and Flammability*, (1975), 6, 311-325.
13. Shafizadeh, F., Chin, P.P.S., and DeGroot, W.F., *Forest Science*, in press.

6

Effect of Fire-Retardant Treatments on Performance Properties of Wood

CARLTON A. HOLMES

Forest Products Laboratory, Forest Service, U.S. Dept. of Agriculture, P.O. Box 5160, Madison, Wis. 53705

The one million fires in buildings in the United States account for about two-thirds of the 12,000 people who die each year in fires. The property loss in building fires is about 85 percent of the total annual $3 billion property loss in fires (1). Building contents are often a primary source of fire and are usually responsible for fire-related deaths before structural members become involved. Nevertheless, wood and wood-base products, extensively used both as structural members and as interior finish in housing and buildings, can be contributors to fire destruction.

To reduce the contribution of wood to fire losses, much research through the years has gone into development of fire-retardant treatments for wood. A total of 21.3 million pounds of fire-retardant chemicals were reported used in 1974 to treat 5.7 million cubic feet of wood products (2). The amount of wood treated was about one tenth of 1 percent of the total domestic production of lumber and plywood and has increased ninefold in 20 years.

How does our research stand in rendering wood fire retardant? What is the effect of fire-retardant treatments on the fire performance properties of wood and on the physical and mechanical properties of wood that are important to its utility? Discussion will be limited to fire retardancy obtained by pressure impregnation, which is currently the most effective method. Fire-retardant coatings, wood-plastic combinations, and chemical modifications of wood will not be considered.

Fire-Retardant Chemicals

Past research on fire retardants, including those for wood, from about 1900 to 1968 is reviewed in John W. Lyons' comprehensive reference book, "The Chemistry and Uses of Fire Retardants" (3). A more recent review by Goldstein (4) gives additional information on fire-retardant chemicals and treatment systems for wood and also discusses some of the topics of this present paper more thoroughly. These two references, together with the older review by Browne (5), are recommended to the reader as basic reviews on selection and chemistry of fire retardants for wood. The chemistry of synergistic effects between chemicals in a fire-retardant system is presented by Lyons (3); it was also discussed more recently by Juneja (6,7) and recommended by him as an area of needed investigation.

Fire-retardant chemicals used by the commercial wood-treating industry are limited almost exclusively to mono- and diammonium phosphate, ammonium sulfate, borax, boric acid, and zinc chloride (4,8). It is believed that some use is also made of the liquid ammonium polyphosphates (9). Some additives such as sodium dichromate as a corrosion inhibitor are also used. Aqueous fire-retardant treatment solutions are usually formulated from two or more of these chemicals to obtain the desired properties and cost advantages. For leach-resistant type treatments, the literature shows that some or all of the following are used: urea, melamine, dicyandiamide, phosphoric acid, and formaldehyde (10-12).

Effect of Fire-Retardant Treatment on Fire Performance Properties

What are the fire performance properties of untreated wood and how are these properties altered by fire-retardant treatments?

Ignition

Wood, like all organic materials, chemically decomposes--pyrolyzes--when subjected to high temperatures, and produces char and pyrolysate vapors or gases. When these gases escape to and from the wood surface and are mixed with air, they may ignite, with or without a pilot flame, depending on temperature. Ignition--the initiation of combustion--is evidenced by glowing on the wood surface or by presence of flames above the surface.

The temperature of ignition is influenced by many factors related to the wood under thermal exposure and the conditions of its environment (5,13,14). Factors include species, density, moisture content, thickness and surface area, surface absorptivity, pyrolysis characteristics, thermoconductivity, specific heat, and extractives content. Environmental conditions affecting temperature of ignition include duration and uniformity of exposure, heating rate, oxygen supply, air circulation and ventilation, degree of confinement or space geometry surrounding the exposed wood element or member, temperature and characteristics of an adjacent or contacting material, and amount of radiant energy present.

Reviews covering ignition of cellulosic solids by Kanury (13), Beall and Eickner (15), Browne (5), and Matson *et al.* (14) report a wide range of ignition temperatures obtained on wood, and dependence on radiant or convective nature of heat. For radiant heating of cellulosic solids, Kanury (13) reports spontaneous transient ignition at a critical temperature of 600°C with piloted transient ignition at 300°C to 410°C. Persistent flaming ignition is reported at a temperature greater than about 320°C.

With convective heating of wood under laboratory conditions, spontaneous ignition is reported as low as 270°C and as high as 470°C (5,14,15). Spontaneous ignition of wood charcoal, which has excellent absorption of oxygen and radiant heat, occurs between 150°C and 250°C (5). In one experiment on ignition, oven-dried sticks of nine different species were ignited by pilot flame in 14.3 to 40 minutes when held at 180°C, in 4 to 9.5 minutes when held at 250°C and in 0.3 to 0.5 minutes when held at 430°C (16).

Many field reports collected by Underwriters' Laboratories, Inc. (UL) (14) show ignition occurring at or near 212°F (100°C) on wood next to steam pipes or other hot materials. Laboratory experiments have not been able to confirm these low ignition temperatures (5,14,16,17). To provide a margin of safety, Underwriters' Laboratories, Inc. suggests that wood not be exposed for long periods of time at temperatures greater than 90°F (32°C) above room temperature or 170°F (77°C). The National Fire Protection Association handbook (18) gives 200°C as the ignition temperature of wood most commonly quoted, but gives 66°C as the highest temperature to which wood can be continually exposed without risk of ignition. McGuire (17) of the National Research Council of Canada suggests that 100°C would be a satisfactory choice of an upper limiting temperature for wood exposure.

Usually the fire-retardant treatment of wood slightly increases the temperature at which ignition will take place. There is evidence, however, that wood treated with some chemical retardants at low retention levels will ignite (flame) or start glowing combustion at slightly lower temperatures or irradiance levels than does untreated wood (19,20), though sustained combustion is usually prevented or hindered.

Thermal Degradation

An extensive review of the literature to 1958 on thermal degradation of wood is given by Browne (5). Beall and Eickner (15) and Goldstein (4) add additional review information on this complex subject. Shafizadeh's (21) review of the pyrolysis and combustion chemistry of cellulose gives a basis for understanding these processes in wood and the effect of fire-retardant treatment on these processes.

Browne (5) described the pyrolysis reactions and events which occur in each of four temperature zones or ranges when solid wood of appreciable thickness is exposed to heat in absence of air. Zone A is below 200°C; Zone B, 200° to 280°C; Zone C, 280° to 500°C; and Zone D, above 500°C. These zones may be present simultaneously. When wood is heated in air, events occurring in these temperature zones include oxidation reactions and, after ignition, combustion of the pyrolysis and oxidation products.

Goldstein (4) more simply divided the thermal degradation processes into those occurring at low temperatures, below 200°C, and those at high temperatures, above 200°C. Decomposition of wood exposed to temperatures below 200°C is slow but measurable (22,23). For example, the average loss in weight of 11 species of wood was 2.7 percent in 1 year at 93°C and 21.4 percent in 102 hours at 167°C (22). Sound wood will not generally ignite below 200°C since products evolved are mostly carbon dioxide and water vapor.

High temperature degradation processes above 200°C include rapid pyrolysis of the wood components, combustion of flammable gases and tars, glowing of the char residue, and evolution of unburned gases, vapors, and smoke.

The most widely accepted theory of the mechanism of fire-retardant chemicals in reducing flaming combustion of wood is that the chemicals alter the pyrolysis reactions with formation of less flammable gases and tars and more char and water (4,5,8,21,24-29). Some fire retardants start and end the chemical decomposition at lower temperatures. Heat of combustion of the volatiles is reduced. Shafizadeh (21) suggests that a primary function of fire retardants is to promote dehydration and charring of cellulose. The normal degradation of cellulose to the flammable tar, levoglucosan, is reduced and the charring of this compound is promoted. Shafizadeh and coworkers used thermogravimetric (TG) and thermal evolution analysis (TEA) data, to confirm two different mechanisms involved in flameproofing cellulosic materials: 1) directing the pyrolysis reactions to produce char, water, and carbon dioxide in place of flammable volatiles, and 2) preventing the flaming combustion of these volatiles (27).

Fire Penetration

The property of a wood material or assembly to resist the penetration of fire or to continue to perform a given structural function, or both, is commonly termed fire resistance. The measure of elapsed time that a material or assembly will exhibit fire resistance under the specified conditions of test and performance is called fire endurance. Large furnaces are used to measure fire endurance of walls, floors, roofs, doors, columns, and beams under the standard ASTM E119 (30) time-temperature exposure conditions.

Wood has excellent natural resistance to fire penetration due to its low thermal conductivity and to the characteristic of forming an insulating layer of charcoal while burning. The wood beneath the char still retains most of its original strength properties.

In wood charring studies by Schaffer at the Forest Products Laboratory (FPL) (31), 3-inch-thick pieces of wood were vertically exposed to fire on one surface. Rate of char development at three constant fire exposure temperatures, 1,000°F (538°C) 1,500°F (816°C), and 1,700°F (927°C), was described by an equation with an Arrhenius temperature-dependent rate constant. When specimens were exposed to the uniformly increasing fire temperatures of ASTM E119 (earlier linear portion of time-temperature curve) (30), the rate of char development was constant, after the more rapidly developed first 1/4 inch of char. Under the standard ASTM fire exposure, temperatures 1/4 inch from the specimen surface reached 1,400°F (760°C) at 15 minutes, 1,700°F (927°C) at 1 hour, and 1,850°F (1,010°C) at 2 hours (31). When wood is exposed to these conditions, the first visual effect of thermal degradation (Figure 1) is indicated by browning of the wood at about 350°F to 400°F (175°C to 200°C). The temperature which characterized the base of the char layer was 550°F (288°C). After the first 1/4 inch of char development, the rate that this char layer moved into the solid wood--the rate of fire penetration--was about 38 millimeters per hour (1-1/2 in/hr).

Schaffer (31) found some differences in char development rate in the three species studied, Douglas-fir, southern pine, and white oak. Charring rate decreased with increase in dry specific gravity and with increase in moisture content. He also found that growth-ring orientation parallel to the exposed face resulted in higher charring rates than when orientation was perpendicular to the exposed face. In studies at the Joint Fire Research Organization in Great Britain (32) on rate of burning, increased permeability along the grain was found to increase rate of char.

Schaffer (33) found that impregnations of southern pine with certain fire-retardant and other chemicals did not significantly change the rate of charring. Boric acid, borax, ammonium sulfate, monosodium phosphate, potassium carbonate, and sodium hydroxide variously reduced the rate of charring after 20 minutes of fire exposure by about 20 percent over untreated wood. Only polyethylene glycol 1,000 reduced the rate of charring over the entire period of fire exposure by about 25 percent over untreated wood. Tetrakis (hydroxymethyl) phosphonium chloride with urea, dicyandiamide with phosphoric acid, monoammonium phosphate, zinc chloride, and sodium chloride had no effect on charring rate.

Commercial fire-retardant treatments generally do not add significantly to the fire endurance of assemblies. It is often more advantageous from the cost standpoint, either to use thicker wood members or to select species with lower charring rates, than to add the cost of the fire-retardant treatment. In some assemblies, however, it has been found worthwhile to use some fire-retardant-treated components in order to gain the extra time which will bring the fire endurance time up to the goal desired. For example, treated wood studs in walls and treated rails, stiles, and cross bands in solid wood doors have been used.

Flame Spread

In the ASTM E84 25-foot tunnel furnace test (34) for measuring flame spread of building materials, an igniting pilot flame is applied to the underside of a horizontally mounted specimen. The flame heats the combustible material to pyrolysis, and the flammable gases given off are ignited by the pilot flame. If the pyrolysis-combustion process becomes exothermic, the flaming on the specimen becomes self-propagating. A flame-spread classification or rating number is calculated from the time-distance progress of the flame along the length of the specimen surface.

The flame-spread number is derived relative to red oak (with an arbitrary flame-spread rating of 100) and to asbestos-cement board (rated zero). Natural wood products (1-inch lumber) usually have flame-spread ratings of 100 to 150 in the test furnace of Underwriters' Laboratories, Inc. (35). Some exceptions are poplar (170-185), western hemlock (60-75), redwood (70), and northern spruce (65).

Wood well treated with current commercial fire-retardant impregnation treatments will have flame-spread ratings of 25 or less. Many treated wood products have obtained a special marking or designation "FR-S" from UL (36) for having a flame-spread, fuel-contributed, and smoke-developed classification of not over 25 and no evidence of significant progressive combustion in an extended 30-minute ASTM E84 (34) test procedure. The fuel-contributed and smoke-developed classifications are also calculated relative to performance of red oak and asbestos-cement board.

Eickner and Schaffer (10) found that monoammonium phosphate (Figure 2) was the most effective of different fire-retardant chemicals in reducing the flame-spread index of Douglas-fir plywood. They used the 8-foot tunnel furnace of ASTM E286 (37). The untreated plywood had a flame-spread index of 115. This was reduced to about 55 at a chemical retention of 2 pounds per cubic foot, to 35 at 3 pounds, 20 at 4 pounds, and to about 15 at retentions of 4.5 pounds and higher. Zinc chloride was next in effectiveness but required higher retention levels to reduce the flame-spread index values equivalent to monoammonium phosphate. It required 5.5 pounds of zinc chloride to reduce flame spread to 35, and 7 pounds to reduce flame spread to 25. Ammonium sulfate and borates were as effective as zinc chloride at retentions of about 4.5 pounds per cubic foot and lower but not as effective at higher retention levels. Boric acid had some effectiveness in reducing flame spread. It was equivalent to zinc chloride, ammonium sulfate, and the borates at a retention of about 2 pounds per cubic foot, but much less effective at high retention levels. A retention of 6 pounds per cubic foot reduced the flame-spread index of the plywood to only 60.

In many laboratories, flame-spread tests of different types have consistently shown that the current acceptable treatments will

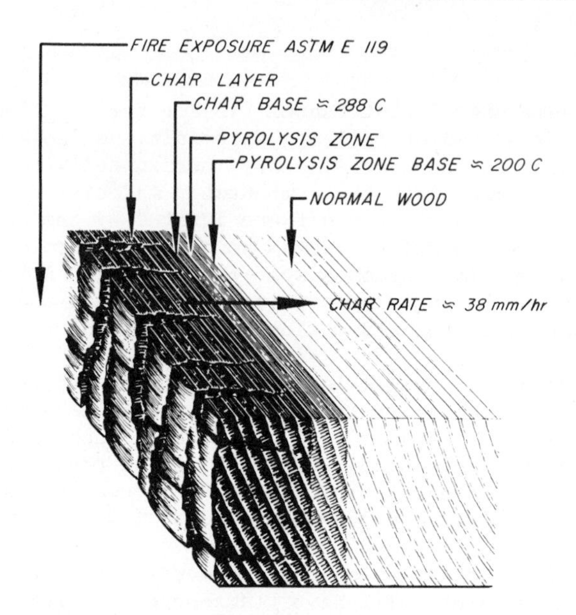

Figure 1. Fire penetration into wood and formation of char layer under the fire exposure conditions of ASTM E119

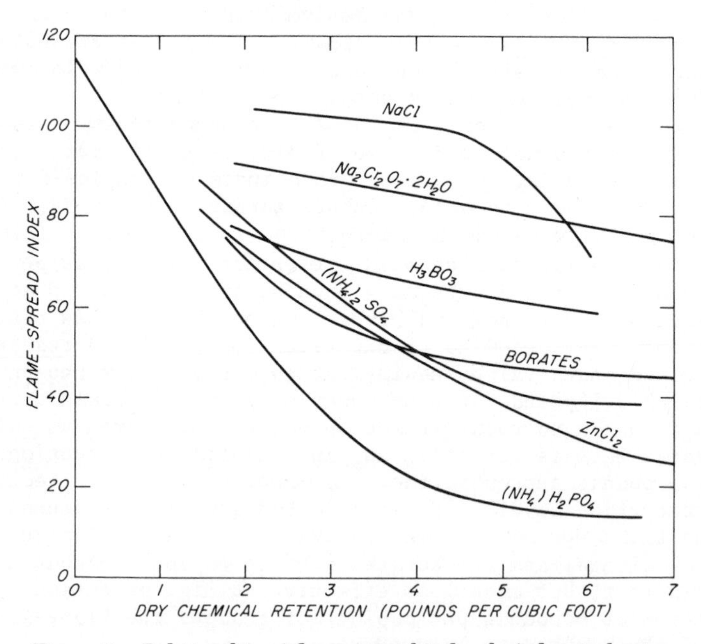

Figure 2. Relationship of flame spread to level of chemical retention in 3/8-in. Douglas-fir plywood evaluated by the 8-ft tunnel furnace method

prevent flaming combustion of the wood and prevent spread of fire over the surface. Wood, properly treated, will be self-extinguishing of both flaming and glowing once the primary source of heat and fire is removed or exhausted.

Glowing

Glowing is the visual evidence of combustion of the carbon in the char layer of the burning wood. If flaming of the released combustible gases has ceased, the glowing of the char is usually termed afterglow.

Of the several possible oxidation reactions in glowing combustion, both Browne (5) and Lyons (3) in their reviews show one possibility to be a two-stage reaction:

(1) $C + \frac{1}{2}O_2 \rightarrow CO + 26.43$ kilocalories per mole

(2) $CO + \frac{1}{2}O_2 \rightarrow CO_2 + 67.96$ kilocalories per mole

The first reaction occurs on the surface of the char and the second is a gas-phase reaction.

Wood that has been effectively treated should not exhibit any afterglowing. Reviews (3,5) covering the subject of glowing point out that the mechanism involved in glow retardance is not clear. Both physical and chemical theories have been suggested. Physical methods include the exclusion of oxygen from the carbonaceous char by formation of coatings of the fire retardant during the combustion process or by a cooling effect due to the fire retardant. The chemical theory with the most supporting evidence indicates that effective glow retardance increases the ratio of CO to CO_2. If the reaction can be directed mostly to the monoxide, step (1) above, the heat liberated is only 28 percent of that given off when the reaction continues to the dioxide. Thus glowing may be eliminated by an insufficient amount of heat to continue combustion.

Effective glow retardants for wood are the ammonium phosphates, ammonium borates, boric acid, phosphoric acid, and compounds that yield phosphoric acid during pyrolysis (3,5). Some chemicals that are reported to stimulate glowing are chromates, molybdates, halides of chromium, manganese, cobalt and copper, and ferric and stannic oxides (5,10). Chemicals found to be ineffective in retarding afterglow in a limited study were ammonium sulfate and sodium borates (10).

Combustion Products

The combustion products of burning wood--smoke and gases--are becoming of increased importance. Code and building officials, builders, producers of building materials and furnishings, and all engaged in fire research are being directed by public interest and scrutiny toward a greater concern for the real hazard to life safety of building materials in a fire situation. A study conducted by the National Fire Protection Association (18) of 311 fatal fires in one- and two-family dwellings, including mobile homes and recreational vehicles, revealed that 73.6 percent of the deaths were caused by products of combustion resulting in asphyxiation or anoxia. In a study of fire fatalities that occurred in the state of Maryland, a Johns Hopkins University group (38) found that carbon monoxide was not only the predominant factor in hindering escape from the fire scene but was also the primary agent in the cause of death in 50 percent of the 129 cases. It was also a major contributor to death in another 30 percent of the cases in combination with heart disease, alcohol in the blood, and burns.

Visibility in a burning building is extremely important. Smoke can obscure vision and exits, thereby retarding escape and resulting in panic. It also hinders the work of firefighters. The particular fraction of smoke, exclusive of any combustion gases, acts as an irritant to the respiratory system and may also result in hypoxia and collapse (39).

Smoke from untreated wood.--For research purposes, there are several methods used for measuring smoke developed by burning building materials (40). These tests generally measure the visible smoke products. One method of smoke determination being used for building code purposes is the 25-foot tunnel furnace method of ASTM E84 which yields a smoke-developed rating relative to red oak. Because the test is conducted under a strong flame, the results are not always indicative of performance in a building fire where materials may have some high-temperature exposure without the presence of flames.

During the last decade, the National Bureau of Standards (NBS) and other laboratories worked to develop a meaningful test method for measuring the smoke development potential of burning wood and other building materials. The method developed at NBS is now extensively used (41,42). This method thermally exposes a small sample in a closed chamber and supplies a specific optical density based on light transmission, light path length, burning area, and volume of enclosure. It is intended to relate to light obscuration and the hindrance in finding exits. This method has been accepted as a standard by the National Fire Protection Association (43) and is expected to be accepted by others and more widely used for rating building materials for regulatory purposes.

The chemical makeup of the combustion products, including aerosols and particulates, will change with burning conditions and the complex processes result in complex mixtures of products (28). More smoke is produced under nonflaming combustion than under flaming combustion. The complexity of the smoke is indicated by the fact that over 200 compounds have been found in the destructive distillation of wood by Goos (45).

In complete combustion, the products from burning wood are carbon dioxide, water, and ash. Other gases and vapors that may be present due to incomplete combustion include carbon monoxide, methane, formic acid, acetic acid, glyoxal, and saturated and unsaturated hydrocarbons (46). The aerosols can also contain various liquids such as levoglucosan and complex mixtures. The solids can consist of unburned carbon particles and high-molecular-weight tars.

There is no standardized test method for determining the combustion products given off from wood or other materials during a real fire situation. The gases and products obtained and their estimated hazard to life will depend on the experimental conditions of any test method selected. Most studies on the toxicity of combustion products show that the dominant hazardous gas from burning wood is carbon monoxide followed by carbon dioxide and the resulting oxygen depletion (46-50).

Considerable research is underway by various institutions and agencies and by industry on the physiological and toxicological effects of smoke and gaseous products. Of particular note are the extensive research programs at the University of Utah under the direction on I. N. Einhorn (44,48), at Johns Hopkins University under R. M. Fristrom (38,50), and at the National Bureau of Standards under M. M. Birky (51).

Smoke from treated wood.--Fire-retardant-treated wood also produces smoke and gaseous combustion products when burned. Many commercial fire-retardant-treated wood products showed greatly reduced smoke development when tested in the 25-foot tunnel furnace used for rating purposes by UL (36). This test, however, does involve flaming exposure, regardless of the flammability of the specimen. In a recent study at the FPL (52), results of tests with the NBS smoke chamber show that plywood treated with specific fire-retardant chemicals may give off more or less smoke than untreated wood depending on the chemicals employed and the conditions of burning.

Eickner and Schaffer (10) examined the effects of various individual fire-retardant chemicals on fire performance of Douglas-fir plywood (Figure 3). Using the 8-foot tunnel furnace test method (37), they found that monoammonium phosphate and zinc chloride greatly increased the smoke density index values for the plywood when treatment levels were above 2.0 pounds per cubic foot. Boric acid, at retentions above 5 pounds per cubic foot, also increased smoke development. Sodium borates and sodium dichromate considerably reduced smoke development. At low retention levels of about 1 to 3-1/2 pounds per cubic foot, ammonium sulfate was also found effective in reducing smoke. In the 8-foot furnace, effective fire retardants produced a low-flaming combustion and this condition generally resulted in more smoke development than in the flaming combustion of untreated wood.

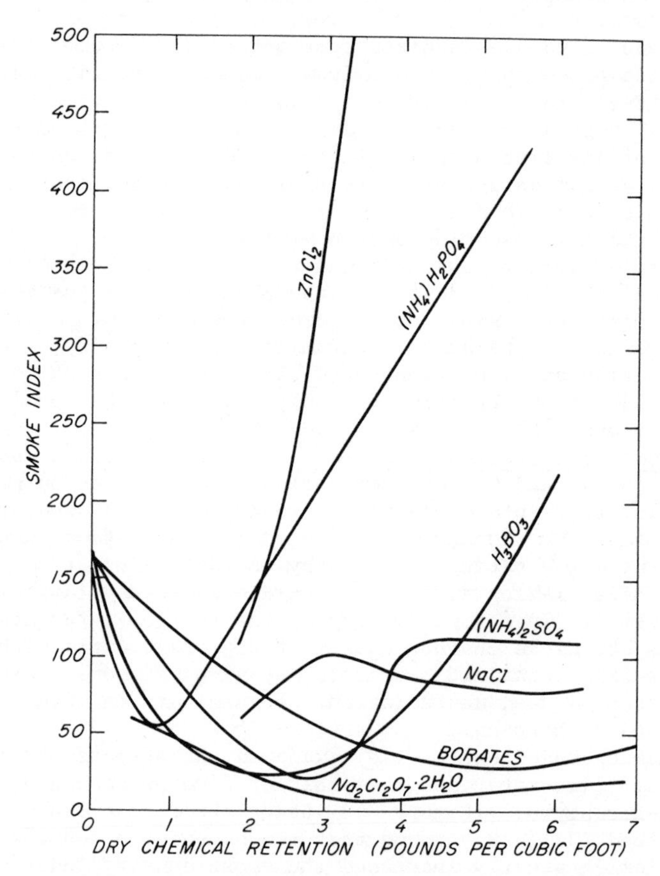

Figure 3. Relationship of smoke density to level of chemical retention in 3/8-in. Douglas-fir plywood evaluated by the 8-ft tunnel furnace method

Satonaka and Ito (53) obtained reduced smoke from fir and oak treated with either ammonium sulfate or diammonium phosphate, or with the commercially used formulations pyresote or minalith. (Pyresote consists of zinc chloride 35 percent, ammonium sulfate 35 percent, boric acid 25 percent, and sodium dichromate 5 percent. Minalith consists of ammonium sulfate 60 percent, diammonium phosphate 10 percent, boric acid 20 percent, and borax 10 percent.) They also obtained reductions in carbon monoxide and carbon dioxide levels compared to the untreated wood with each of the four treatments at the two pyrolysis temperatures employed, 400°C and 700°C.

The possibility of toxic gas formation can occasionally be predicted from the chemical composition of fire-retardant formulations. Chemicals containing chlorine may produce chlorine gas, hydrogen chloride, or other chlorinated products. Ammonia gas may also be a noxious gas from ammonia-containing compounds. The trend in recent years has been toward increased investigation and use of organic compounds as fire retardants for wood. The thermal decomposition products from wood treated with these compounds is not clearly understood, particularly in regard to their toxicological and physiological effects. Information on research in this area is lacking.

Heat Contribution

At some time after the initial exposure of wood to heat and flames in a fire situation, the burning process becomes exothermic and heat is contributed to the surroundings. The total heat of combustion of wood varies with species and is affected by resin content. It varies from about 7,000 to 9,000 Btu per pound, but not all of this potential heat is released during a fire. The degree to which the total available heat is released depends on the type of fire exposure and the completeness of combustion.

During the initial stages of a fire, fire-retardant-treated wood contributes less heat than does untreated wood, especially from the flammable volatiles (8,26). This means that the spread of fire to nearby combustibles is slow. The fire tends to be confined to the primary source. In the ASTM E84 test for building materials, treated specimens produce about 75 percent less heat than untreated red oak. In a total combustion test, however, such as the National Bureau of Standards "potential heat" method (54), both treated and untreated wood release about the same total heat.

Heat release rate is another relevant measure of the combustibility of a material along with ease of ignition and flame spread. Smith (55) points out that the release rate data, obtained under different test exposures, will be useful in predicting the performance in actual fires under different fuel loading. Release rate data can thus be used--along with other fire performance characteristics--for specifying materials and products in a particular location in an occupancy with a given fuel load rating.

The rate of heat release during the initial stages of fire exposure is considerably less, however, for treated wood than for untreated wood. Brenden (56), using the FPL rate of heat release method, obtained a maximum heat release rate of 611 Btu per square foot per minute for untreated 3/4-inch Douglas-fir plywood, with an average release rate of 308 Btu per square foot per minute for the first 10 minutes. Fire-retardant-treated Douglas-fir plywood, with 3.6 pounds per cubic foot dry chemical and a reported flame spread of 25 to 28, had a maximum heat release rate of 132 Btu per square foot per minute at 42 minutes and an average rate of 16 Btu per square foot per minute for the first 10 minutes.

Treatment-Related Properties of Fire-Retardant-Treated Wood

Strength

Gerhards (57) reviewed the results of 12 separate studies on strength properties of fire-retardant-treated wood conducted at the FPL and other laboratories. He concluded that modulus of rupture (MOR) is consistently lower and modulus of elasticity (MOE) and work to maximum load are generally lower for fire-retardant-treated wood than for untreated wood if fire-retardant treatment is followed by kiln drying. The effect may be less or negligible if the fire-retardant-treated wood is air dried instead of kiln dried. The most significant loss was in work to maximum load, a measure of shock resistance or brashness, which averaged 34 percent reduction.

The losses from treating and kiln drying for small clear specimens averaged about 13 percent for MOR and 5 percent for MOE. Losses in structural sizes were about 14 percent for MOR and 1 percent for MOE (57). Losses due to high temperature kiln drying, above 65°C may be considerably greater (58).

The National Forest Products Association recommends that the allowable stresses for fire-retardant-treated wood for design purposes be reduced by 10 percent as compared to untreated wood; the allowable loads for fasteners are also reduced 10 percent (59). The 10 percent reduction in design stresses was confirmed providing the swelling of the wood resulting from treatment is taken into account (57,60). Treated wood is slightly more hygroscopic than untreated, therefore the density of equivalent cross sections of the treated test samples was slightly lower.

Brazier and Laidlaw (58) at the Princes Risborough Laboratory have written that, until more research is done in this area, it is wise to assume a loss in bending strength of 15 to 20 percent for treated wood dried at 65°C.

In addition to strength loss due to kiln drying at high temperatures (above 65°C), progressive loss of strength in treated wood members can be caused by acidic degradation of the wood by some treatment chemicals (58). There is evidence that phosphate and sulfate salts may be broken down to acidic residues within the wood. The degradation and resultant loss in strength may continue at a slower rate under use at normal temperatures.

The fire-retardant treatment of large structural members for applications where strength is a predominant factor is usually not recommended. The adverse effects of treatment chemicals on strength and other properties such as hygroscopicity outweigh any benefit obtained by the treatment. Large wood members have good fire resistance and if treatment is required for reducing flame spread, it would be better to use fire-retardant coatings or other protection.

Hygroscopicity

Wood that has not been treated will absorb moisture from the surrounding air until its moisture content reaches an equilibrium condition. The hygroscopicity of wood treated with inorganic fire-retardant chemicals is usually greater than that of the untreated wood and is dependent on size and species of wood, temperature, relative humidity, and type and amount of chemicals used (8,60). The increase in equilibrium moisture content is negligible at 27°C and 30 to 50 percent relative humidity. A 2 to 8 percent increase in moisture content occurs in the treated wood at 27°C and 65 percent relative humidity, and at 80 percent relative humidity the moisture content may increase 5 to 15 percent and cause exudation of the chemical solution from the wood.

In an unpublished study at the U.S. Forest Products Laboratory, the moisture content of wood treated with two commercial formulations reached 48 to 58 percent (based on ovendry weight of treated sample) in 4 weeks' exposure at 27°C and 90 percent relative humidity. Continuous exposure of wood treated with water-soluble salts to conditions above 80 percent relative humidity can result in loss of chemicals and in adverse effects on dimensional stability and paint coatings. Corrosion of some metals in contact with the wood will also occur.

Zinc chloride will add considerably to the equilibrium moisture content of wood in the range of 30 to 80 percent relative humidity (8). Ammonium sulfate will add at relative humidities exceeding 65 percent, and borax and boric acid will attract water at lower humidities. Phosphate salts affect hygroscopicity mostly when relative humidity exceeds 80 percent (58).

Most commercial treatment formulations are developed for use under conditions not greater than 80 percent relative humidity. An exterior type, leach-resistant treatment that is not hygroscopic is available (61).

Gluing

Generally, the bonding obtainable with fire-retardant-treated wood is satisfactory for decorative purposes. Treated wood members can be bonded into structural assemblies with specially formulated adhesives under optimum bonding conditions (8). However, the quality of bonds is not usually equal to that obtainable with untreated wood, particularly in evaluation after exposure to cyclic wetting and drying (62).

Corrosivity

Current treatment solutions containing corrosive inorganic salts usually also contain corrosion inhibitors such as sodium dichromate or ammonium thiocyanate or are formulated to a more neutral pH (60). However, soluble-salt-treated wood in contact with metals should not be exposed to high relative humidities for prolonged periods. The treatment chemicals can attack and deteriorate metal fasteners. The corrosion products in turn deteriorate the wood. For example, under humid conditions, ammonium sulfate will attack the zinc and iron of galvanized punched-steel nail plates used in trusses (58). Alkaline and acidic areas are developed in the wood next to the attacked metal fastener, and cause degradation of the wood (58,63).

Paintability

Paintability is generally not a problem under dry normal conditions. Unusually high relative humidity conditions can affect adhesion of the paint film or cause chemical crystal blooming on the paint surface due to the increased moisture content of the wood. Natural or clear finishes are generally not used for treated wood because the chemicals may cause darkening or irregular staining.

Machining

The abrasive effect of treatments with inorganic salt crystals can reduce tool life. Where machining is necessary, this can be minimized by using tools of abrasive-resistant alloys.

Durability

Fire-retardant-treated wood is durable and stable under normal exposure conditions. Treatments using inorganic water-soluble salts, however, are not recommended for exterior exposures to rain and weathering unless the treatment can be adequately protected by water-repellent coating. Exterior-type treatments in which the chemicals are "fixed" in the wood in some manner are leach resistant and nonhygroscopic.

Current Research

The latest "Directory of Fire Research in the United States 1971 to 1973," by the National Research Council (64), shows that only a few of the listed federal, university, private, and industrial laboratories are doing research involving fire-retardant impregnation treatments for wood. Published research indicates that the current effort is in the development of leach-resistant types of fire-retardant treatments for both exterior and interior uses. Major emphasis is on reduction of flame spread as determined by ASTM E84 (34), and reduction of flaming and fire penetration as determined by ASTM E108 (65). Development is also directed toward enhanced properties of the treated wood in nonhygroscopicity, gluability, paintability, strength, and preservation against biodegradation. Some attention--but not enough--is being given to reduction of smoke and noxious gases. The current research emphasis on the toxicological and physiological effects from the combustion products of natural and synthetic polymers is expected to eventually include fire-retardant-treated wood.

FPL Research

At the U.S. Forest Products Laboratory, many of the research programs involve fire-retardant-treated wood. This has included extensive basic study of pyrolysis and combustion reactions of wood and its components and the effects of chemical additives on these reactions (15,24-26,28,29,66). A cooperative study (9) with the Division of Chemical Development of the Tennessee Valley Authority, showed the effectiveness of liquid ammonium polyphosphate fertilizers as fire retardants for wood. The commercial use of these products, made from electric furnace superphosphoric acid, has been shown to be economically feasible. Work has been completed by Schaffer (33) on the rate of fire penetration in wood treated with different types of chemicals. Some results of this study are reported elsewhere in this paper.

Studies are currently being conducted on smoke development and heat release rate from treated and untreated wood and wood products (52,56). An evaluation of the available treatment systems for wood shingles and shakes was completed using artificial weathering (11). A further development from this work was a new ASTM Standard Method D2898 (67,68) for testing durability of fire-retardant treatment of wood.

Other Institutions

Using full-scale fire test facilities of the Illinois Institute of Technology-Research Institute (IITRI), Christian and Waterman (69) studied fire and smoke behavior of interior finish materials including fire-retardant-treated wood products. The authors found that the materials performed according to a "relative hazard" position, but that the tunnel test flame-spread number does not quite place them in the proper order. They state that "attempts to distinguish between hazards of materials whose tunnel test flame-spread numbers differ by 25 or less do not seem justified." These same IITRI researchers (70) also found that in some situations a significant amount of material with a flame spread of 90 can be safely used on walls of corridors wider than 6 feet when the ceiling material has a rating of 0 to 25.

Effective fire-retardant treatments for wood for exterior uses under conditions of leaching and weathering have been needed for many years. For wood shingle or shake roofing, a commercial treatment system has been developed (61) in the United States that meets acceptance requirements of Underwriters' Laboratories, Inc. Lumber and plywood are also available with this exterior-type treatment.

The success of this treatment system indicated a breakthrough in the development of a commercially successful system whereby fire-retardant chemicals are pressure impregnated into the wood and fixed or converted to a leach-resistant state without serious impairment of the desirable natural wood properties. This development has stimulated research with leach-resistant type treatments. Chemicals employed usually involve organic phosphates and compounds that can react with phosphorous-containing chemicals or with the wood cellulose structure to give permanence of treatment.

The Eastern Forest Products Laboratory (12,71) at Ottawa, Ontario, has been active in development of leach-resistant treatments using melamine or urea with dicyandiamide, formaldehyde, and phosphoric acid. Decay resistance is also shown for a urea-based treatment (72). One stystem has met the requirements for Class C wood roofing under ASTM E108 by Underwriters' Laboratories of Canada (12,73). This treatment, or one similar, is expected to be introduced into the United States within the year as an approved exterior-type leach-resistant treatment.

McCarthy and coworkers (74) at the Australian Forest Products Laboratory reported that a pressure treatment for pine posts with zinc-copper-chromium-arsenic-phosphorus preservative produced a leach-resistant treatment having both fire retardancy and preservation against decay. This treatment system is reported to have commercial application in Australia.

Basic research on the chemistry of cellulosic fires is being studied by Shafizadeh at the University of Montana (75,76). Working with model compounds, he has shown how thermal reactions affect the cleavage of the glycosidic bond with breakdown of the sugar units through a transglycosylation mechanism which eventually results in formation of combustible tar and volatile pyrolysis products. Interference of the transglycosylation process by acidic additions, amino groups, and phosphate and halogen derivatives has been demonstrated to retard combustion by producing more water and char.

One area of continuing research at the Stanford Research Institute (SRI) is concerned with the effects of flame retardants on thermal degradation of cellulose (77,78). The results of a recent study (78) for treating wood showed that existing wood roofs can be given a self-help fire-retardant treatment equivalent to a Class C (65) rating for a 5-year period. To obtain adequate depth of penetration, the treatment is effective only on weathered shingle or shake roofs at least 5 years old. The treatment consists of a spray application of a 20 percent aqueous solution of diammonium phosphate, followed by a 20 percent aqueous solution of magnesium sulfate to form the water-insoluble magnesium ammonium phosphate.

Of particular interest is the application of the Parker-Lipska model for selecting fire retardants (77). This model of pyrolysis processes predicts the efficiency of candidate chemical fire retardants based on increased char yield and elimination of flaming. Efficient retardants will be those that have high oxygen content per molecule and contain phosphorus or boron to prevent afterglow. In addition, the studies at SRI have shown that the optimal add-on weight of a chemical retardant is about 10^{-4} mole per gram of cellulose.

Areas of Needed Research

From the viewpoint of life safety, the most urgent area of fire-retardant research is the development of treatments for wood that will reduce not only flame spread but also smoke and noxious gases. The treatments should not add or create new noxious gases. Basic research on the combustion of wood being conducted in many laboratories should be studied and carefully gleaned for clues on treatment chemicals or other means to alter the cellulose and lignin structure to reduce smoke and harmful gases. Because combustion products have been shown to be the primary cause of death in fires, research on the reduction of smoke and gases should take precedence over reduction of flame spread.

Another area of necessary research is development of treatments that will increase resistance of wood to fire penetration. The work done by Schaffer (31,33) and others in this field should be carried further. The slow rate of fire penetration in thick wood members is one of the basic assets of wood and has been accepted and utilized for many years in heavy timber construction. But thin wood members and paneling have a considerably higher fire penetration rate than thick wood members under severe fire conditions. A fire-retardant system that will give slower fire penetration means more available safety time for fire fighting personnel and for evacuation of occupants from a burning building.

Continuing basic research is needed in the pyrolysis, combustion, and fire chemistry of wood leading toward the selection of fire-retardant chemicals and their more efficient application to wood.

The high loading required (2 to 6 pounds of dry chemical per cubic foot of wood) for chemicals in present use puts a severe limitation on cost of usable treatments. A higher cost treatment could be tolerated if it proved more efficient. A large part of the cost of treated wood to the consumer is the full-cell pressure process required by present-day formulations. A less costly method of getting the chemical into the wood is needed. We need not limit the choice of chemical candidates only to those that can be used in a water-treating solution. Application with hydrocarbon solvents or liquified gases with subsequent recovery of the carrier may prove practicable.

Further research should be directed toward the development of test methods to properly evaluate fire-retardant-treated wood. Current methods have been criticized (79) for not giving a true hazard evaluation of materials on their potential performance in a real fire. The limitations of small-scale test methods should be understood as adequate only for products research and development. Even the 25-foot rating furnace of ASTM E84 (34) has been criticized regarding its correlation with full-scale fires and the meaning of the numbers it produces for flame spread and smoke density (40,69,80). There is a trend toward more full-scale fire testing with the objective of relating the results of smaller scale tests including the 25-foot furnace to performance in real fires. Full-scale tests are too expensive, of course, to prove out building products on a routine basis. Perhaps the corner-wall test (80-82) is adequately realistic and could be used in conjunction with small-scale tests for determining product performance and fire hazard.

As new criteria are developed for defining combustibility, a method is needed to realistically indicate heat release rate of wood products exposed to building fires instead of dependence on "total heat values."

Continuing research must yield information on the treatment-related properties of fire-retardant-treated wood and methods for their improvement. The properties of the conventional salt treatments which need improvement especially are hygroscopicity, strength properties, gluing, and finishing.

Literature Cited

1. National Commission on Fire Prevention and Control
"America burning: The report of the national commission on fire prevention and control." 53 p. Washington, DC. 1973.
2. American Wood Preservers' Association
"Wood preservation statistics 1974." Compiled by Ernst & Ernst, In Am. Wood Preserv. Assoc. Proc. 71:225-263. AWPA, Washington, DC. 1975.
3. Lyons, J. W.
"The chemistry and uses of fire retardants." Wiley-Interscience Div., John Wiley and Sons, New York. 1970.
4. Goldstein, I. S.
Degradation and protection of wood from thermal attack. In "Wood Deterioration and Its Prevention by Preservative Treatment. Vol 1. Degradation and Protection of Wood," p. 307-339. D. D. Nicholas, ed. Syracuse Univ. Press. Syracuse, N.Y. 1973.
5. Browne, F. L.
U.S.D.A. For. Prod. Lab. Rep. No. 2136, Madison, Wis. 1958.
6. Juneja, S. C.
Synergism and fire retardance of wood and other cellulosic materials. In "Advances in Fire Retardants, Part 2." V. M. Bhatnagar, ed. Technomic, Westport, Conn. 1973.
7. Juneja, S. C.
Wood Sci. 7(3):201-208. 1975.
8. Eickner, H. W.
J. Mater. 1(3):625-644. 1966.
9. Eickner, H. W., J. M. Stinson, and J. E. Jordan
Am. Wood Preserv. Assoc. Proc. 65:260-271. 1969.
10. Eickner, H. W., and E. L. Schaffer
Fire Tech. 3(2):90-104. 1967.
11. Holmes, C. A.
Evaluation of fire-retardant treatments for wood shingles, Ch. 2. In "Advances in Fire Retardants, Part 1." V. M. Bhatnagar, ed. Technomic, Westport, Conn. 1971.
12. Juneja, S. C., and L. R. Richardson
For. Prod. J. 24(5):19-23. 1974.
13. Kanury, A. M.
Fire Res. Abstr. and Rev. 14(1):24-52. 1972.
14. Matson, A. F., R. E. Dufour, and J. F. Breen
Bull. of Res. No. 51, p. 269-295. Underwriters' Lab. Inc., Northbrook, Ill. 1959.
15. Beall, F. C., and H. W. Eickner
U.S.D.A. For. Serv. Res. Pap. FPL 130, For. Prod. Lab., Madison, Wis. 1970.
16. U.S. Forest Products Laboratory.
U.S.D.A. For. Prod. Lab. Rep. No. 1464 (rev.), Madison, Wis. 1958.

17. McGuire, J. H.
Fire Technol. 5(3):237-241. 1969.
18. National Fire Protection Association
"Fire protection handbook." 13th ed. pp. 1-7, 5-7. Natl. Fire Prot. Assoc. Boston, Mass. 1969.
19. Broido, A.
Chem. Tech. 3(1):14-17. 1973.
20. Prince, R. E.
"Tests on the inflammability of untreated wood and of wood treated with fire-retarding compounds". In Publ. Proc. of Annu. Meet. Natl. Fire Prot. Assoc. Boston, Mass. 1915.
21. Shafizadeh, F.
Pryolysis and combustion of cellulosic materials. In "Advances in Carbohydrate Chemistry." Vol. 23, pp. 419-474. M. L. Wolfrom, and R. S. Tipson, eds. Academic press, New York. 1968.
22. MacLean, J. D.
Am. Wood Preserv. Assoc. Proc. 47:155-168. 1951.
23. Stamm, A. J.
Ind. and Eng. Chem. 48(3):413-417. 1956.
24. Beall, F. C.
Wood Sci. 5(2):102-108. 1972.
25. Brenden, J. J.
U.S.D.A. For. Serv. Res. Pap. FPL 80, For. Prod. Lab., Madison, Wis. 1967.
26. Browne, F. L., and J. J. Brenden
U.S. For. Serv. Res. Pap. FPL 19, For. Prod. Lab., Madison, Wis. 1964.
27. Shafizadeh, F., P. Chin, and W. DeGroot
J. Fire & Flammability/Fire Retardant Chem. 2:195-203. Aug. 1975.
28. Tang, W. K.
U.S. For. Serv. Res. Pap. FPL 71, For. Prod. Lab., Madison, Wis. 1967.
29. Tang, W. K., and H. W. Eickner
U.S. For. Serv. Res. Pap. FPL 82, For. Prod. Lab., Madison, Wis. 1968.
30. American Society for Testing and Materials
ASTM Design. E119-73. Philadelphia, Pa. 1973.
31. Schaffer, E. L.
U.S. For. Serv. Res. Pap. FPL 69, For. Prod. Lab., Madison, Wis. 1967.
32. Great Britain Ministry of Technology and Fire Offices' Committee, Joint Fire Research Organization
"Fire research 1964." Rep. of the Fire Res. Board with the Rep. of the Dir. of Fire Res. p. 12. London. 1965.
33. Schaffer, E. L.
J. Fire & Flammability, Fire Retardant Chem. Suppl. 1:96-109. April 1974.

34. American Society for Testing and Materials
ASTM Desig. E84-75. Philadelphia, Pa. 1975.
35. Underwriters' Laboratories, Inc.
"Wood--fire hazard classification, card data service." Serial No. UL527. Underwriters' Laboratories, Inc., Northbrook, Ill. 1971.
36. Underwriters' Laboratories, Inc.
"Building materials directory Part I." Build. Mater. List. Underwriters' Laboratories, Inc., Northbrook, Ill. Jan. 1976.
37. American Society for Testing and Materials
ASTM Design. E286-69 (reapproved 1975). Philadelphia, Pa. 1969.
38. Halpin, B. M., E. P. Radford, R. Fisher, and Y. Caplan
Fire J. 69(3):11-13, and 98-99. 1975.
39. Thomas, D. M.
Fire Command 38(4):23-27. 1971.
40. Yuill, C. H., et. al.
Task Group of Subcomm. IV of ASTM Comm. E-5 on Fire Tests of Mater. and Constr., Mater. Res. and Stand., MTRSA 11(4):16-23, 42. 1971.
41. Brenden, J. J.
For. Prod. J. 21(12):22-28. 1971.
42. Lee, T. G.
"The smoke density chamber method for evaluating the potential smoke generation of building materials." U.S. Dep. Commer., Natl. Bur. Stand., Tech. Note No. 757. 1973.
43. National Fire Protection Association
Natl. Fire Prot. Assoc. NFPA No. 258. Boston, Mass. 1976.
44. Einhorn, I. N., D. A. Chatfield, J. H. Futrell, R. W. Mickelson, K. J. Voorhees, F. D. Hileman, and P. W. Ryan
"Methodology for the analysis of combustion products." UTEC 75-073, FRC/UU49. Univ. of Utah, Salt Lake City, Utah. 1975.
45. Goos, A. W.
The thermal decomposition of wood, Ch. 20. In "Wood Chemistry." L.E. Wise and E. C. Jahn, eds. 2nd ed. Reinhold, New York. 1952.
46. Wagner, J. P.
Fire Res. Abstr. and Rev. 14(1):1-23. 1972.
47. Birky, M. M.
Polymer Prepr. 14(2):1011-1015. 1973.
48. Einhorn, I. N., M. M. Birky, M. L. Grunnet, S. C. Packham, J. H. Petajan, and J. D. Seader
"The physiological and toxicological aspects of smoke produced during the combustion of polymeric materials." Annu. rep. 1973-1974. UTEC-MSE 74-060, FRC/UU26. Univ. of Utah, Salt Lake City, Utah. 1974.

49. O'mara, M. M.
J. Fire & Flammability 5(1):34-53. 1974.
50. Robison, M. M., P. E. Wagner, R. M. Fristrom, and A. G. Schulz
Fire Technol. 8(4):278-290. 1972.
51. Birky, M. M.
"Review of smoke and toxic gas hazards in fire environment." Int. Symp. Fire Safety of Combust. Mater., Univ. of Edinburgh, Scotland. Oct., 1975.
52. Brenden, J. J.
U.S.D.A. For. Serv. Res. Pap. FPL 249, For. Prod. Lab., Madison, Wis. 1975.
53. Satonaka, S., and K. Ito
J. Jap. Wood Res. Soc. 21(11):611-617. 1975.
54. Loftus, J. J., D. Gross, and A. F. Robertson
"Potential heat of materials in building fires." U.S. Dep. Commer., Natl. Bur. Stand., Tech. News Bull. 1962.
55. Smith, E. E.
Fire Technol. 12(1):49-54. 1976.
56. Brenden, J. J.
U.S.D.A. For. Serv. Res. Pap. FPL 230, For. Prod. Lab., Madison, Wis. 1975.
57. Gerhards, C. C.
U.S.D.A. For. Serv. Res. Pap. FPL 145, For. Prod. Lab., Madison, Wis. 1970.
58. Brazier, J. D., and R. A. Laidlaw
"The implications of using inorganic salt flame-retardant treatments with timber." BRE Information IS 13/74. In News of Timber Res. Princes Risborough Lab., Aylesbury, Buckinghamshire. Dec. 1974.
59. National Forest Products Association
"National design specifications for stress-grade lumber and its fastenings." Natl. For. Prod. Assoc., Washington, DC. 1973.
60. U.S. Forest Products Laboratory.
"Wood handbook: Wood as an engineering material." pp. 15-10, 15-11. U.S.D.A. Agric. Handb. No. 72, (rev)., 1974.
61. Shunk, B. H.
For. Prod. J. 22(2):12-15. 1972.
62. Schaeffer, R. E.
U.S.D. A. For. Serv. Res. Note FPL-0160, For. Prod. Lab., Madison, Wis. 1968.
63. Baker, A. J.
U.S.D.A. For. Serv. Res. Pap. FPL 229, For. Prod. Lab., Madison, Wis. 1974.
64. National Research Council
"Directory of fire research in the United States 1971-1973." Comm. on Fire Res., Div. Engr., Natl. Res. Coun., Natl. Acad. Sci., Washington, DC. 1975.

65. American Society for Testing and Materials
ASTM Desig. E108-75. Philadelphia, Pa. 1975.
66. Browne, F. L., and W. H. Tang
U.S. For. Serv. Res. Pap. FPL 6, For. Prod. Lab., Madison, Wis. 1963.
67. American Society for Testing and Materials
ASTM Desig. D2898-72. Philadelphia, Pa. 1972.
68. Holmes, C. A.
U.S.D.A. For. Serv. Res. Pap. FPL 194, For. Prod. Lab. Madison, Wis. 1973.
69. Christian, W. J., and T. E. Waterman
Fire. Tech. 6(3):165-178, 188. 1970.
70. Christian, W. J., and T. E. Waterman
Fire J. 65(4):25-32. 1971.
71. King, F. W., and S. C. Juneja
For. Prod. J. 24(2):18-23. 1974.
72. Juneja, S. C., and J. K.Shields
For. Prod. J. 23(5):47-49. 1973.
73. Fung, D.P.C., E. E. Doyle, and S. C. Juneja
Inf. Rep. OP-X-68, Dep. of the Environ., Canadian For. Serv., Eastern For. Prod. Lab., Ottawa, Ont. 1973.
74. McCarthy, D. F., W. G. Seaman, E.W.B. DaCosta, and L.D. Bezemer
J. Inst. Wood Sci. 6(1):24-31. 1972.
75. Shafizadeh, F.
Appl. Polymer Symp. 28, Proc. of the Eighth Cellulosic Conf., Part I, pp. 153-174. T. E. Timell, ed. John Wiley and Sons, N.Y. 1975.
76. Shafizadeh, F.
J. of Appl. Polymer Sci. 12(1):139-152. 1976.
77. Amaro, A. J., and A. E. Lipska
"Development and evaluation of practical self-help fire retardants. Annual Report." SRI Proj. No. PYU-8150. Stanford Res. Inst., Menlo Park, Calif. 1973.
78. Lipska, A. E., and A. J. Amaro
"Development and evaluation of practical and self-help fire retardants. Final Report." SRI Proj. No. PYU-8150. U.S. Dep. Commer., NTIS No. AD-A014 492. Stanford Res. Inst., Menlo Park, Calif. 1975.
79. Yuill, C. H.
Am. Soc. Test. Mater. Stand. News, STDNA 1(6):26-28, 47. 1973.
80. Williamson, R. B., and F. M. Baron
J. Fire & Flammability 4(2):99-105. 1973.
81. Underwriters' Laboratories, Inc.
"Flammability studies of cellular plastics and other building materials used in interior finishes." Subj. 723. Underwriters' Laboratories, Northbrook, Ill. 1975.

82. U.S. Forest Products Laboratory.
U.S.D.A. For. Serv. Res. Note FPL-0167. For. Prod. Lab., Madison, Wis. 1967.

7

Properties of Wood during Carbonization under Fire Conditions

F. C. BEALL

Faculty of Forestry and Landscape Architecture, University of Toronto, 203 College St., Toronto, M5S 1A1 Canada

Most of the knowledge concerning the behavior of wood under high temperature conditions is limited to such externally-measured quantities as mass loss or gas generation. Two general approaches have been used to predict the pattern of wood deterioration during fire exposure. The more elegant approach is through the differential form of the unidirectional heat-flow equation. Thermal diffusivity, which is the equation "constant", has a very complex behavior with temperature, precluding a formal solution to the equation. However, numerical solutions, such as can be developed from finite-element techniques, permit a means of evaluating this "constant" for small temperature increments (1). No serious effort has been made in past research to fully evaluate thermal diffusivity of wood as it changes with temperature. Such information is elementary for designing methods of fire retardation, which in the past, have been almost exclusively trial and error. It is also evident that the physical changes in wood during fire exposure should be examined in an effort to modify thermal diffusivity. A second approach involves the evaluation or creation of empirical equations based on experimental data. The most widely-used method relates the rate of charring to the change of wood density (2). However, just as the density factor in thermal diffusivity is relatively undefined, so is the density change from wood to char. Fragmentary information has been published on the density change (3, 4), but no systematic study has been done.

The purpose of this study was to clarify the change of density with species, heating rate, and temperature under oxygen-deficient conditions. The major constraint was an arbitrary specimen size (10-mm cube), based on the observation that thicknesses equal to or greater than about 6 mm (approximately the half-thickness of the cubes) produce consistent charring rates (5). The effect of thickness on density changes is currently being studied.

Experimental Procedure

Sample Preparation. The six wood species (Table I) were

Table I. Mass, density, and shrinkage relationships at 600°C among wood species heated at 1°C/min.

Species	$\frac{m_c}{m_o}$[a]	ρ_o (g/cm^3)	ρ_c (g/cm^3)	$\Delta L/L_o$	$\Delta T/T_o$	$\Delta T/\Delta R$		$\Delta V/V_o$	
						char	moisture[b]	char	moisture[b]
White oak	0.31	0.80	0.58	0.175	0.326	1.44	1.87	0.57	0.16
Hard Maple	0.25	0.71	0.46	0.176	0.361	1.38	2.06	0.61	0.15
Southern pine	0.25	0.55	0.35	0.188	0.323	1.16	1.54	0.64	0.12
Douglas fir	0.24	0.45	0.26	0.176	0.327	1.33	1.58	0.58	0.12
Basswood	0.22	0.39	0.24	0.184	0.367	1.17	1.41	0.65	0.16
Redwood	0.29	0.37	0.23	0.180	0.262	1.25	1.69	0.52	0.07

[a] Subscript c = char; o = ovendry control value

[b] Wood Handbook (6).

selected for a wide range of density, structure, and shrinkage (from moisture loss) behavior. All specimens were selected from flat-sawn, kiln-dried boards which were free from visible defects. Several lengths of nominal 10 by 10 mm stock from each species with good radial and tangential faces provided sufficient specimen material, either end-matched or containing virtually the same growth rings. After the stock was crosscut, final dimensions of the 10 mm cubes were obtained by sanding. All specimens were ovendried and stored over desiccant until measured.

Sample Measurements. Mass was determined using an analytical balance of 10^{-5}g sensitivity. Prior research on carbonized wood has shown that its spongy and/or fragile nature could cause errors in mechanical measurement of dimensions. Therefore, dimensions were obtained using a camera mounted on an incident light microscope to photograph specimens with a calibrated grid (micrometer eyepiece) resting on each of the faces. The developed negatives were mounted in slides and projected to directly measure each face. Two widths were determined at uniform spacing in each axis, for a total of 4 measurements per face, or 24 per cube. The paired measurements were later averaged. After carbonization, the cube faces were photographed in an identical sequence to permit shrinkage analysis of individual faces. Specimen exposure to atmospheric conditions was minimized.

Sample Runs. A block diagram of the system is shown in Figure 1. Three specimens were placed in a nickel boat and positioned in the Vycor furnace tube. The system was initially flushed with nitrogen at 0.8 ℓt/min to remove oxygen, and reduced to 0.2 ℓt/min during the run. Heating rate (1, 10, 50°C/min) was preset on the temperature programmer which maintained a linear rate using a platinum resistance sensing element directly below the furnace tube and having a proportionating output voltage to the furnace. The sample temperature was monitored with a CR/AL thermocouple adjacent to the face of one sample. Thermocouple EMF was fed through an electronic reference junction to a strip-chart recorder with a calibrated span. When the desired final temperature was reached, the heating was stopped and the split furnace element opened to expedite cooling.

Results and Discussion

A typical mass loss curve for the three heating rates is shown in Figure 2. The fractional residual (char) mass at 600°C is given for each species in Table I. Considerable data are available from the literature, particularly from thermogravimetry studies, on the influence of specimen and heating parameters on mass loss characteristics.

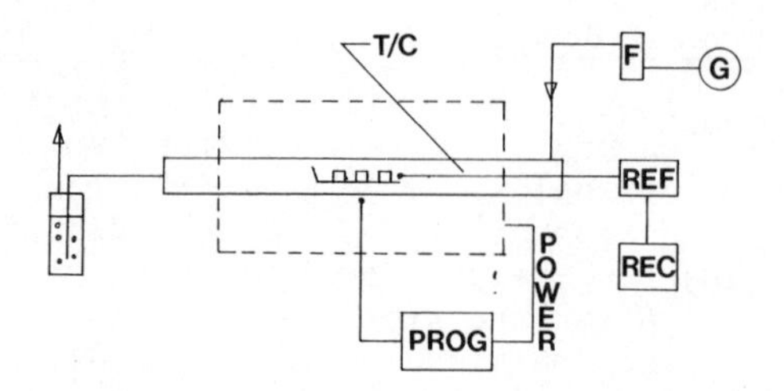

Figure 1. Block diagram of heating system. G = gas; F = flowmeter; T/C = thermocouple; REF = reference junction; REC = recorder. Dashed line indicates boundary of furnace.

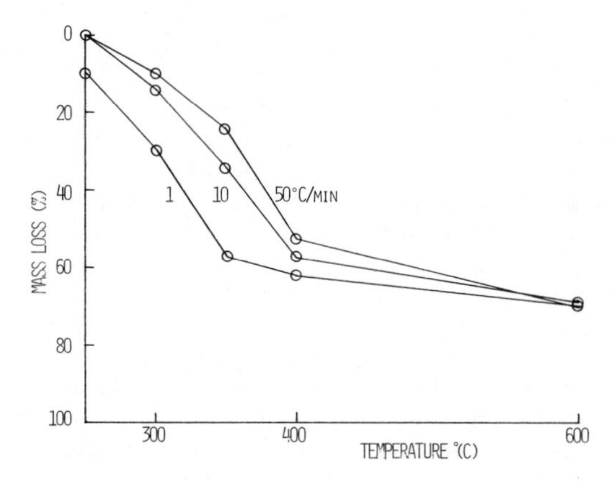

Figure 2. Mass loss of redwood at three heating rates to end temperatures of 250°, 300°, 350°, 400°, and 600°C

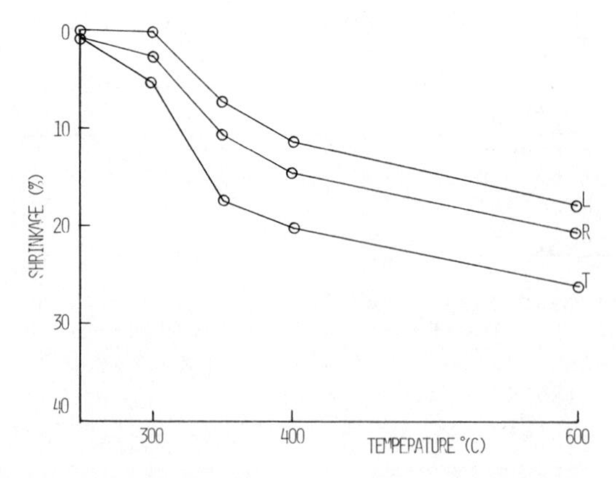

Figure 3. Longitudinal, radial, and tangential shrinkage of redwood heated at 1°C/min

General Shrinkage Behavior. Shrinkage in the major axes is shown in Figure 3 for a typical species heated at 1°C/min. Certain features were common for all species: similar shrinkage rates in all axes starting at about 350°C, greater tangential than radial shrinkage at all temperatures, a lag in onset and lowest value for longitudinal shrinkage, and practically identical longitudinal shrinkage behavior and values for all species (Table I). The T/R shrinkage ratios from carbonization show a similarity to those reported for moisture-loss shrinkage (6), however, measurements must be made on controls before the relationship can be established. In general, there is a tendency toward isotropism in transverse shrinkage (T, R), particularly for basswood and southern pine. Redwood, because of its relatively low tangential shrinkage, behaves more isotropically than the others. The analogy between char and moisture is less clear for volumetric shrinkage, although a trend is obvious. As with the T/R analogy, measurements must be made on controls to clarify any relationship.

Density Changes. The variation of density with temperature was the major relationship sought in this study. From the char density values obtained at 600°C, it was possible to establish the following regression equations:

(1) $\rho_c = -0.078 + 0.79\ \rho_o$ $(\dot{T} = 1\ ^\circ C/min)$ $r^2 = 0.98$

(2) $\rho_c = -0.049 + 0.71\ \rho_o$ $(\dot{T} = 10\ ^\circ C/min)$ $r^2 = 0.99$

(3) $\rho_c = -0.006 + 0.56\ \rho_o$ $(\dot{T} = 50\ ^\circ C/min)$ $r^2 = 0.97$

All of these relationships apply to conditions of an oxygen-deficient atmosphere and rapid escape of volatiles. The more complex dependence of density on both temperature and heating rate is shown in Figure 4. Higher rates of heating delay the density change until about 400°C, where the curves cross and show a direct dependence between density change and heating rate.

Longitudinal Shrinkage. Despite the possible analogies between carbonization and moisture loss for transverse shrinkage of wood, longitudinal shrinkage at 600°C did not vary significantly among species. The mean value of longitudinal shrinkage for the six species was 18.0% with a standard deviation of 0.5%. Calculations by Bacon and Tang (7) show that the reduction in length of cellibiose, if it were transformed into graphite, would be 17.3%. This close agreement supports the concept of *in situ* cellulose losing oxygen and forming a graphitic-type structure during contraction of the chains. Additionally, the three heating rates produced different shrinkage paths, but the same (statistically) endpoint at 600°C (Figure 5). A further difference is obvious between longitudinal and transverse shrinkage when these are plotted against mass loss (Figure 6) instead of temperature (Figure 3). The lag in longitudinal

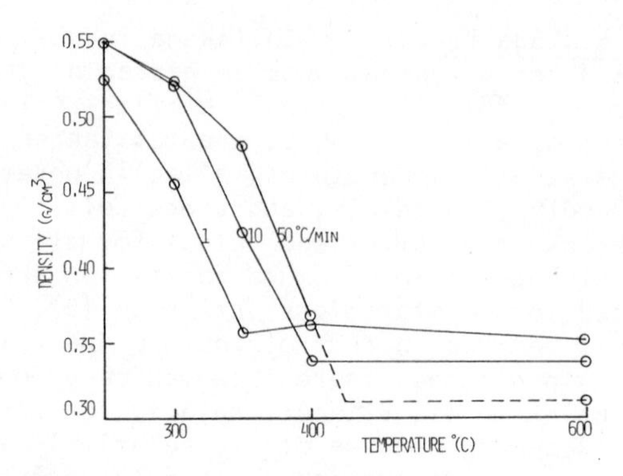

Figure 4. Density of southern pine as affected by heating rate and temperature. Dashed line is the probable path between points at 400° and 600°C.

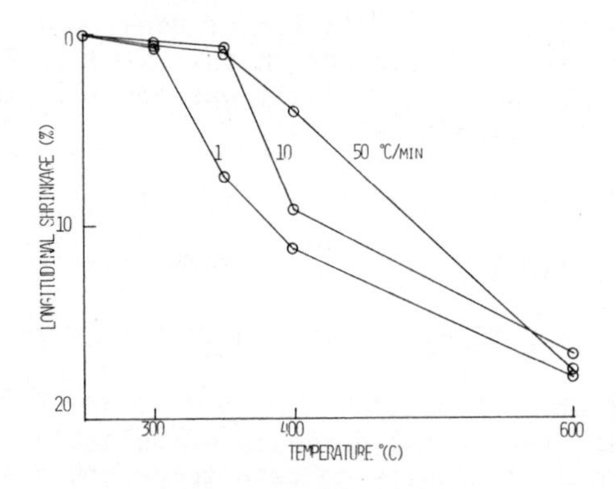

Figure 5. Effect of heating rate on longitudinal shrinkage of redwood

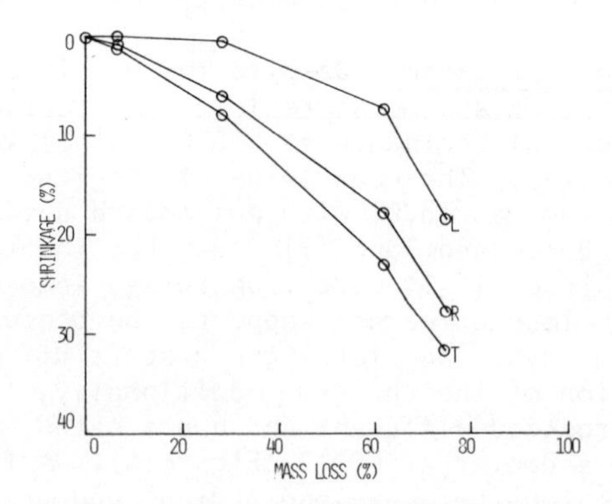

Figure 6. The relationship of shrinkage in the major axes to mass loss of southern pine

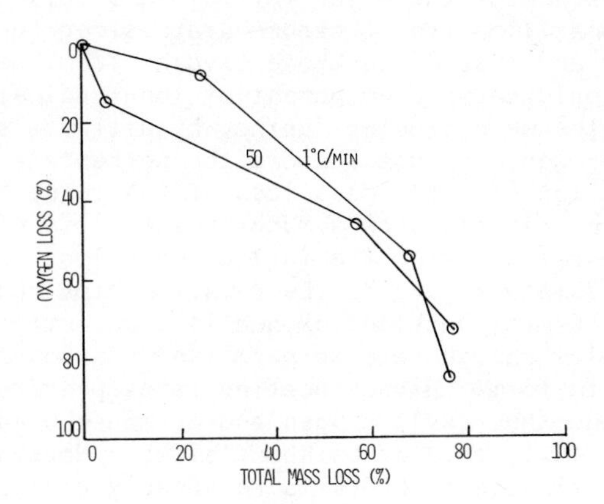

Figure 7. Effect of heating rate on loss of oxygen from Douglas fir

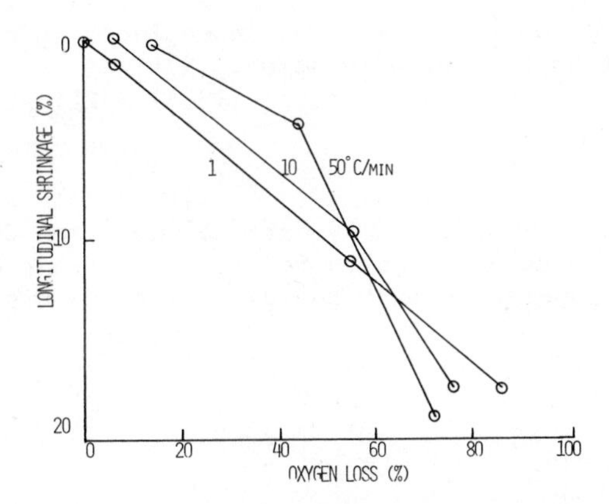

Figure 8. Effect of heating rate on the relationship between longitudinal shrinkage and loss of oxygen

shrinkage is much more pronounced, particularly at 300°C, which strongly suggests that the major initial mass loss (to about 30%) occurs from volatilization of carbohydrate sidegroups. Shrinkage from lignin or any loss of backbone oxygen from the polysaccharides should cause a component of longitudinal shrinkage. The relationships were studied further by ultimate analysis for oxygen and hydrogen. Figure 7 shows the percentage of oxygen loss as a function of total mass loss at the three rates of heating. The curves clearly show a much greater loss of oxygen at higher heating rates during the initial mass loss. By combining Figures 6 and 7 into Figure 8, the relationships appear much clearer. At 1°C/min, backbone oxygen is apparently lost at a sufficiently slow enough rate to permit new carbon-to-carbon valence bonds to form. Higher heating rates preferentially remove sidegroup (hydroxyl) oxygen and/or cause a lag in C-C bonding at the newly-created backbone sites. However, the longitudinal shrinkage at 600°C is reasonably constant for all species and heating rates.

Literature Cited

1. Knudson, R.M. and A.P. Schniewind. For. Prod. J. (1975) 25 (2):23-32.
2. Schaffer, E.L. U.S. For. Serv. Res. Note FPL-0145. 1966.
3. Beall, F.C. P.R. Blankenhorn, and G.R. Moore. Wood Science (1975) 6 (3):212-219.
4. McGinnes, E.A., S.A. Kandeel, and P.S. Szopa. Wood and Fiber (1971) 3 (2):77-83.
5. Akita, K. Report Fire Res. Inst. Japan 9 (1, 2). 1959.
6. Anon. "Wood Handbook - wood as an engineering material". Govt. Printing Office. Washington. 1974.
7. Bacon, R. and M.M. Tang. Carbon (1964) 2:221-225.

This research was funded by National Research Council Canada and the Canadian Forestry Service. Portions of the study were in cooperation with the University of Missouri.

8

Dimensional Changes of Wood and Their Control

ALFRED J. STAMM

School of Forest Resources, Department of Wood and Paper Science,
North Carolina State University, P.O. Box 5516, Raleigh, N.C. 27607

Wood, like all other plant materials, is laid down from aqueous solution. The cellulose, hemicellulose, and lignin polymers formed are no longer soluble in water, but water still dissolves in them to form solid solutions on the polar hydroxyl groups. Water is held within the cell wall structure by hydrogen bonding (1, 2). Sorption is of the polymolecular sigmoid type (1, pg 146). Each adsorption site, consisting primarily of hydroxyl groups, can take up 5 to 7 molecules of water (1, pg 162) as shown from adsorption isotherms using the equation of Brunauer, Emmett, and Teller (3). The energy of adsorption decreases rapidly as molecules beyond monomolecular are taken up (1, pg. 208). The final molecule adsorbed on any particular site is taken up by a force just exceeding that required to open up the structure to accommodate it.

The take up of water or other liquids within the cell walls of wood involve the take up of a molecule at a time and its movement from one adsorption site to another (molecular jump phenomenon) under a concentration gradient. This is distinct from flow of bulk liquids into the coarse capillary structure under a capillary force or pressure gradient.

Fundamentals of Shrinking and Swelling

Adsorbed water virtually adds its volume to that of the dry cell walls of wood, causing swelling (1, Chapter 13). This is the case because of the fact that the dry cell walls are virtually free of voids that could fill with water without swelling occurring. The volumetric swelling of wood substance, with a specific volume of 0.685 (1, Chapter 3) and a fiber saturation point of 30%, ignoring any adsorption compression, is 43.7%. If wood swelled like a stress strainless gel, the voids in wood would increase in volume the same amount as the wood substance (1, Chapter 13). Fortunately, natural voids, such as the lumen of fibers and the vessels of hardwood change only slightly in volume with changes in volume of the wood substance because of the internal restraining action by fibril wrappings.

External dimensional shrinking and swelling of wood is roughly proportional to the specific gravity of the wood (1, Chapter 13). Swelling of wood can be forced to be almost entirely internal by applying strong external restraint (1, pg. 233).

Wood is anisotropic, that is it swells and shrinks differently in the three structural directions. Shrinking in the fiber direction is usually only 0.1 to 0.3%, varying with the slope of the fibril wrappings of the S_2 layer of the cell walls and with the slope of the grain (4, 2, pg 100). The tangential shrinkage is usually 1.5 to 2.5 times that in the radial direction. Tangential shrinkage of commercial woods grown in the United States ranges from 7 to 11% and radial shrinkage from 3 to 7% (5). Greater tangential than radial shrinking has been explained on the basis of ray cell restraint; greater tangential shrinkage of the denser latewood than of the earlywood forcing increased earlywood shrinkage in the tangential direction; greater slope of fibrils on radial faces of fibers than on the tangential faces due to concentration of pits on the radial faces; and greater amounts of middle lamella in the radial walls than in the tangential walls (2, pg. 106-118). All of these effects may be operative to various extents. Further, the ratio of tangential to radial swelling increases with an increase in moisture content above about 20% (6, 7). Swelling in aqueous solutions beyond the swelling in water is almost entirely in the tangential direction (1, pg. 251).

Dimensional Stabilization

There are five known methods by which the shrinking and swelling of wood can be materially reduced in rate or in final magnitude. They are: 1. applying mechanical restraint by cross-laminating, 2. applying external or internal water resistant coatings, 3. reducing the hygroscopicity of the wood components, 4. chemically cross-linking the structural components of the wood, 5. bulking the cell walls of wood with chemicals.

Cross-laminating. Wood because of its anisotropic nature, swells thirty to one hundred times as much transversely as longitudinally (1, Chapter 13, 2). When veneer is made up into plywood, the lateral external swelling of each ply is mechanically restrained from being its normal amount due to the much smaller longitudinal swelling of the adjacent plies. Swelling of plywood in the two sheet directions is only slightly greater than the longitudinal swelling of the unassembled plies. The mechanical restraint reduces the hygroscopicity by several percent (8) but cannot alone account for the large reduction in external swelling. The chief effects are a relief of the stresses by an increased swelling in the thickness direction of the sheets and an internal swelling into the lumen of the fibers

(1, pg 233).

This simple method for obtaining dimensional stability of plywood, in the important sheet directions, has the shortcoming that it promotes face checking. Plywood is known to face check as a result of the restraining stresses set up under alternate swelling and shrinking considerably more than in normal wood or in parallel laminates (9). It will be shown later that this face checking can be greatly reduced by subjecting the face plys of plywood to a fiber bulking treatment before assembly.

External Coatings. Applying water resistant coatings or finishes to wood will appreciably reduce the rate of adsorption of liquid water or adsorption of water vapor and thus reduce the rate of swelling and face checking, but has only a minor effect upon equilibrium swelling. The effectiveness of coatings varies with the nature of the coating and the exposure conditions. Unfortunately all known coatings that adhere to wood are somewhat permeable to water. Applying aluminum foil to all surfaces of small wood panels, with curved edges and corners, between coats of varnish or oil base paints gave moisture excluding efficiencies of 99% (weight gain of the uncoated control minus the weight gain of the coated specimen divided by that of the control when exposed to a relative humidity of 97% for one week) (10, 11). Aluminum powder dispersed in varnish or oil base paint gave values ranging from 90 to 95%. Two coats of pigmented oil base paint over a primer gave values ranging from 60 to 90%. These measurements were made prior to the advent of water bomb emulsion paints so they were not included in the study. They undoubtedly would have given still lower values. Two coats of varnishes, enamels, or cellulose nitrate laquers gave values ranging from 50 to 85%. Five coats of linseed oil followed by two coats of wax gave values of only about 8%.

The moisture excluding efficiency of coatings decreases rapidly with time, relative humidity cycling, and weathering exposure. When cyclic or weathering tests are extended for periods of a year or more moisture exclusion is practically eliminated.

Internal Coatings. Impregnating wood with water-resisting materials dissolved in a volatile solvent has the advantage of not being weathered away or degraded by ultraviolet light. Experience has shown, however, that less perfect coatings are obtained in this way. Internal coating with water repellents (natural resins,waxes or drying oils dissolved in volatile hydrocarbon solvents containing a toxic agent such as pentachlorophenol) are used to some extent to give temporary protection to millwork, especially against adsorption of liquid water (12, 13). They are usually applied to dry millwork by a simple three minute dip technique. Penetration is chiefly confined to end penetration. This superficial treatment imparts some decay resistance to wood and reduces face checking and grain raising, but has little or no effect on alternate seasonal shrinking and swelling.

Reduction in Hygroscopicity. Obviously any treatment or chemical change in wood that reduces its affinity for water will reduce its tendency to swell. Replacing polar hydroxyl groups with less polar groups should accomplish this. An ideal case would be to replace all hydroxyl groups accessable to water by hydrogen. Unfortunately all known hydrogenation procedures break down both cellulose and lignin (14, Chapter 17). Wood can, however, be acetylated without chemical break down of the structure. This would be expected to reduce the hygroscopicity and swelling and shrinking to about half of normal. It actually caused a greater reduction due to bulking of the fibers. Acetylation will hence be considered under bulking. The only presently known dimension stabilizing method for wood that results from a loss in hygroscopicity alone is heat stabilization.

Heat Stabilization. When wood is heated, preferably in the absence of oxygen, under temperature-time conditions that cause some loss of water of constitution and other minor breakdown products, swelling and shrinking are appreciably reduced (1, pg. 304). Figure 1 is a plot of the logarithm of heating time against heating temperature for three different softwood species having different thicknesses that were heated beneath the surface of a low fusion Woods metal (15) to minimize oxidation and cause rapid heat transfer. Linear plots result for the three different reductions in shrinking and swelling. Reductions of 40% are obtained by heating at 315°C for one minute, 255°C for one hour, 210°C for one day, 180°C for one week, 160°C for one month, and 120°C for one year. Unfortunately, this simple means of obtaining dimensional stability of wood is accompanied by relatively large strength losses, especially toughness, and abrasion resistance. Abrasives actually gouge out entire fibers rather than abrading away parts of fibers. Table I gives data for the effect of heating wood for 10 minutes at three different temperatures upon four different strength properties (15). Heat stabilization imparts considerable decay resistance to wood. Heating to attain a dimensional stabilization of 40% gave a negligible weight loss due to decay when subjected to block culture tests with *Trametes serialis* for two months (17). The corresponding weight loss of the unheated controls was 28.4%.

Heat stabilization was at first believed to be due to the formation of ether linkages between adjacent cellulose chains as a result of splitting out of water between two hydroxyl groups (18). It was later shown that heat stabilized wood swells to a

Table I. Weight and strength losses accompanying heat stabilization of dry softwoods heated beneath the surface of molten Wood's metal for ten minutes at three different temperatures (15)

Temp. °C	Weight Loss %	Modulus of rupture loss %	Hardness loss %	Toughness loss 1/ %	Abrasion resistance loss 2/ %	Reduction in swelling and shrinking %
210	0.5	2.0	5.0	4.6	40.	10
245	3.0	5.0	12.5	20.0	80.	25
280	8.0	17.0	21.0	40.0	92.	40

1/ Forest Products Lab. toughness test (16)

2/ Heated in air

greater extent than unheated wood in concentrated sodium hydroxide solutions and in pyridine (19). As neither of these chemicals break ether bonds, another explanation for heat stabilization was sought.

Hemicellulose, the most hygroscopic component of wood, is also the most subject to thermal degradation (20) to furfural and various sugar break-down products which polymerize under heat to water insoluble polymers, thus reducing the hygroscopicity. These polymers are presumably soluble or at least swell in concentrated sodium hydroxide solution or pyridine thus accounting for the increased swelling in these media. This also accounts for the extremely low abrasion resistance of heat stabilized wood. In normal wood the fibers are at least partially held together by hemicellulose chains that pass through the middle lamella (1 pg. 319). If these chains are severed by heat, complete fibers can be separated by abrasion. Hardboards, made from steam hydrolyzed wood chips, when heat tempered or stabilized lose little if any strength as the hemicelluloses are removed in the hydrolysis step and they are no longer needed for bonding.

Any applied use of the simple heat stabilization technique to wood will be limited by the large loss in abrasion resistance and toughness.

Cross Linking. Tieing together of the structural units of wood with stable molecular cross-links should greatly reduce its tendency to swell. This is illustrated by the fact that incorporating only small amounts of divinyl benzene in the vinyl benzene used in making polystyrene, converts the polymer from a benzene soluble to a benzene insoluble material (21), with single cross-links per several thousand carbon atoms in each polymer chain.

Formaldehyde has long been known to act as a cross-linking agent for cellulose (22) and is used as a crease resistant treatment for cotton fabrics (23). Cotton fabrics are soaked in a formalin solution containing a low concentration of a mildly acidic salt, followed by drying. Tarkow and Stamm (24) applying the treatment to wood, showed that dimensional stabilization does not occur until the wood is almost dry and then only when the acidity was quite high. It thus seemed desirable to treat the wood with formaldehyde in the vapor phase over heated paraformaldehyde (25). Appreciable permanent dimensional stabilization occurred only in the presence of strong mineral acids such as hydrochloric or nitric acid. Permanent weight gains of 4 to 5% were accompanied by dimensional stabilizations of up to 70%, expressed as antishrink efficiencies,

$$A.S.E. = 1 - \frac{(S_s - S_t)}{S_c} \times 100$$

where S_c is the shrinkage of the control and S_t that of the treated specimen. Optimum ASE values were obtained when the wood had a moisture content of 5 to 10% at the time of treatment (24). When formic or acetic acids were used as catalysts anti-shrink efficiencies of less than 10% resulted. Unfortunately cross-linking for high dimensional stability of wood requires a catalyst pH of 1.0 or less, in contrast to the much lower acidity that is adequate for obtaining crease resistance in cotton fabrics. Table II shows the drastic effect of the reaction on the two most adversely affected strength properties of wood. These losses are largely due to acid hydrolysis of the hemicelluloses and cellulose, as they occur when wood is treated with the catalysts without formaldehyde present. Paper can be cross-linked with formaldehyde to give good dimensional stability with less acidic catalysts, and considerably smaller permanent weight increase (26, 27, 28).

The formaldehyde reaction with wood is undoubtedly one of cross-linking as it is accomplished with a much smaller weight increase than in the case of the bulking treatments and reactions, to be considered in the following section. Further, dimensional stabilization is attained by reducing swelling rather than by a reduction in shrinkage, which is the case for bulking treatments. Formaldehyde reacted wood, unlike heat stabilized wood, swells only slightly in concentrated sodium hydroxide solutions and in pyridine, which would be expected if the reaction involved cross-linking (24).

Other aldehydes than formaldehyde have been tested as to their cross-linking ability (24). None gave as good dimensional stability as formaldehyde or proved as permanent,and all required the high concentrations of embrittling acids to catalyze the reaction. Chloral required no addition of acid but it developed its own embrittling acidity on heating. Other types of cross-linking agents have been sought that do not require the high acidity needed to attain high dimensional stability by the formaldehyde reaction. Although these efforts have not as yet met with success they should be continued because of the smaller amount of short cross-linking reactant needed compared to bulking reactions.

Bulking Treatment with Water Soluble Non-Reacting Chemicals

When chemicals are either deposited in or chemically reacted with the cell walls of wood so as to increase the volume of the dry cell walls, the external volumetric shrinkage of the wood is materially decreased as a result of bulking of the fibers. This principle was first observed when thin cross sectional wafers of softwoods were swollen in concentrated salt solutions followed by drying to equilibrium with various decreasing relative vapor pressures at which the tangential and radial dimensions of the wafers were measured. Figure 2 is a plot of the external cross-sectional shrinkage against the relative vapor pressure over saturated solutions of the following salts and their fraction-

Table II Critical strength losses caused by formaldehyde cross-linking of softwoods to various permanent weight gains and antishrink efficiencies, A.S.E.

Weight increase of dry wood	A.S.E.	Toughness loss 1/	Abrasion Resistance loss
%	%	%	%
0.10	10	27	60
0.55	25	45	80
2.20	50	70	91
4.20	70	84	95

1/ Forest Products Lab. Toughness Test (16)

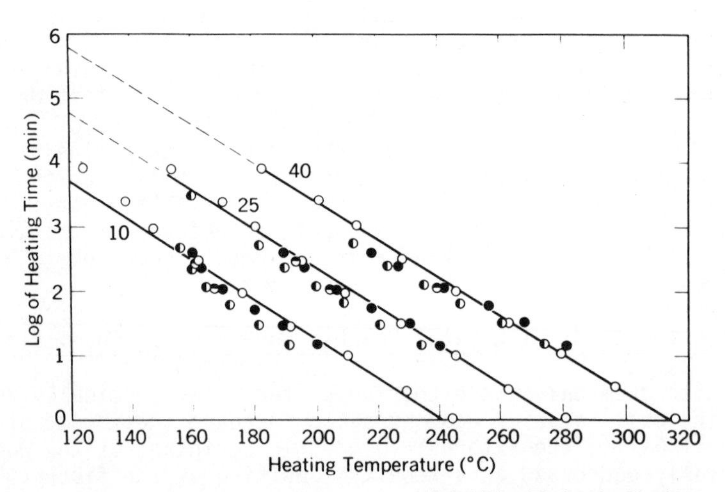

Figure 1. Logarithm of heating time vs. temperature required to give three different reductions in swelling and shrinking when the heating was done beneath the surface of a molten metal to exclude oxygen (15). ○, 1/16-in. thick Sitka spruce veneer; ●, 1/2-in. thick cross sections of western white pine; ○, 3/8-in. flat sawn western white pine; ○, 15/16-in. thick eastern pine boards. Numbers on plot indicate antishrink efficiency (A.S.E.) in percent.

al reduction in vapor pressure: barium chloride, 0.916; sodium chloride, 0.758; manganese chloride, 0.543; magnesium chloride, 0.331; and lithium chloride, 0.117 at 25°C (29). The plot shows that no shrinkage occurs until the relative vapor pressure falls below that in equilibrium with a saturated solution of the salt in the wood. The shrinkage to the final oven dry condition was in each case reduced by the volume of salt finally attained within the cell walls. Figure 3 is a plot of the shrinkage versus the moisture content, giving virtually parallel straight lines. This indicates that shrinkage, in all cases, is the same function of the volume of water lost below the fiber saturation point. Water thus virtually adds its volume to that of the cell walls, further indicating that the extent of voids in the dry cell walls must be virtually negligible. Shrinkage due to bulking is reduced merely because there is less moisture to be lost.

Reducing the relative vapor pressure at which shrinkage begins has no advantage in attaining dimension control, as the wood in equilibrium with higher relative vapor pressure values is always damp. It is, however, advantageous in so called salt seasoning by reducing drying stresses (30).

The ideal bulking agent for wood would be a non-corrosive non-volatile solid, approaching infinite solubility in water, that does not materially reduce the vapor pressure of water. These conditions are more nearly approached with sugars than with salts, as shown in Figure 4 (31). Treatment of wood with aqueous sugar solutions containing a toxic agent was commercially practiced in England for a short period (32). The chief shortcoming was that the wood became damp at relative humidities above 80% and that adhesion of wood finishes was reduced.

Polyethylene glycols proved to be considerably better bulking agents than sugars, as shown in Figure 5 (33). Sitka spruce cross sections saturated with 25% solutions of polyethylene glycols with molecular weights of 1000 and less gave almost complete replacement of the solution by the polymer on slow drying. Thus, the wood approaches having an antishrink efficiency of 100%. This can occur only as the solubility of the polymer in water approaches 100%. The higher molecular weight polyethylene glycols are less effective bulking agents because of their lesser solubility in water and the finding that fractionated polyethylene glycols with molecular weights exceeding about 3500 cannot, because of their bulk, penetrate the cell walls of wood (34). The fact that presumably higher molecular weight polymer entered the cell walls of wood (see Figure 5) can be explained on the basis that depolymerization occurred during boiling to put them into solution and that the commercial polymers had an appreciable spread in molecular weights. Figure 5 shows that a slight swelling occurs during the initial stages of drying in the case of the low molecular weight polymers. This is due to the fact that swelling in aqueous solutions of hygroscopic chemicals increases

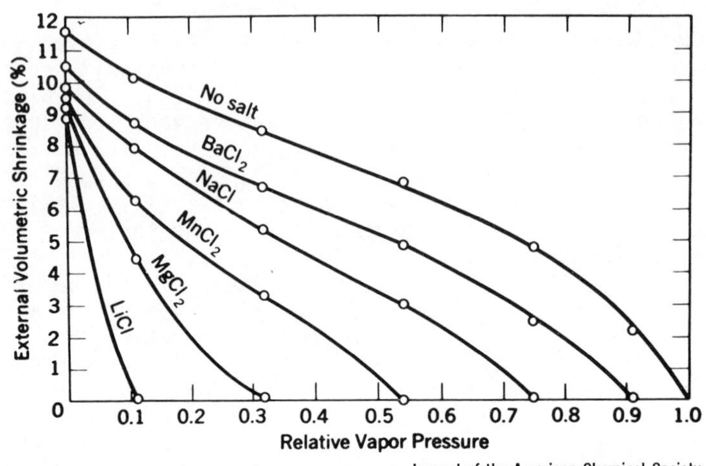

Figure 2. External volumetric shrinkage vs. relative vapor pressure for thin Sitka spruce cross sections containing originally different quarter-saturated salt solutions (29)

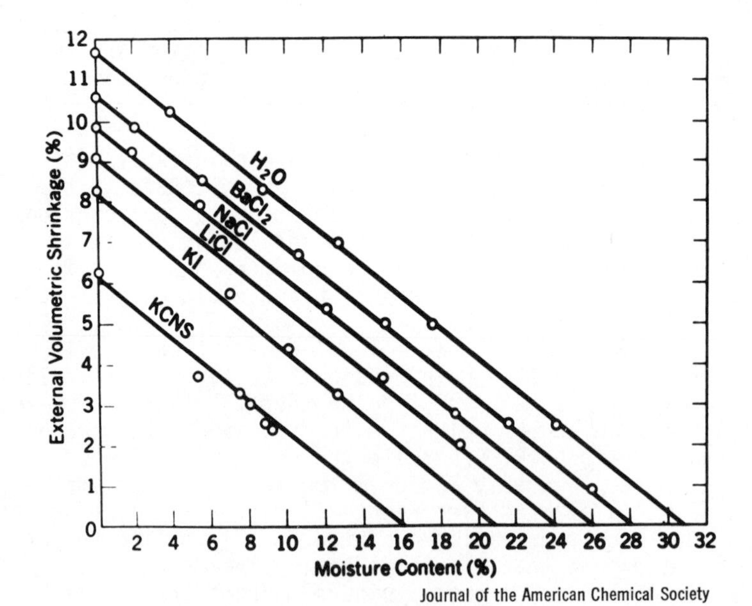

Figure 3. External volumetric shrinkage vs. moisture constant for thin Sitka spruce cross sections containing originally different quarter-saturated salt soltuions (29)

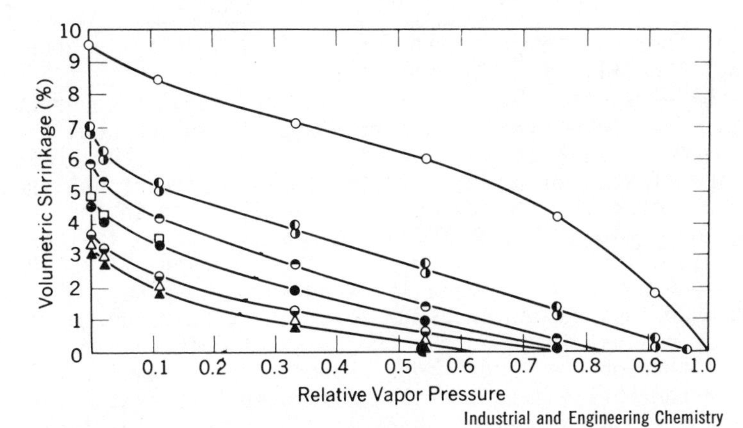

Figure 4. External volumetric shrinkage vs. relative vapor pressure for thin white pine cross sections presoaked in different concentrations of sucrose or invert sugar (31). ○, water only; ◐, 6.25% sucrose; ⊖, 12.5% sucrose; □, 25.0% sucrose; ●, 80.0% sucrose; ◑, 12.5 invert sugar; △, 25.0% invert sugar; ▲, 50.0% invert sugar.

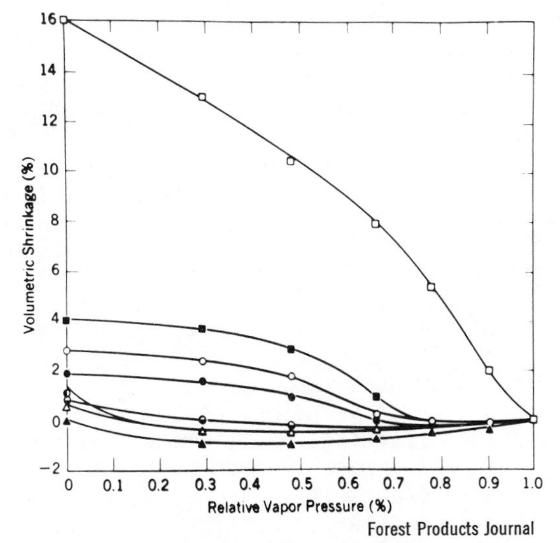

Figure 5. External volumetric shrinkage vs. relative vapor pressure for thin Sitka spruce cross sections presoaked in 25% by weight aqueous solutions of glycerine and polyethylene glycol having the average molecular weights given in parenthesis in the legend (33). □, water only; ◑, glycerine; ▲, polyethylene glycol (200 and 400); △, polyethylene glycol (600); ⊖, polyethylene glycol (1,000); ●, polyethylene glycol (1,540); ○, polyethylene glycol (4,000); ■, polyethylene glycol (6,000).

with an increase in concentration of the solute up to about a 50% solution (1, pg. 249).

Polyethylene glycol treatment is best applied to green wood. The simplest technique is to merely soak the green wood in a 30% by weight aqueous solution of polyethylene glycol - 1000. The time of soaking varies with the permeability of the species and the amount of end grain exposed as diffusion in the fiber direction is about ten to fifteen times as fast as in the transverse directions. Figure 6 is a photograph of two adjacent cross-sections of an originally green loblolly pine tree 3 cm thick. One was soaked in a 30% solution of polyethylene glycol-1000 for one day followed by air drying of both specimens. The control developed a large wedge shaped check extending from the pith to the bark due to stresses developed because the tangential shrinkage was about twice the radial shrinkage. The treated specimen shrank so little that it developed a minimum of damaging stresses. This was accomplished with only a 16% take up of the polymer (35).

An alternate method for treating green wood is to apply several liberal coats of molten polyethylene glycol-1000 a day apart to all surfaces,and storing the specimens in sealed polyethylene bags between coatings to avoid drying. This should be repeated for a week to a month, depending on the permeability and size of the specimens, followed by air drying.

Checking of the face plies of plywood resulting from relative humidity cycling can be virtually eliminated by pretreatment of the face plies with polyethylene glycol-1000 so as to attain approximately a 25% dry weight increase. The treated plys can be assembled with untreated core plies using any type of glue. To insure a good bond the face plies should be oven dried just prior to assembly to reduce the surface moisture content (35). Drying just prior to the application of finishes is also desirable. A surface treatment of loblolly pine house siding with polyethylene glycol improved the weathering properties of applied alkyd emulsion paints and that of two-can clear polyurethane finishes (36, 37).

Wood treated with polyethylene glycol has considerable decay resistance under non leaching conditions in spite of it's non toxicity (17). This is probably due to the fact that there is insufficient water present within the cell walls to support decay.

The strength properties of polyethylene glycol treated wood are virtually those of the swollen wood. This is not surprising as the polymer tends to maintain green wood dimensions. Unlike heat stabilized and formaldehyde cross-linked wood and wood bulked by resin forming polymers within the cell walls (to be considered later), the toughness of the wood is not adversely affected by polyethylene glycol treatment (35).

Green tree cross sections, with bark intact, are being treated, on a limited scale, with polyethylene glycol for table and stand tops and decorative plaques to prevent checking (38). Green,

Forest Products Journal

Figure 6. Adjacent originally green loblolly pine tree cross sections (27-cm diameter and 3 cm thick). Left, soaked in a 30% by weight solution of polyethylene glycol (1000) for 24 hours; right, soaked in water for 24 hours, both followed by air drying. The treated specimen, on the left, developed no periferal radial checks. The control, on the right, developed a large periferal radial check extending almost to the pith. The treated specimen took up on the average 16% of the polyethylene on a dry weight basis (35).

roughed out decorative carving blanks are also being treated by diffusion with polyethylene glycol. The green treated wood carves more easily than dry wood. Treatment is more complete where there is a maximum of end grain, just the parts of the carving that need treatment most to prevent checking on final drying. Artists can set their own pace in carving, as between carving sessions the carving should be immersed in a 25 to 30% solution of polyethylene glycol or stored in a polyethylene bag following application of the molten polymer to all surfaces. Gunstocks of exotic woods are being commercially treated with polyethylene glycol to give them dimensional stability and to avoid face checking (39).

Merely dipping thin fancy crotch face veneer in a solution of polyethylene glycol gives sufficient take up of the polymer to plasticize the sheets so that they dry flat, thus avoiding breaking and checking when assembled with core plies.

Wood artifacts, recovered in the water logged condition, are treated with polyethylene glycol to prevent serious break down of the structure on drying. A notable example is the treatment of the Swedish wooden battleship Vasa, which was sunk in the harbor of Stockholm in 1628, and recovered in 1961 (40). The most remarkable recovery case is that of a pine log hermetically sealed in a bog in a glacial moraine in Northern Wisconsin. Radioactive dating technique showed that the log was buried for a period of 31,000 years. Air drying of a section of the log resulted in serious break down of the specimen to a pile of chips as a result of drying stresses. Other sections of the log were soaked in increasing concentrations of polyethylene glycol-1000 from 10 to 30% for several weeks. The specimens remained perfectly sound upon air drying, with no additional checking. The slight shrinkage that did occur was presumably sufficiently great to allow hydrogen bonds to replace broken covalent bonds.

Recent experiments to determine the dimension stabilizing efficiency of water soluble fire retardent chemicals (41) showed ammonium sulfamate to be superior to phosphate salts, giving antishrink efficiencies of 51 to 66% compared to polyethylene glycol-1000 values of 63 to 77%. Sodium silicate, because of its alkalinity, caused collapse of the wood that resulted in negative antishrink efficiencies. Strongly alkaline systems should hence be avoided.

Bulking Treatment with Water Insoluble Chemicals. The chief shortcomings of dimensional stabilization of wood with polyethylene glycol are that it can be leached from the wood and that the wood feels damp when held for prolonged periods of time at relative humidities of 80% and above. It thus appears desirable to deposit water insoluble materials within the cell walls of wood. This can be done by a replacement process with waxes (42). Water in green wood is replaced by Cellosolve (ethylene glycol monoethyl ether) by soaking the wood in this chemical, followed by slowly distilling off the water which has a lower boiling point than the Cellosolve. No shrinkage occurs during this first stage of replacement, if carried out slowly. The specimens are then immersed in a molten wax or natural resin and the Cellosolve slowly distilled off. This step invariably involves some shrinkage. Antishrink efficiencies of 80% are, however, obtainable in this way with mixtures of beeswax and rosin. This treatment appears suitable for the preservation of wood artifacts. Christensen (43) has treated wood artifacts by replacing the water with tertiary butanol and this with polyethylene glycol-4000.

A simpler approach for depositing water insoluble chemicals within the cell walls of wood is to impregnate the wood with solvent soluble resin forming chemicals containing a catalyst that penetrate the cell walls followed by evaporation of the solvent and then heating to polymerize the resin. This has been accomplished with the following water soluble resin forming systems: phenol, resorcinol, melamine and urea-formaldehydes, phenol-furfural, furfuryl-aniline and furfuryl alcohol (44). The most successful of these has been phenol-formaldehyde (45). It is cheaper than resorcinol and melamine-formaldehydes and gives higher dimensional stability and is more weather resistant than urea-formaldehyde (46). Further, less chemical is lost on drying and polymerizing than in the case of furfural-aniline and furfuryl alcohol when slightly prepolymerized but still water soluble "A" stage phenol-formaldehyde slightly alkaline resin is used. A number of suitable "A" stage resins are commercially available (47). Their aqueous solid resin contents range from 33% to 70%, pH from 6.9 to 8.7 and relative viscosities in 33% solutions from 3.5 to 4.7. Wood treated with these resins is called Impreg.

Difficulty was encountered in adequately distributing "A" stage resins in sizable pieces of solid wood. Limited amounts of Impreg have been made by impregnating easily treated solid woods such as ponderosa pine and basswood. Most of the Impreg presently made is laminated from treated veneer. Predried fancy face veneer, 1/32 inch or less is thickness, can be adequately treated merely by soaking in a 30 to 60% solid content "A" stage resin for a few minutes up to an hour or two depending upon the thickness and the amount of cross grain. Cross grain accentuates capillary absorption which is followed by diffusion into the cell walls. The rate of diffusion into the cell wall varies inversely with the square of the thickness. Straight grain veneer, 1/16 inch or more in thickness, requires excessive soaking time for the take up of 25 to 30% of resin forming chemical. Veneer having a low to medium specific gravity, in thickness up to 1/8 inch and moisture contents of 20 to 30%, can be readily treated using compression roll equipment (48). The veneer is passed between compression rolls beneath the surface of the solution where it is compressed to about half of its original thickness. On emerging from between the rolls, the veneer tends to recover its original thickness and in doing so sucks in the treating solution.

The chief method for treating air dry thicker veneer is by pressure impregnation in a treating cylinder. The usual procedure is to immerse one sheet of veneer at a time in a tank filled with the treating solution to insure wetting of the faces of each ply, making close piling possible without fear of forming dry pockets. The sheets of veneer are then held down in the solution with metal weights. The height of the treating solution is adjusted so that following impregnation the top sheet is still submerged. The tank is then rolled into the treating cylinder and 20 to 200 psi of air pressure is applied for ten minutes to six hours, depending on the wood species, whether sapwood or heartwood and the thickness of the veneer. Heartwood of basswood or cottonwood veneer 1/16 inch thick will take up its own weight of 30% solids content solution in 15 minutes at 30 to 40 psi. Birch heartwood veneer 1/16 inch thick will require a pressure of 75 psi for two to six hours to attain the same take up (45).

The treated veneer should then be close piled for one to two days, with a water proof cover over it, to allow for equalization of the resin content by diffusion. The veneer can then be dried and the resin polymerized in a continuous veneer drier or in a dry kiln. Real fast initial drying should be avoided to prevent excessive migration of the as yet uncured resin to the surfaces. The treated veneer is then laminated into panels of any desired thickness in a hot press using phenolic glue (45).

The dimensional stability of Impreg made in the aforegoing way increases with an increase in the resin content of the veneer up to about 70% antishrink efficiency at a resin content of 30 to 35%. This ASE value is less than that obtainable with polyethylene glycol because of loss of water and subsequent contraction of the resin forming chemicals within the cell walls as polymerization occurs.

Face checking of plywood and parallel laminates, with phenolic resin treated faces, is practically eliminated on indoor exposure. Under out-of-doors weathering conditions face checking and erosion are materially reduced (9).

Phenolic resin treatment imparts considerable decay resistance to wood as do other dimension stabilization treatments (17). The treatment increases the electrical resistance materially (49). It also gives wood considerable acid resistance (45) and heat resistance (50). Treated specimens have been subjected to cyclic heating to 205°C followed by cooling more than 50 times without visual harm, whereas untreated controls charred and disintegrated badly after a few heating cycles. Phenolic resin treatment, however, does not impart true fire resistance to wood, but it does improve the integrity of the char, thus cutting down on fire spread (45).

Phenolic resin treatment causes a slight loss in tensile strength properties of wood and a considerable increase in compressive properties and hardness. Flexural properties are increased slightly. Shear parallel to the grain is decreased. Toughness is, however, reduced to about one-third of normal (51, 1, pg. 131).

Impreg is used for automobile die models of all of the body surfaces (50). Parallel laminates of phenolic resin treated cativo veneer are hot pressed to one inch thick panels and these are glued together to the desired thickness with cold setting glue, followed by carving. Impreg is also used for various shell molding dies (50) where its excellent heat resistance is utilized. The mold is imbedded in sand containing a heat setting resin, heat cured, cooled, and the mold removed. The Impreg mold can be reused up to 50 times.

Compreg is similar to Impreg except that the treated veneer prior to heat curing is appreciably compressed. Phenolic resin, still in the "A" stage, is an excellent plasticizer for wood. Pressures of 1000 psi or less at 275 to 300°F are sufficient to compress most species to dry volume specific gravities of 1.2 to 1.35, thus approaching the specific gravity(1.46)of the wood substance (52). Drying of treated veneer without cure of the resin can be accomplished by kiln drying for five to eight hours at 140 to 150°F, (52). Compressed parallel laminates can be made from dried but uncured veneer containing at least 30% of resin forming solids without the use of a laminating glue when compressed to a specific gravity of at least 1.3 at about 300°F as sufficient resin exudes from the plies to form a good bond. When the plies are crossed, contain less than 30% of resin forming chemicals, and are compressed to less than a specific gravity of 1.3, additional hot press phenolic bonding resin must be used. It is important to predry the treated plies to a moisture content of 2 to 4% prior to application of a waterborne laminating glue as this tends to introduce excessive moisture in the panel which is trapped on compression. It is further desirable to again dry the glue spread veneers to this low moisture content before assembly. Failure to do this may result in checking of the panels as they slowly dry to this reduced equilibrium moisture content.

Treated plies fortunately respond to compression under heat and pressure more rapidly than they cure even at 280°F. This makes possible molding of Compreg by a so called expansion molding technique. Single sheets of dry uncured resin treated veneer with a surface coat of bonding resin are rapidly preheated to 220 to 240°F and then compressed in a fast operating cold press at about 1500 psi. The plies respond rapidly to compression. The contained resin does not set in a thermosetting sense but sets in a thermoplastic sense as the veneer is cooled. The sheets of veneer can be kept in this compressed condition for weeks at room temperature and low relative humidity without springback. They are cut to template sizes layed up in proper sequence in a split mold to completely fill the mold. The mold is firmly locked in a closed position and then heated to about 270°F. The resin loses its thermoplastic set. The plies tend to lose their compression and exert a pressure on the mold approaching that at which they were compressed. As heating continues the resin sets in a thermosetting sense (53).

Compreg swells in the thickness direction two to three times as much as Impreg on the basis of its compressed dimensions but the swelling is extremely slow and the panels do not recover from compression as do untreated compressed wood panels (52). It has a golden to dark brown color, depending on the species. It has a natural lustrous finish that can be restored by merely sanding and buffing when cut or scratched. It can be readily cut or turned using metal working tools operated at reduced speeds. Compreg can be glued to Compreg or normal wood with both hot press phenolic and room temperature setting resorcinol glues (52).

Compreg is highly resistant to decay and attack by termites and marine borers (52). Its electrical and acid resistances are also real high.

The strength properties of Compreg are in general increased over those of the wood from which it was made about in proportion to the increase in specific gravity except for the hardness which is increased by ten to twenty fold (54) and the toughness which is reduced to 0.75 of that for the original wood (51). The toughness is improved if the Compreg is made with a spirit soluble phenolic resin rather than an "A" stage water soluble resin (55), but the dimensional stability is not so good.

Compreg was used during World War II largely for the roots of wooden airplane propellers, for ship screw bearings and experimental aircraft landing surfaces of aircraft carriers. More recent uses have been for forming dies and jigs, weaving shuttles, knife handles, glass door pulls and railroad track connectors where electrical resistance is needed for automatic signaling systems.

Furfuryl Alcohol Resin has been successfully formed in the cell walls of wood to give the wood high dimensional stability (ASE values of 65 to 75%) and high alkali as well as acid resistance (56). Anhydrous furfuryl alcohol swells dry wood very slowly. Only about 5% of water present either in the furfuryl alcohol or the wood makes it a good swelling agent for wood. The reaction requires an acid catalyst (57). The use of strong mineral acids should be avoided as the polymerization may proceed even at room temperature with explosive violence. The shelf life of furfuryl alcohol with catalyst present (90% furfuryl alcohol 5% water and 5% catalyst) was tested by determining the time at room temperature beyond which the viscosity of the solution increased significantly. Of a series of acid salts and di-and tri-basic organic acids tested only zinc chloride and citric and malic acids gave shelf lives over one month. These systems on heating at 100°C for 24 hours gave resin yields of

72 to 75%.

Furfuryl alcohol resin treated wood varies in color from dark brown to black depending on the resin content. At high resin contents, a high degree of polish is attained by sanding and buffing. Hardness and crushing strength perpendicular to the grain are increased materially. Toughness, as in the case of phenolic resin treated wood, is decreased (in terms of the Charpy impact test from 70 to 30 inch - lb.) (56). Relative Forest Products Lab toughness values obtained by the author ranged from 0.3 to 0.67 using different acid catalysts and varying concentrations.

Drastic alkali resistance tests consisting of heating wood specimens in boiling 10% NaOH for 16 days reduced the crushing strength at the elastic limit for untreated southern yellow pine from 620 to 80 psi and for the wood containing 71% furfuryl alcohol resin from 2650 to 890 psi (56).

Vinyl Resin Treatment. Considerable interest has developed in recent years in polymerizing various vinyl resins in cellulosic materials. Most of the vinyl monomers, with the exception of acrylonitrile, swell wood only slightly,(58, 59) and hence would not be expected to be good dimension stabilizing bulking agents except when a non-aqueous fiber penetrating solvent is used to aid in the fiber penetration. Normally the liquid monomers are impregnated into solid wood and polymerized either by gamma ray irradiation which generate free radicals that act as excitation sites in the system (60) ,by free radicals generated by thermal break down of a peroxide catalyst such as benzoyl peroxide (61), or by Vazo , a DuPont catalyst that breaks down on heating to two free radicals and a nitrogen molecule (62).

Distribution of monomer was found to be good only at high loading which resulted in the polymer being mostly in the homopolymer form in the void structure (63, 64, 65, 66, 67, 68). These modified woods are being made to a limited extent largely to take advantage of the improved mechanical properties, especially hardness and abrasion resistance (68). Good dimensional stability, 60-70% ASE, is obtained only with acrylonitrile and its combination with other vinyl monomers and then only at high loadings (69).

Vinyl resin treated wood, at high loadings has a natural lustrous appearance as does Compreg. Its advantages over Compreg for flooring are its greater toughness, abrasion resistance and undarkened color. Because of the much higher resin content it should be potentially considerably more expensive than Compreg.

The step of impregnating with vinyl monomers could be greatly simplified and made more uniform if veneer was treated as in the case of Impreg and Compreg. In this case a low volatility monomer, such as tributyl styrene (70) dissolved in a volatile wood swelling solvent such as methyl alcohol should be the impregnant to avoid depletion of resin at the surfaces of the plies after evaporating off the solvent and thus insuring dimensional stability. Curing of the resin with benzoyl peroxide or Vazo and heat could be carried out in a press simultaneously with assembly of the plies.

Chemical Reactants. The bulking agents for wood thus far considered depend merely upon deposition of chemicals within the cell walls. Forming of resins within the cell walls may or may not involve some chemical reaction with the wood. Even if the resin cannot be leached from the wood with resin solvents there is no assurance that it is chemically attached at the wood. If polymers are formed within the cell walls with molecular weights exceeding about 3500 they may be merely mechanically entrapped as homopolymers (34). There is, however, one group of bulking agents that definitely react with the available hydroxyl groups within the cell walls of wood. Acetylation has proved to be the most successful of these reactions.

Acetylation of cellulose to the triacetate has been carried out without breaking down of the structure with acetic anhydride containing pyridine to help open up the cell wall structure and to act as a catalyst (71). This led Stamm and Tarkow (72) to test the liquid phase reaction on wood. High dimensional stabilization without break down of the structure was obtained, but excessive amounts of chemical were used. They hence devised a vapor phase method at atmospheric pressure that proved suitable for treating veneer up to thicknesses of 1/8 inch. Acetic anhydride pyridine vapors generated by heating an 80-20% mixture of the liquids were circulated around sheets of veneer suspended in a box lined with sheet stainless steel. Hardwood veneer, 1/16 inch thick, required about a 6 hour exposure at 90°C to obtain an acetyl content of 18 to 20% and an ASE of 70%. Softwood veneer required an acetyl content of 25% to obtain the same ASE value and an exposure time of 10 to 12 hours. Clermont and Bender (73) showed that dimethyl formamide can be substituted for pyridine as the swelling agent and catalyst for acetylation of wood.

Goldstein et al. (74) showed that acetylation of wood can be carried out in the liquid phase with acetic anhydride without the addition of a catalyst. Only one acetyl group of the anhydride molecule reacts with the wood, the other forming acetic acid. Following surface reaction on the wood the acetic acid formed presumably helps open up the structure and promote further reaction. These investigators also devised a means of avoiding the use of excessive amounts of acetic anhydride by dissolving just the needed amount in an aromatic or chlorinated hydrocarbon, impregnating solid wood with this solution under pressure of about 150 psi in a treated cylinder, heating while still under pressure to 100 to 130°C for 8 to 16 hours to promote the reaction followed by draining the cylinder and evacuation to remove any excess of acetic anhydride, the solvent and formed acetic acid. The drained reactants were found to be reusable for several subsequent impregnations. The liquid phase reaction with acetic anhydride alone and also when diluted to 25% with xylene at 125°C gave ASE values for twelve species of wood ranging from 70 to 80% (74).

This process was carried out on pilot plant scale for several years by the Koppers Co., Pittsburgh, Pa. It was never converted to a large scale process for economic reasons.

Baird, (75) showed that the vapor phase reaction can also be carried out without a catalyst to attain acetyl contents of 20% in 2 hours at 130°C with white pine cross sections. The addition of 15% of dimethylformamide gave an acetyl content of 25% under the same conditions. The presence of catalyst was found helpful only in attaining the higher levels of acetylation.

Acetylated wood is highly stable. Ten cycles of relative humidity change between 30 and 90% at 27°C over a period of four months gave no loss in anti-shrink efficiency (ASE) (76). Soaking in a 9% aqueous sulfuric acid solution for 18 hours at 25°C had no effect on the subsequent ASE. When the temperature was increased to 40°C the ASE dropped only from 75 to 65%. Exposure of acetylated birch panels in the warm salty water of the Gulf of Mexico for a year showed no attack by *Teredo* and no loss in ASE whereas the untreated controls were badly attacked. Acetylated birch stakes inserted in termite infected soil showed no sign of attack in 5 years. Acetylated Sitka spruce with an ASE of 70% when exposed to *Lenzites trabea* in a 3 month soil-block culture test showed a negligable loss in weight compared to 47% for the controls (17). Similar results were obtained by Goldstein et al., (74), using six different cultures.

Douglas fir plywood with acetylated faces, when exposed to the weather on a test fence for two years without a surface finish developed only a slight roughening and checking whereas the controls weathered and checked badly (76). The weathering of exterior paints on panels with acetylated faces were considerably better than on the controls. Presurface acetylation also seemed to improve the weathering properties of painted wood (36, 37).

Acetylation in general causes a slight bleaching of the wood. It causes little change in the specific gravity of wood, the weight increase being virtually offset by the bulking. Acetylation causes virtually no change or a small increase in most of the strength properties of wood (72, 74, 75, 76) including toughness which is adversely affected by all resin forming bulking treatments.

Other Reactants. Vapor phase reactions of isosyanates with wood have been studied as a means of obtaining dimensional stability (75). Isocyanates are poor swelling agents for wood. It was thus necessary to use an accompanying swelling agent such as dimethyl formamide to open up the structure. The most suitable isocyanate, butyl, gave ASE values up to 78% for a weight increase of 49% when heated for two hours at 130°C. Toughness and abrasion resistance were, however, reduced to 72 and 75% of the values for the untreated controls.

Another bulking reaction of interest is with ethylene oxide with trimethyl amine present to open up the structure and serve as a catalyst (77). Small wood specimens were evacuated at 95°C in an autoclave. Trimethyl amine at 65°C was admitted to a pressure of 1 psi absolute. Ethylene oxide was then introduced into the system under a pressure of 50 psi and held until the desired extent of weight increase of 20 to 30% due to reaction was attained, to give ASE values up to 65%.

Recently Rowell and Gutzmer (78) have shown that good dimensional stability can be imparted to wood by reactions with other alkylene oxides namely propylene and butylene oxides and epichlorohydrin catalyzed with triethylamine. All of these chemicals are liquids at room temperature so that complicated gas handling equipment is not needed. Optimum ASE values of 66 to 68% for Southern yellow pine reacted with propylene oxide were obtained when the add on weight ranged from 28 to 34%. Higher add on values evidently resulted in rupture of the fiber with an appreciable increase in swelling. The optimum ASE values were obtained when the wood was impregnated under a pressure of 150 psi with 95 parts of propylene oxide and five parts of the triethylamine and heated for one hour at 110 to 120°C. Epichlorohydrin gave similar ASE values with a slightly broader range of weight increases. Epichlorohydrin treatment gave excellent decay resistance as shown by block culture tests.

Conclusions

The most effective dimension stabilizing treatments for wood thus far devised that introduce a minimum of accompanying detremental properties are all of the bulking types. The best all around treatment is acetylation. It has the least effect on the appearance and specific gravity of wood. It is the only treatment other than with polyethylene glycol that does not reduce the toughness of wood. It gives ASE values as high as 75% with weight increases of only 18 to 20% for hardwoods and 26 to 28% for softwoods. It is highly stable and gives the optimum resistance to organisms. The reaction can be carried out simply in the vapor phase on veneer up to 1/8 inch thick or in the liquid phase on solid wood when the reactant is dissolved in a hydrocarbon solvent.

Other bulking treatments have their special applications. Phenolic resin treatment, the first to be developed, gives high permanent dimensional stability, decay, heat, acid, and electrical resistance. When compressed prior to setting of the resin, it gives the hardest treated wood known, hardness increases up to

20 fold. There is, however, a loss in toughness. Furfuryl alcohol resin treatment imparts alkali as well as acid resistance to wood, making it suitable for chemical processing equipment. The chief improved property of vinyl resin treated wood is its high abrasion resistance making it suitable for floor surfaces.

Polyethylene glycol treatment is suitable for the treatment of green wood, especially water swollen artifacts as it materially reduces the shrinkage that occurs on drying, and the accompanying degrade. The treatment is also highly useful in carving green wood and avoiding degrade on drying.

The newest treatment with alkylene oxides shows promise of being developed into a commercial process.

Literature Cited

(1) Stamm, A. J. "Wood and Cellulose Science" Ronald Press Co., New York (1964).
(2) Skaar, C. "Water in Wood" Syracuse Univ. Press., Syracuse, N. Y. (1972).
(3) Brunauer, S., Emett, P. H. and Teller, E., J. Am. Chem. Soc. (1938) 60, 309.
(4) Koehler, A. "Longitudinal Shrinkage of Wood", U. S. Dept. Agr. For. Prod. Lab. Report 1093 (1946).
(5) Markwardt, L. J. and Wilson, T. R. C. "Strength and Related Properties of Woods Grown in the U. S.", U. S. Dept. Agr. Tech. Bull. 479. (1935).
(6) Keylwerth, R., Holz Roh Werkstoff, (1962) 20 (7) 252-259.
(7) Hittmeier, M. E. Wood Sci. and Tech. (1967) 1 (2),109-121.
(8) Barkas, W. W. "A Discussion of the Swelling Stresses and Sorption Hysteresis of Plastic Gels", Great Brit. Dept. Sci. Ind. Research, Forest Products Special Report No. 6 (1947).
(9) Lloyd, R. A. and Stamm, A. J., For. Prod. J. (1958) 8 (8), 230-234.
(10) Hunt, G. M. "Effectiveness of Moisture-Excluding Coatings on Wood", U. S. Dept. Agr. Circular No. 128 (1930).
(11) Browne, F. L., Ind. Eng. Chem. (1933) 25, 835-842.
(12) Browne, F. L., Architectural Record, (1949), Mar: 131-133.
(13) Browne, F. L. and Downs, L. E. "A Survey of the Properties of Commercial Water Repellants and Related Products" U. S. For. Prod. Lab. Mimeo R1495. (1945).
(14) Stamm, A. J. and Harris, E. E., "Chemical Processing of Wood", Chem. Pub. Co., N. Y. (1953).
(15) Stamm, A. J., Burr, H. K., and Kline, A. A., Ind. Eng. Chem. (1946) 38:630-637.
(16) Forest Products Lab. "Toughness Testing Machine", U. S. For. Prod. Lab. Report 1308 (1956).
(17) Stamm, A. J. and Baechler, R. H., For. Prod. J. (1960) 10 (1):22-26.
(18) Stamm, A. J. and Hansen, L. A., Ind. Eng. Chem. (1937) 29: 931-938.

(19) Seborg, R. M., Tarkow, H., and Stamm, A. J., J. For. Prod. Research Soc. (1953) 3 (3):59-67.
(20) Stamm, A. J. Ind. Eng. Chem. (1956) 48 413-417.
(21) Staudinger, H., Trans. Faraday Soc. (1936) 32:323-335.
(22) Eschalier, X., French Patent No. 374, 724 additions 8422 (1906); 9904 (1908); 9905 (1908); 10760 (1909).
(23) Gruntfest, I. J. and Gagliardi, D. D. Textile Research J. (1948) 18 643-649.
(24) Tarkow, H., and Stamm, A. J., J. For. Prod. Research Soc. (1953) 3:33-37.
(25) Walker, J. F. "Formaldehyde", Reinhold Pub. Corp. New York, (1944).
(26) Cohen, W. E., Stamm, A. J. and Fahey, D. J.,TAPPI (1959), 42, 934-940.
(27) Stamm, A. J., TAPPI, (1959) 42:44-50.
(28) Stamm, A. J., TAPPI, (1959) 42:39-44.
(29) Stamm, A. J., J. Am. Chem. Soc., (1934), 56:1195-1204.
(30) Loughboruogh, W. K., Southern Lumberman, (1939), Dec. pg. 137.
(31) Stamm, A. J., Ind. Eng. Chem. (1937), 29:833-836.
(32) Batson, B. A., Chem. Trade J. (1939) 105 (8) (2724):93, 98.
(33) Stamm, A. J., For. Prod. J. (1956) 6 (5):201-204.
(34) Tarkow, H., Feist, W. C. and Southerland, C. F., For. Prod. J. (1966) 16 (10):61-65.
(35) Stamm, A. J., For. Prod. J. (1959) 9 (10):375-381.
(36) Campbell, G. G. "An Investigation of Improving the Durability of Exterior Finishes on Wood", M. S. Thesis, Dept. Wood and Paper Sci. North Carolina State Univ, Raleigh, N. C. (1966).
(37) Campbell, G. G. "The Effect of Weathering on the Adhesion of Selected Exterior Coatings to Wood", PhD Thesis, Dept. Wood and Paper Sci., North Carolina State Univ. Raleigh, N. C. (1970).
(38) Mitchell, H. L. and Iverson, E. S., For. Prod. J. (1961) 11 (1), 6-7.
(39) Mitchell, H. L. and Wahlgren, H. E., For. Prod. J. (1959) 9 (12), 437-441.
(40) Franzen, A. National Geographic Mag. (1962) 121 (1) 42-57.
(41) Stamm, A. J., Wood Sci. and Tech. (1974) 8:300-306.
(42) Stamm, A. J. and Hansen, L. A., Ind. Eng. Chem. (1935) 27: 148-152.
(43) Christensen, B. B., "The Conservation of Waterlogged Wood in the National Museum of Denmark", National Museum of Denmark Copenhagen (1970).
(44) Stamm, A. J., and Seborg, R. M., Ind. Eng. Chem. (1936) 28: 1164-1170.
(45) Stamm, A. J. and Seborg, R. M., Ind. Eng. Chem. (1939) 31: 897-902.
(46) Millett, M. A. and Stamm, A. J., Modern Plastics (1946) 24, 150-153.

(47) Burr, H. K., and Stamm, A. J., "Comparison of Commercial Water-Soluble Phenol-Formaldehyde Resinoids for Wood Impregnation", U. S. For. Prod. Lab.Mimeo 1384 (1943).
(48) Stamm, A. J., "Wood Impregnation", U. S. Patent No. 2350135. (1944).
(49) Weatherwax, R. C., and Stamm, A. J., Elect. Eng. Trans. (1945) 64:833-839.
(50) Seborg, R. M. and Vallier, A. E., J. For. Prod. Research Soc. (1954), 4 (5):305-312.
(51) Erickson, E. C. O. "Mechanical Properties of Laminated Modified Wood", U. S. For. Prod. Lab. Mimeo No. 1639 Revised. (1958).
(52) Stamm, A. J. and Seborg, R. M. Trans. Am. Inst. Chem. Eng. (1941) 37:385-397.
(53) Stamm, A. J. and Turner, H. D. "Method of Molding", U. S. Patent No. 2391489 (1954).
(54) Weatherwax, R. C., Erickson, E. C. O., and Stamm, A. J., "Modulus of Hardness Test," Am. Soc. Testing Materials, Bull No. 153 (1948).
(55) Findley, W. H., Werley, W. J., and Kacatieff, C. D., Trans. Am. Soc. Mech. Eng. (1946) 68, 317-325.
(56) Goldstein, I. S., For. Prod. J. (1955) 5 (4) 265=267.
(57) Goldstein, I. S., and Dreher, W. A., Ind. Eng. Chem. (1960) 52, 57-58.
(58) Siau, J. F., Wood Sci. (1969) 1 (4):250-253.
(59) Loos, W. E., and Robinson, G. L., For. Prod. J. (1968) 18 (9); 109-112.
(60) Chapiro, A., and Stannett, V. T., International J. Applied Radiation and Isotopes (1960) 8, 164-167.
(61) Meyer, J. A., For. Prod. J. (1965) 15 (9):362-364.
(62) DuPont Co. "DuPont Vazo 64 Vinyl Polymerization Catalyst" "Product Information" (1974).
(63) Kenaga, D. L., Fennessey, J. P. and Stannett, V. T., For. Prod. J. (1962), 12 (4), 161-168.
(64) Kent, J. A., Winston, A., and Boyle, W. R., "Preparation of Wood - Plastic Combinations using Gamma Radiation to Induce Polymerization", U. S. Atomic Energy Commission Report O. R. O. - 600 and 612 (1962).
(65) Loos, W. E., Walters, R. E. and Kent, J. A., For. Prod. J. (1967) 17 (5): 40-49.
(66) Ramlingham, K. V., Werezak, G. N. and Hodgins, J. W., J. Polymer Sci. (1963) Part C Polymer Symposium No. 2: 153-167.
(67) Siau, J. F., Meyer, J. A. and Skaar, C., For. Prod. J. (1965) 15 (10):426-434.
(68) Ellwood, E., Gilmore, R., Merrill, J. A. and Poole, W. K., "An Investigation of Certain Physical and Mechanical Properties of Wood-Plastic Combinations", U. S. Atomic Energy Commission Report ORO-638 (RTI-2513-T13) (1969).
(69) Ellwood, E., Gilmore, R., and Stamm, A. J. Wood Sci. (1972) 4 (3) 137-141.

(70) Kenaga, D. L., Wood and Fiber (1970) 2 (1), 40-51.
(71) Hess, K., Ber., (1928) 61, 1460.
(72) Stamm, A. J., and Tarkow, H., J. Phys. and Colloid Chem. (1947) 31: 493-505.
(73) Clermont, L. P. and Bender, E., For. Prod. J. (1957) 7 (5), 167-170.
(74) Goldstein, I. S., Jeroski, F. B., Lund, A. E., Nielson, J. F., and Weaver, J. W., For. Prod. J. (1961) 11 (8), 363-370.
(75) Baird, B. R., Wood and Fiber (1969) 1 (1) 54-63.
(76) Tarkow, H., and Stamm, A. J. and Erickson, E. C. O. "Acetylated Wood", U. S. Dept. Agr. For. Prod. Lab. Mimeo No. 1593 (1955).
(77) McMillin, C. W., For. Prod. J. (1963) 13 (2), 56-61.
(78) Rowell, R. M., and Gutzmer, D. I., Wood Sci. (1975) 7 240-246.

9

Dimensional Stabilization of Wood with Furfuryl Alcohol Resin

ALFRED J. STAMM

School of Forest Resources, Department of Wood and Paper Science, North Carolina State University, P.O. Box 5516, Raleigh, N.C. 27607

A number of methods have been developed in the last forty years for the dimensional stabilization of wood (1, 2). Unfortunately none of these have proved to be economically successful on a large scale. One of the methods, involving the treatment with furfuryl alcohol (3) was shown by Goldstein (4, 5) to impart both alkali as well as acid resistance to wood, giving it a distinctive black color. This treatment is of renewed interest at the present time because furfuryl alcohol is made from the renewable resource, corn cobs, and can also be made from the hydrolizate of hardwood waste (6).

Goldstein and Dreher (7), have shown that five parts of the catalyst zinc chloride or organic di-or tri-basic acids, such as citric acid, gave solutions in 5 parts of water added to 90 parts of furfuryl alcohol that remain stable for a month or more at room temperature without a significant amount of polymerization, whereas they polymerize to give high yields of resin when heated for a day at 100°C. This makes possible the impregnation of wood with this liquid in conventional treating cylinders, draining off the excess liquid for reuse, and then heat curing the resin within the wood. In this way an antishrink efficiency

$$\left(1 - \frac{\%\ \text{swelling of treated wood}}{\%\ \text{swelling of untreated wood}} \times 100\right)$$

of 63 was obtained at 48% resin content and 70 at 120% resin content with Idaho pine cross sections (4, 7). The toughness of southern pine sticks (0.5 by 0.5 by 4 inch span) as shown by the Charpy impact test at 68% resin content was, however, reduced from 69 to 27 foot pounds (relative toughness 0.39). It thus appeared desirable to determine if the use of less catalyst would improve the toughness and if the long tie up of an oven or dry kiln for the cure of the resin could be avoided.

Experimental

Douglas fir, loblolly pine, and Engelmann spruce specimens 4.5 by 4.5 cm in the radial and tangential directions were cut into a series of end matched cross sections 3 mm thick for treatment. Yellow poplar sticks 1 by 1 cm in the radial and tangential directions by 40 cm long were also treated. All specimens were stress relieved by swelling in water followed by air and then oven drying. Part of the sticks were cut into four end matched sticks 9.5 cm long for treatment and toughness tests.

Three different catalysts were used. Five parts by weight of zinc chloride or of citric acid were dissolved in 5 parts of water and added to 90 parts by weight of furfuryl alcohol, giving 5% catalyst concentrations. Zinc chloride was also used in one fifth and one twenty fifth of the former concentration. Formic acid was used in 5% by weight concentrations with no water present. The presence of the small amounts of water in the case of the two solid catalysts aid in their solution and cause almost as rapid swelling of wood as in water alone, whereas the swelling in water-free furfuryl alcohol is extremely slow. Formic acid accelerates the rate of swelling of wood in furfuryl alcohol, similar to the effect of water.

The highly permeable loblolly pine and Engelmann spruce cross sections were treated by merely immersing the air dry specimens in the treating solutions for 5 seconds. The much denser Douglas fir heartwood cross sections were treated by pulling a vacuum for 30 seconds over the solution immersed specimens. The 9.5 cm long yellow poplar sticks were treated by immersing them for 10 minutes under vacuum in the treating solutions. The 40 cm long yellow poplar sticks were treated in glass tubes by pulling a vacuum of 0.1 mm of mercury on the oven dry specimens, running in the treating solution under vacuum and holding for one minute (8).

The treated specimens were wrapped in aluminum foil and held at room temperature for one day to allow for equilization of the solution through the structure by capilarity and diffusion. The specimens were then weighed and the radial and tangential dimensions determined. They were again wrapped in aluminum foil and heat cured at 120°C for either 18 or 6 hours. The specimens were weighed and measured, oven dried for 2 hours to remove unpolymerized volatiles and again weighed and measured. The specimens were then immersed in distilled water for at least two days, measured, air dried followed by oven drying, weighing and measuring. Specimens that were well cured lost little weight and dimensions between the heat cured condition and the first and second oven drying.

Anti-Shrink and Polymerization Efficiencies

Antishrink efficiencies (ASE) for the treated specimens were calculated from the changes in cross sections between the original untreated water swollen and oven dry conditions and the treated water swollen and second oven dry condition. Figure 1 is a plot of the ASE for the three species of cross sections and the yellow poplar sticks versus the resin content. The ASE values increase approximately linearly with an increase in resin content as resin is formed within the cell walls to bulk the fiber. Above about 40% resin content by weight, resin is deposited within the natural void structure with little increase in ASE. This point corresponds to an optimum bulked volume of 32.5% as the specific gravity of cast furfuryl resin was found to be 1.23 by the suspension method in an aqueous zinc chloride solution (1 pg 55). This value is only slightly greater than the fiber saturation point or optimum bulking of untreated wood by water, 30%. The furfuryl alcohol - catalyst solutions swelled the wood 6 to 8% beyond the swelling in water. The bulking by the solutions was thus 31.6 to 32.4%. This indicates a real high efficiency of diffusion of resin forming chemicals into the cell walls of wood under the conditions used.

The efficiency with which resin is formed from the weight of resin forming material taken up within the wood was calculated on the basis of one mole of water being lost for each mole of furfuryl alcohol polymerized (0.815) and all catalyst and water lost as vapor or from leaching subsequent to the final water soak. The efficiency of polymerization is then the final resin content of the wood in weight percent, divided by the original solution content of the wood before polymerization times one minus the initial fractional moisture content of the wood times the fraction of the weight of the treating solution that was furfuryl alcohol times 0.815. These efficiencies together with the antishrink efficiencies are given in Table I for the matched yellow poplar sticks 9.5 cm long and in Table II for the Douglas fir and Engelman spruce cross sections. Under conditions where polymerization was virtually complete (18 hr. cure) the efficiencies were 90% or better using both zinc chloride and citric acid catalysts. When formic acid was used as the catalyst, the efficiency was significantly lower. When the curing time was reduced to 6 hours high efficiency of polymerization resulted only at zinc chloride concentrations of 1% and 5%.

Mechanical Properties

Static bending tests were made on nine treated yellow poplar sticks (40 x 1 x 1 cm) (three each with 2.5% zinc chloride, 2.5% citric acid and 5% formic acid catalyst cured for 18 hr. at 120°C), and four control sticks. Loading was in the tangential direction. The resin contents ranged from 40 to 72%, ASE values ranged from 72 to 77% (av. 74%). The average relative stress to proportional limit was 1.3, the average relative modulus of rupture was 0.89 and the average relative modulus of elasticity was 2.0. The increases in the stress to proportional limit and the modulus of elasticity are due to stiffening of the fibers. The loss in modulus of rupture resulted from acid embrittlement of the fibers. There was no significant difference due to curing with the different catalysts except in the case of the relative modulus of rupture which averaged 0.99 for the specimens

Table I. Efficiency of Furfuryl Alcohol Resin Treatment of End Matched Yellow Poplar Sticks (9.5 x 1.0 x 1.0 cm) Cured at 120°C in Aluminum Foil, The Antishrink Efficiency (ASE) and the Relative Toughness

Catalyst	No. Spec.	Av. Solution Content Wt.%	Cure Time hr.	Av. Resin Content Wt.%	Theoretical Resin Content Wt.%	Efficiency %	ASE %	Av. Relative Toughness[1/]
5% formic acid	9	91.5	18	51.2	66.5	77.1	70.6	0.78
1% citric acid	9	102.3	18	67.1	74.3	90.3	73.6	.55
1% zinc chloride	6	91.8	18	55.9	59.5	90.7	73.4	.57
0.2% zinc chloride	6	73.8	6	20.7	57.2	36.2	27.7	.74
1% zinc chloride	6	74.4	6	49.7	57.6	86.3	54.6	.58
5% zinc chloride	6	71.2	6	53.7	53.2	97.3	68.8	.33

1/ U. S. Forest Products Lab. toughness test (9) 3 inch span top weight. 30° angle, pull in tangential direction.

Table II. Efficiency of Furfuryl Alcohol Resin Treatment of Douglas fir and Engelmann Spruce cross Sections Cured at 120°C in Aluminum Foil, the Antishrink Efficiency (ASE) and the Relative Abrasion Resistance.

Catalyst	No. Spec.	Av. Solution Content	Cure Time	Av. Resin Content	Theoretical Resin Content	Efficiency	ASE	Relative Abrasion Resistance
		wt. %	hr.	wt. %	wt. %	%	%	
		1/						2/
Douglas Fir								
5% formic acid	8	81.5	18	47.7	59.4	80.5	73.0	0.71
1% citric acid	4	62.9	18	43.3	45.8	94.5	77.2	.64
1% zinc chloride	4	77.8	18	55.3	56.6	97.5	74.1	.63
Engelmann Spruce								
5% formic acid	4	113.3	18	62.7	82.5	76.3	69.2	.66
1% citric acid	4	103.9	18	67.5	75.7	89.5	72.5	.63
1% zinc chloride	4	124.1	18	88.1	90.3	97.3	73.6	.63
0.2% zinc chloride	4	143.0	6	66.7	104.2	64.0	61.7	.83
1% zinc chloride	4	141.0	6	98.2	102.8	95.5	72.0	.80
5% zinc chloride	4	149.0	6	106.0	108.5	97.9	66.7	.73

1/ Original moisture content of wood before treatment, 6%
2/ Loos Abrader (10)

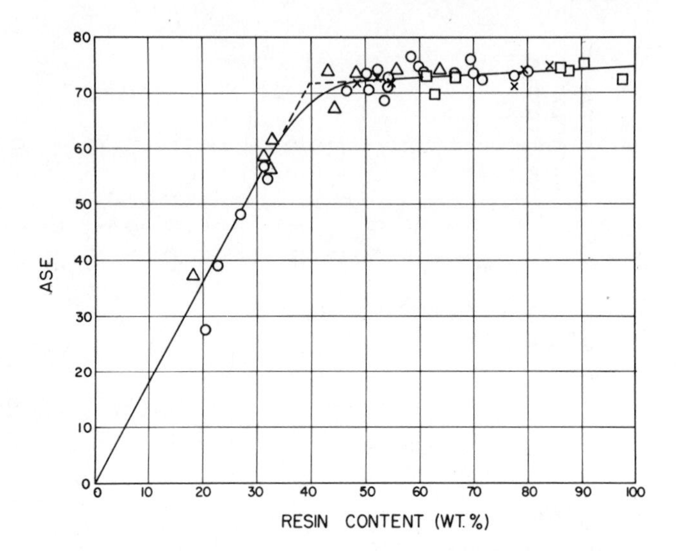

Figure 1. Antishrink efficiency (ASE) vs. furfuryl resin content. ○, yellow poplar sticks; △, Douglas fir cross sections; □, Engelmann spruce cross sections; ×, loblolly pine cross sections

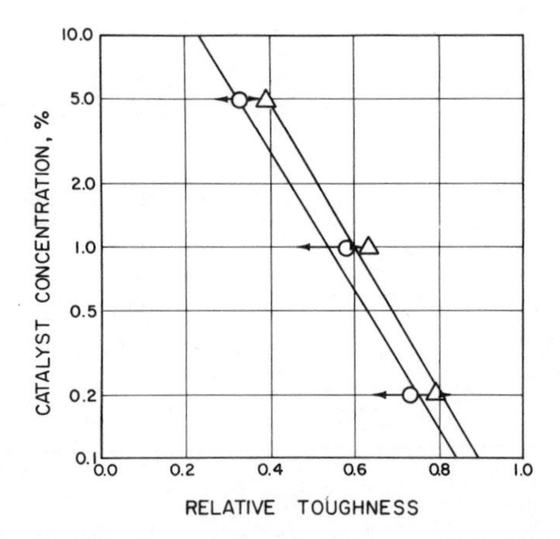

Figure 2. Relative toughness vs. logarithm of the zinc chloride catalyst concentration for end-matched furfuryl alcohol resin treated yellow poplar sticks. ○, average relative toughness of resin-treated specimens. △, average relative toughness of specimens impregnated with the same aqueous catalyst concentration and subjected to the same heating and time cycle with no furfuryl alcohol present. Arrows indicate range of relative toughness values for the resin-treated specimens.

cured with the formic acid catalyst. This higher value is probably due to somewhat less embrittlement of the fibers by formic acid.

Toughness determinations were made on the matched 9.5 cm long yellow poplar sticks using a U.S. Forest Products Laboratory toughness tester (9). One stick of each end matched four sticks served as an untreated control. The other three sticks were treated identically or with three different concentrations of zinc chloride catalyst. The results are given in Table I. Relative toughnesses ranged from 0.33 to 0.78. Formic acid catalyst embrittled the wood less than zinc chloride and citric acid but with some loss in both polymerization and antishrink efficiences. Figure 2 shows that the relative toughness is approximately inversely proportional to the logarithim of the zinc chloride catalyst concentration. Most of the toughness loss is due to acid hydrolysis under the curing conditions used. This is shown in Figure 2 by data for sticks containing the same abundance of zinc chloride catalyst but no furfuryl alcohol that were subjected to the same heating conditions as those in which the resin was formed.

Edge abrasion resistances were determined on 3 mm thick by 4.5 cm by 4.5 cm cross sections of treated and untreated Douglas fir and Engelmann spruce with a simple rotating disc - rocker arm mounted abrader developed by Loos (10) especially for tests on small thin cross sections. Sandpaper (120 grit) was mounted on the under surface of the rotating disc that was pressed against the radial edge of the cross sections with a pressure of one Kg. The disc was rotated at a speed of 100 revolutions per minute for 100 revolutions, determined with a revolution counter. Loss of tangential dimension was determined with a dial gauge to one thousandth of an inch. As the abrasiveness of the sandpaper decreased with use only slightly between the first 100 revolutions and 1000 revolutions, measurements were alternately made on treated and untreated control specimens repeated five times. Relative abrasion resistances (loss in tangential dimension for the controls divided by loss for the treated specimens) were averaged for tests 2 through test 5 for each set of specimens. Fresh sandpaper was used for each new set of specimens. Table II shows that furfuryl alcohol resin treatment reduces the abrasion resistance significantly but far less than for heat treatment and formaldehyde cross linking (3). There is a tendency for specimens catalized with formic acid to give a slightly higher abrasion resistance than these catalized with citric acid and with zinc chloride but the difference is hardly significant. The loss in abrasion resistance, like the toughness, is probably due to a combination of the embrittling effect by the catalyst and the stiffening effect by the resin within the cell walls. The abrasion resistance, like the toughness, is roughly inversely proportional to the logarithm of the catalyst concentration (see Figure 3).

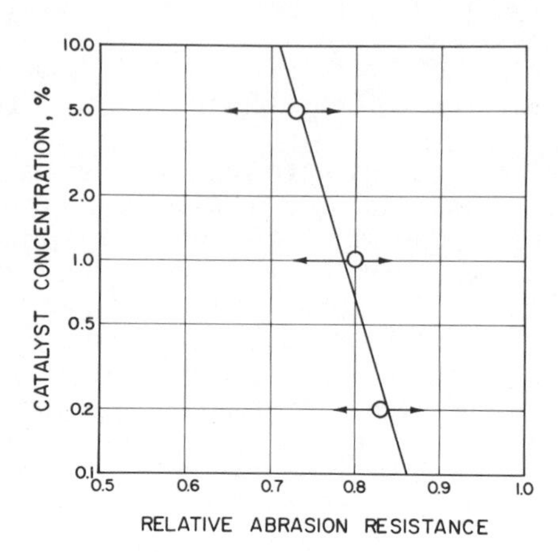

Figure 3. Relative abrasion resistance vs. logarithm of the zinc chloride catalyst concentration for end-matched cross sections of furfuryl alcohol resin-treated Engelmann spruce. Arrows indicate range of relative abrasion resistance values.

Table III Effect of furfuryl alcohol resin treatment of wood on the weight loss due to decay under ASTM soil-block tests for 12 weeks. 1/

Culture	Species	Resin content %	weight loss 2/ %
Lenzitestrabea	Yellow Poplar	0	40.50
		50	1.62
		81	0.02
	Loblolly Pine	0	36.90
		34	0.12
		66	0.10
Polyporus-versicolor	Yellow Poplar	0	51.30
		50	1.30
		81	0.02

1/ Thanks are extended to Professor Ellis Cowling and Gerald Preston of the Plant Pathology Dept., N. C. State University for making these tests.

2/ Average of four replications

These tests show that furfuryl resin treatment of wood imparts to it a high degree of decay resistance.

Abstract

Douglas fir, Engelman spruce, and loblolly pine cross sections and Yellow poplar sticks were treated with furfuryl alcohol resin using zinc chloride, citric acid, and formic acid catalysts giving the wood a distinctive black color. Antishrink efficiencies of 69 to 75% were obtained with 1% zinc chloride and with 1% citric acid catalysts when cured at 120°C for 18 hours, with a treatment efficiency of 90% or better. Formic acid (5% concentration) gave similar antishrink efficiencies with significantly lower polymerization efficiency but a significantly higher modulus of rupture and toughness and a slightly higher abrasion resistance. Curing time can be reduced to 6 hours using 1% zinc chloride without a significant loss in properties.

Literature Cited

(1) Stamm, A. J., "Wood and Cellulose Science", Ronald Press Co. N. Y. (1964).

(2) Stamm, A. J. "Solid Modified Woods", Chapter 2, Vol. 2 "Principles of Wood Science and Technology" by Kollmann, F. F. P. Kinzie, E., and Stamm, A. J., Springer-Verlag, Berlin, (1975).

(3) Stamm, A. J. and Seborg, R. M., "Minimizing Wood Shrinking and Swelling: Treating with Synthetic Resin-Forming Materials", Ind. Eng. Chem. (1936) 28 1164-1170.

(4) Goldstein, I. S. "Impregnations of Wood to Impart Resistence to Alkali and Acid". Forest Products J. (1955) 5, 265-267.

(5) Goldstein, I. S. "Impregnating Solutions and Method of Impregnation Therewith" (1959) U. S. Patent 2,909,450.

(6) Stamm, A. J. and Harris, E. E. "Chemical Processing of Wood", Chemical Publishing Co., Inc. N. Y. pg. 511 (1953).

(7) Goldstein, I. S. and Dreher, W. A. "Stable Furfuryl Alcohol Impregnating Solutions" Ind. Eng. Chem. (1960) 52, 57-58.

(8) Stamm, A. J. "Penetration of Hardwoods by Liquids", Wood Sci. and Tech. 7 (1973) 285-296.

(9) U. S. Forest Products Lab. "Forest Products Lab. Toughness Testing Machine", Report No. 1308. (1941).

(10) Loos, W. E., "Modification of Wood and Paper Properties by B^{10} (n,α) Li^{7} Initiated Graft Copolymerization", PhD thesis in Wood Sci. and Tech. North Carolina State Univ., Raleigh, N. C. (1966) pg. 33-35.

10

Adhesion to Wood Substrates

J. D. WELLONS

Forest Products Department, Oregon State University, Corvallis, Ore. 97331

Glued wood products continue to capture an increasing proportion of the market. The world population is expanding while the average living standard also is improving. Both strain our wood resources to the point that we must use wood as a scarce commodity. We are forced to use smaller trees, previously unused species, or residues from other manufacturing processes to make our products. Using wood from these new sources usually requires that smaller pieces of wood be glued to obtain products of the desired size. This trend likely will continue for many years, ever increasing the need for research on wood gluing processes and the potential for application of its results. Thus, wood gluing research is a major activity in many organizations.

This paper reviews the status of the art of adhesion to wood. The term "adhesion" here means the forming or the result of a durable interface, or zone of "intimate" contact, between one piece of wood and a second material, whether it be adhesive, coating, or another piece of wood. The review will focus primarily on solid wood, but will refer to selected literature on wood fiber or fiber-wall components if the concepts presented apply to solid wood. Several specific topics will be considered in detail: mechanisms of adhesion to wood; techniques for predicting whether or not adequate adhesion will occur or has occurred; wood properties affecting adhesion; and, finally, techniques for enhancing adhesion.

Adhesion is a necessary part of wood gluing, but it is only a part of that subject. Many topics in wood gluing are omitted from this review to allow more depth to those that are included. Advances in the chemistry and technology of wood adhesives, the rheology of wood adhesive joints, and details of specific industrial gluing processes are all important, but they must be left for others to review. Even within the subject of adhesion to wood, I will limit comment primarily to research since 1970. Marian and Stumbo (1, 2), Jurecic (3), Halligan (4), Patton (5), and Collett (6) have carefully reviewed the earlier contributions to adhesion to wood. Their efforts will not be duplicated except where necessary to provide perspective.

Mechanisms for Adhesion

As the science of adhesion has developed, various theories of adhesion have been advocated for one material or another. With wood as a substrate, mechanical interlocking, interdiffusion of polymers, intermolecular attractive forces, and covalent chemical bonding all have been proposed, either individually or collectively, to explain adhesion. In reality, no experiments reported to date have been able to disprove the existence of any one of these mechanisms, or to quantify their relative importance. A most exasperating feature of research on adhesion to wood is that factors presumed to be independent in experiments are never totally independent.

Mechanical Interlocking and Interdiffusion. The interlocking of microscopic adhesive tendrils in the pore structure of wood long has been considered a minor contributor to adhesion (7). It is based on the premise that spreading, penetration, wetting, and molecular proximity, which are prerequisite to every bonding mechanism, carry the adhesive into the minutest capilliaries where mechanical interlocks occur when the adhesive is fully cured. This mechanism has been discarded by most other material scientists (8). In fact, it has been ignored by most wood scientists of the last 10 years, often to be replaced by the concept of molecular interdiffusion. Horioka (9) did consider mechanical interlocking important in a limited way. He accounted for adhesive bond strength with wood of various densities by presuming that stress was transferred through interlocking adhesive "plugs" in the fiber lumen. This assumption did not eliminate the need to assume also intermolecular attractive forces to transmit stress from those adhesive plugs to the fiber walls, so Horioka considered mechanical interlocking to be secondary in importance to intermolecular attraction.

The interdiffusing of water-soluble adhesive polymer into the fiber walls of wood has been established clearly by recent research, after many years of debate. Tarkow et al. (10) showed that polyethylene glycol molecules up to molecular weight 3,000 (18- to 20-A° radius of gyration in water) could penetrate the fiber wall from aqueous solutions. Collett (11) used scanning electron microscopy (SEM) to examine plywood gluelines containing either lead oxide or rhodamine B dye in the phenolic adhesive. He believed that these techniques indicated adhesive penetration into fiber walls, but had no assurance that the lead oxide or rhodamine B dye had penetrated to the same depth as the adhesive. Smith and Côté (12) examined brominated phenol-formaldehyde resin adhesive on wood with an energy dispersive x-ray spectrometer interfaced with an SEM to confirm that resin molecules did penetrate the fiber wall from aqueous alkali. Their results are illustrated by Figure 1, which includes the SEM view of the fiber walls and lumen with the superposed trace of bromine concentration. Nearn (13) likewise concluded that phenolic resins penetrated the wood fiber

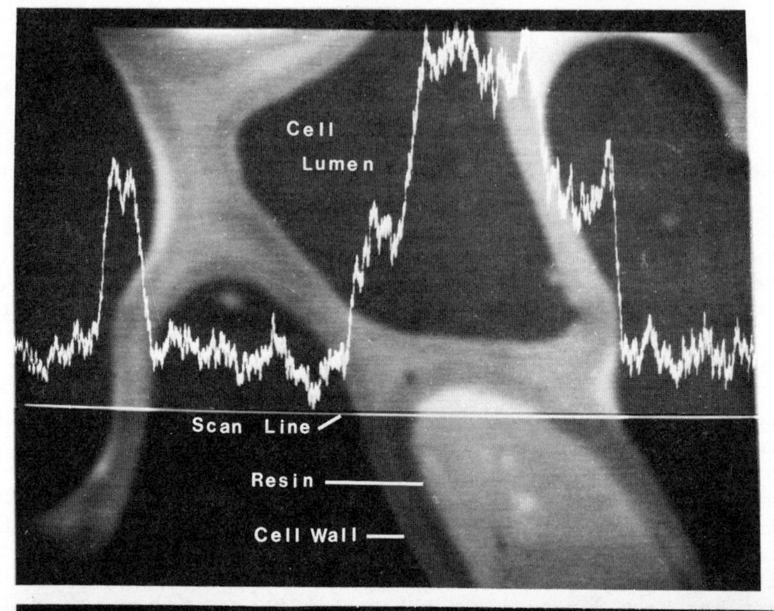

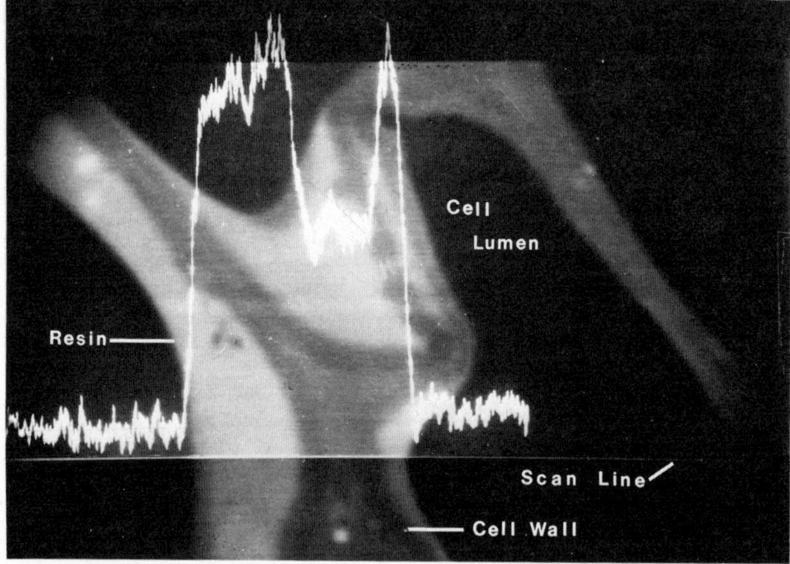

Journal of Paint Technology

Figure 1. SEM view of wood fibers impregnated with brominated phenol-formaldehyde resin. The line trace indicates bromine concentration in the wood structure (12).

wall, based on fluorescent microscopy, autoradiography of phenolic resins synthesized with C^{14}, and transmission electron microscopy of gluelines containing brominated phenolic resin.

The conditions for adhesive or coating polymers in nonaqueous solvents to penetrate the fiber wall are less well established. Schneider (14) reviewed the evidence for penetration of various coating polymers into the fiber wall and concluded that most studies did not allow differentation between solvent penetration and polymer penetration. Furuno et al. (15) claimed that methyl methacrylate did penetrate the fiber wall from methanol solution but not from dioxane. As might be expected, all authors report that swelling of the wood is a necessary prerequisite to penetration of the fiber wall by polymers.

Nearn (13) contends that fiber-wall penetration is essential for durable adhesive bonds to wood. He reached this conclusion by comparing glue-bond performance to fiber-wall penetration. This does not mean necessarily that interdiffusion of adhesive into the fiber wall is essential for adhesion. An adhesive bond could fail prematurely because of microscopic failures in the wood fiber near its surface. Such defects would be repaired by adhesive penetrating the fiber structure, resulting in better bond performance. Also, a zone of interpenetrating wood and adhesive molecules might allow a more gradual transition between the mechanical properties of the adhesive and the wood, also improving bond performance, but not necessarily improving adhesion itself.

Intermolecular Attraction and Adsorption. Specific adhesion, or the physical adsorption of adhesive polymers onto wood by Van der Waal's attractions and hydrogen bonds long has been considered the major mechanism of adhesion between wood and adhesive or coating polymers. This conviction led to the detailed thermodynamic descriptions of the adhesion process and the concepts of wetting and spreading, as reviewed by Marian and Stumbo (2) and Collett (6). The major additions to our knowledge of this important subject during the last decade have to do with quantifying the process of adsorption of polymers onto wood from solution.

Proof that polymers are adsorbed onto wood from solution came from the work of Tarkow and Southerland (16) working with polyvinyl acetate dissolved in carbon tetrachloride and benzene. Okuro (17) extended this understanding to aqueous systems by showing that methylolated phenol was adsorbed onto wood from both benzene and water. He showed further that substantially greater quantities of o-methylolphenol were adsorbed from the water and speculated that the difference resulted from swelling, which allowed interdiffusion of this phenol into the fiber wall. The effectiveness of intermolecular forces at causing adsorption of polymers on wood was determined by Mizumachi and Fujino (18) and Mizumachi and Kamidohzono (19). The first study used dynamic mechanical properties of styrene-butadiene copolymers adsorbed on wood; the second study used dielectric properties of polyvinyl acetate adsorbed on wood components. Both studies verified that the wood immobilized chain segments of the synthetic polymer to a greater degree than in the bulk polymer.

Although none of these studies of polymer adsorption help us determine the relative importance of intermolecular attraction in adhesion, they collectively do establish that this mechanism is exceedingly important as a first step, and in some instances the only step, in the formation of wood-adhesive interfaces.

Covalent Chemical Bonds. The possibility of covalent chemical bonding between wood and adhesive has been established. Although early efforts to resolve this issue failed to distinguish between covalent bonding and entanglement of polymer chains, recent research seems to have avoided this difficulty. Troughton (20) and Troughton and Chow (21) established that the kinetics of acid hydrolysis of melamine-formaldehyde and urea-formaldehyde adhesives depended on whether or not the adhesive was cured in the presence of wood components. Formaldehyde was released from melamine-wood systems faster and at a substantially lower activation energy than from the melamine adhesive. From urea-wood systems, formaldehyde was released slower, but at a slightly higher activation energy, than from the urea adhesive. The same kinetic parameters were obtained when wood meal was replaced with isolated cellulose or lignin. These effects were assumed to mean that wood-adhesive covalent bonds were being broken. The authors proposed (equation I) that the covalent bonds were oxymethylene bridges formed by condensing the methylolated adhesive with aliphatic hydroxyl groups on cellulose or lignin.

$$R\text{-}CH_2OH + \begin{cases} HO - \text{Cellulose} \\ HO - \text{Lignin} \end{cases}$$

$$\Downarrow$$

$$\left.\begin{matrix} R\text{-}CH_2\text{-}O\text{-}\text{Cellulose} \\ \text{or} \\ R\text{-}CH_2\text{-}O\text{-}\text{Lignin} \end{matrix}\right. + H_2O \qquad \text{I}$$

R represents the remainder of the methylolated glue molecule. Ramiah and Troughton (22) reinforced these convictions by differential thermal analysis of melamine- and urea-cellulose glue mixtures. They found peaks in the thermograms of glue-cellulose or glue-cellobiose mixtures that were not present in the thermograms of any of the individual components. Acetylation of the hydroxyl groups in the "cellulosic" material eliminated those unexplained peaks, resulting in thermograms that were the sum of the individual components. The acid hydrolysis and thermal decomposition of phenolic adhesives were too slow to allow interpretation about wood-phenolic covalent bonds.

Polymerization kinetics have been used to question whether or not phenolic adhesives covalently bond to wood. Chow (23) estimated the rate of cure of phenol-formaldehyde in the presence of cellulose, cellobiose, and glucose by measuring the ultraviolet absorbance of the water solubles from the partly cured resin. He found lower activation energy for resin cure when carbohydrates were present and postulated a mechanism analagous to that proposed in equation I. The results of Mizumachi and Morita (24) are less convincing, however. They also examined the cure of phenolic adhesives, but in the presence of wood meal using differential thermal analysis (DTA). Their study indicated that the activation energy of resin cure could be increased, decreased, or unchanged by including wood meal, depending on the species selected. DTA, however, is based on the net thermal effect of not only resin cure but any other thermal responses of extractive and fiber-wall components in the wood.

The best evidence for covalent bonding between wood and phenol-formaldehyde adhesives was provided by Allan and Neogi (25). They measured the quantity of 3,5-dibromo-4-hydroxybenzyl alcohol reacted with α-cellulose and extractive-free lignocellulose fiber. The ortho bromine atoms blocked further polymerization, which reduced the likelihood of physical entanglement. The possibility of adsorption was discounted from the lack of interaction between O-bromophenol and the wood components. Allan and Neogi found little or no reaction with the α-cellulose but a substantial reaction with lignin-containing wood fiber; they proposed (equation II) that the methylolated phenol condensed with lignin at ortho positions unoccupied by methoxyl groups.

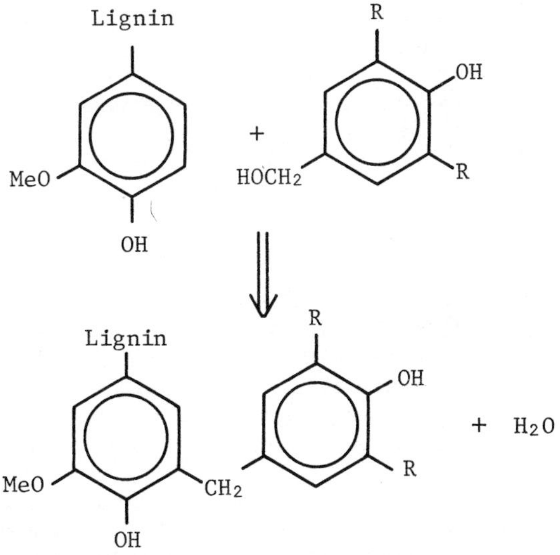

Lignin
MeO
OH
+
R
OH
$HOCH_2$
R

⇓ II

Lignin
MeO
OH
CH_2
R
OH
R
+ H_2O

The lack of reaction between methylolated phenol and cellulose reported by Allan and Neogi seems to contradict the findings of Chow and coworkers. One possible explanation for this disparity could be the difference in available free formaldehyde in their systems. Allan's model phenolic adhesive would have the equivalent of only one mole of formaldehyde per mole of phenol and would not be expected to have significant quantities of free formaldehyde. The resins used by Chow and coworkers had about 2 moles of combined formaldehyde per mole of phenol. Such resins are able to release formaldehyde during cure when condensation occurs between two methylol groups. This formaldehyde might then add at the aliphatic hydroxyls on cellulose or lignin resulting in condensation, as proposed by Chow, between the methylolated wood components and the phenolic resins.

O'Brien and Hartman (26) studied the interface of a model system--epoxy resin, regenerated cellulose fibers--by attenuated total reflectance infrared spectroscopy. They compared spectra of the components to spectra of epoxy cured on cellulose and found for the mixture a diminished hydroxyl absorption (3,350 cm^{-1}) and C-O stretching (1,050 cm^{-1}), and disappearance of the epoxy band (915 cm^{-1}). From this they concluded that covalent bonding does occur between the epoxide groups and cellulose hydroxyls.

In summary, the cumulative evidence seems to point beyond doubt to covalent bonding between wood and adhesive as a reality, especially when formaldehyde-based adhesive resins are used. But is covalent bonding essential to provide water-proof adhesive bonds? Some argue that many polar materials are held together cohesively by nothing more than intermolecular physical attractions, and that because these materials resist water, adsorption is sufficient for water-proof adhesive bonds. Others argue that the swelling of wood and adhesive in water would eventually disrupt even an efficient array of hydrogen bonds, requiring water-impervious covalent bonds to provide durable adhesion. Unequivocal experiments to resolve these arguments are yet to be reported, so we must speculate still about the relative importance of these two mechanisms.

Predicting Adhesion

A major thrust to moving the art of wood gluing more toward a science has been aimed at successful prediction of when adhesion has, or will, occur between adhesive and a woody substrate. Much of this effort has been in wettability and scanning electron microscopic (SEM) studies. Judging from existing knowledge, prediction of adhesion is still empirical.

Wettability. Good wetting or intimate molecular contact is certainly well established as a necessary condition for adhesion. Thus, early research was preoccupied with thermodynamic estimates of the surface free energy of wood. Collett (6) amply reviewed the subject of contact angles, and Iylengar and Erickson (27) additionally considered the value of solubility parameters to predict when the adhesive and wood are compatible. These concepts are certainly important when assessing new adhesive systems, but have serious drawbacks as predictors of bond quality with the aqueous cross-linking resins most commonly used for wood.

One difficulty with some of the contact angles and resulting critical surface tensions reported in the literature for liquids on wood is that they are nonequilibrium values. When water or aqueous solutions are used as the wetting medium, water is absorbed by the wood, resulting in swelling and a change in wettability with time. Chen (28) correlated contact angles to the spreading of adhesive droplets on wood surfaces. He found that wettability increased substantially as moisture content increased. Jordan and Wellons (29) showed that contact angles of aqueous alkali on wood decreased substantially over a 4-minute period (Figure 2). In fact, varying contact time between wood and liquid was more important than wood drying temperatures at influencing contact angle values, even though drying temperature is a well-established factor affecting adhesion.

A second difficulty with many of the contact angles reported for liquids on wood is that they do not match adequately the gluing conditions they are supposed to predict. Many researchers have used distilled water. But Jordan (30) found that distilled water formed droplets of very high and stable contact angles (about 90°) on selected veneers. On these same veneers an aqueous solution of NaOH (pH = 11) formed contact angles of about 40° that diminished to zero with time. For adequate prediction of the performance of an adhesive, the wetting liquid certainly needs to match the adhesive in its ability to solubilize components on the wood surface, swell the wood surface, and so on. The best alternative might be to measure contact angles with the adhesive, except that observed contact angles then also depend on adhesive viscosity (28). Thus, the apparent equilibrium contact angle of an adhesive on wood may reflect resistance to viscous flow rather than thermodynamic equilibrium. In addition, many wood adhesives are used at temperatures of 100 to 150°C. We have no experimental basis for extrapolating wettability data to these higher temperatures.

A final difficulty in interpreting wettability data arises when one attempts to state what factors about wood really control contact angles. Luner and Sandell (31) and Lee and Luner (32) estimated critical surface tensions from contact angles of nonpolar liquids on assorted cellulose, hemicellulose, and lignin films. They found surprisingly little difference in wettability for all fiber-wall components (Table I) and concluded that observed differences in wettability must be related to other wood properties such as extractives, density, porosity, and related differences in roughness, of which extractives would have a far greater effect on adhesion *per se* than the other factors, although all would influence various gluing parameters. The authors acknowledge, however, that carbohydrates and lignin do differ in ability to interact with hydrogen bonding solvents, so lignin likely is somewhat less wettable than carbohydrates when exposed to polar solvents.

Table I. Critical surface tension of wood components (29, 30).

Sample	Critical surface tension, erg/cm^2
Cellulose	35.5 - 42.5
Hardwood xylan	34.0 - 36.5
Softwood xylan	35
Galactoglucomannan	36.5
Arabinogalactan	33
Lignin	33 - 37

Wood Science

These difficulties with wettability data are well illustrated in the research of Hse (33). He found that the contact angle between phenolic adhesives and southern pine veneer decreased with increasing caustic in the adhesive (synonymous with swelling power of the adhesive for wood); that the contact angle increased with increasing formaldehyde content (synonymous with higher molecular weight and degree of branching); and finally he found that glue-line performance increased as contact angle increased--exactly the opposite one would expect if contact angle were the factor controlling adhesion. Under this condition, contact angle seemed to be predicting the ability of the adhesive to penetrate (or over-penetrate) the wood capillary structure.

These comments should not be interpreted to mean that measures of wettability are useless at predicting adhesion. They do seem clearly to indicate that contact angles and critical surface tensions reported for wood are not necessarily thermodynamic quantities or well-defined material parameters. Because most contact angles are dynamic values, they should be interpreted with caution and considered as relative measures of adhesion, for which the absolute scale is yet unknown. Further, we need to keep in mind that although wetting is necessary for adhesion, it may not be the limiting factor in many real situations.

Glueline Microscopy. Another technique for evaluating adhesive bond performance is to examine the glueline with light and electron microscopes. As the resolving power of electron microscopes has improved, some authors have attempted to use them to evaluate adhesion. A major limitation lies in the fact that adhesion occurs over a distance of less than 10 A°, regardless of which mechanisms are acting. No microscopic technique currently is able to resolve such small distances on wood substrates. On the other hand, if the microscope is able to detect interfacial failure at any level of magnification we are assured of no adhesion. The basic techniques of electron microscopy, especially scanning microscopy of gluelines in wood products, have been reviewed by Collett (11), Smith and Côté (34), and more recently Parham (35).

The major focus of glueline microscopy has been: the zone of failure, the continuity of the glueline, and penetration of the adhesive into the wood fiber structure (as discussed previously). Similar glueline characteristics are reported by Fengel and Kumar (36), Koran and Vasishth (37), Borgin (38), and Hare and Kutscha (39). All report few, if any, interfacial glueline failures from mechanical tests. These observations confirm the long-held belief that once adhesion occurs, true adhesion failure is unlikely with most wood adhesives. Only in those limited instances when wood or adhesive properties prevent good adhesion does an interface fail.

Bond Durability. Because we lack an adequate theoretical basis for predicting adhesion, many empirical studies of bond performance have been reported. Some most interesting and potentially applicable studies of adhesive durability have been reported recently (40, 41, 42, 43, 44) but are mentioned only in passing because the thrust of this review is adhesion, not adhesives.

Most empirical estimates of glue-bond quality evaluate adhesion and cohesion simultaneously by applying combinations of thermal, mechanical, or chemical stress to the bonded assembly. Northcott *et al*. (45) reported on a major study comparing the effectiveness of many accelerated-aging tests of plywood glue bonds. One major conclusion of that study was that no one accelerated-aging test was best for detecting all bond inadaquacies. Glue bonds can fail for several altogether different reasons. A second conclusion from that study was that glue-bond degradation could be considered analagous to a chemical reaction in that the logarithm of bond life was linearly related to the inverse of absolute temperature (see Figure 3). Similar studies of bond durability of coatings on wood and fiber substrate have been reported by Nack and Smith (46) and Marck (47).

Using exposure to higher temperature to accelerate bond degradation has been studied in detail by Gillespie and coworkers during the last 10 years. Most recently (48), effect of dry heat was compared to earlier work with wet heat. They concluded that the Arrhenius model does describe the degradation of the bonds,

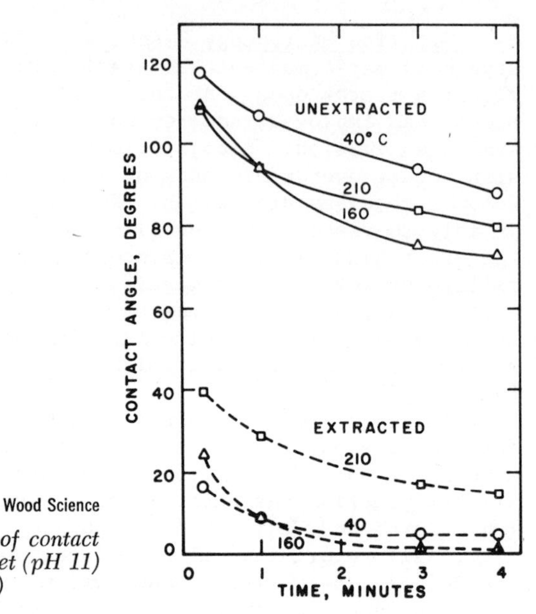

Wood Science

Figure 2. Change with time of contact angle of aqueous NaOH droplet (pH 11) on keruing wood (29)

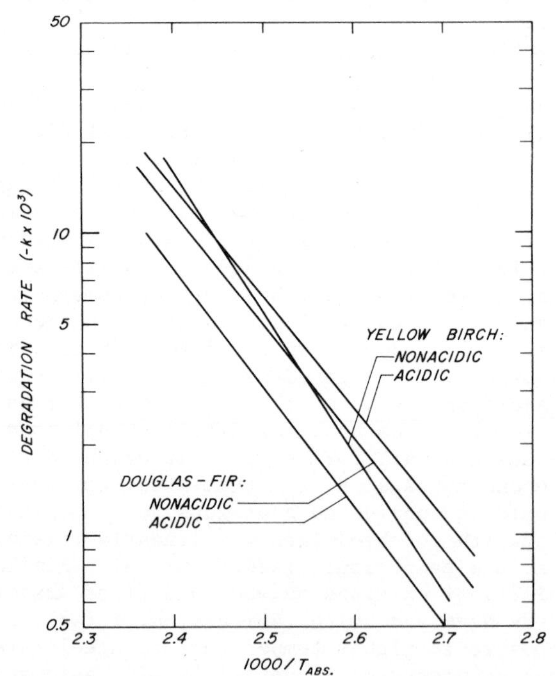

Forest Products Journal

Figure 3. Temperature dependence of rate of wet-strength loss caused by dry heat in joints of yellow birch and Douglas-fir with acidic and nonacidic adhesives (48).

but that the degradation is almost totally cohesive--in the glue with wet heat and hydrolizable adhesives and otherwise in the wood substrate with wet or dry heat. Whether or not these tests would detect marginal adhesion was not certain.

The work of Northcott et al. (45) led the Weyerhaeuser Company to examine in detail boil-chill-dry cyclic tests for wood-bond degradation. Kreibich and Freeman (49) found that 10 minutes boiling in water, 3.75 minutes cooling in air at room temperature, and 60 minutes drying at 107°C efficiently degraded adhesive bonds between parallel laminated wood. Their studies have resulted in a new practice being recommended by the American Society for Testing and Materials (ASTM D-3434-75). Wilkie and Wellons (50) found this cyclic test most efficient with plywood glue bonds, whether the bond was inadequate because of poor adhesion or low cohesive strength of the glue.

Most empirical bond-durability tests require some measure of the proportion of failure occurring in the wood or in the glueline (including both adhesive or cohesive failures). Such estimates are, at best, artibrary and time consuming. Recent studies by Carroll (51) and Carroll and Warren (52) offer hope of simplifying such estimates. They propose determining the proportion of specimens with greater than a prespecified percentage wood failure, rather than determining the average wood-failure area. Their studies show that these estimates are related to each other both theoretically and empirically.

Wood Properties That Affect Adhesion

Many studies through the years have shown that selected properties of both the wood and the adhesive do affect adhesion. Although many wood and adhesive properties have been studied (see, for example, 53, 54, 55), wood extractives and thermal inactivation of the wood surface have received considerable attention recently and will be discussed in detail.

Wood Extractives.

Glue bonds can be affected in many ways by extraneous components on or near the surface of wood. Chen (56) showed that extractive removal did improve bond performance and that the extractives, in part, reduce wettability. This effect is reflected also in Figure 2 from Jordan and Wellons (29). Hergt and Christensen (57) also showed that extractives can slow water adsorption from the adhesive into the wood, thus slowing development of cohesion in the adhesive.

Plomley et al. (58) discussed in detail (Table II) and gave examples of the various ways extractives could interfere with the formation of a glue-wood interface--forming a barrier at the interface that may prevent wetting or cause mechanical weakness, changing adhesive properties and altering the normal flow and spreading of adhesive on the wood surface, or even preventing cure of adhesive at the interface. Wellons et al. (59) gave evidence that both of the last two mechanisms are responsible for poor bonding between phenolic adhesives and wood of the Dryobalanops species. Examples of the effects of extractives on adhesion will likely become more prevalent as more tropical hardwoods are used in glued wood products.

Table II. Possible behavior of surface contaminants, especially extractives, on wood surface (56).

Reaction with adhesive	Location of contaminant		
	Layer between wood and adhesive	Dispersed by adhesive, but concentrated near interface	Thoroughly dispersed in adhesive
No reaction	Contaminant layer determines bond strength	Properties of adhesive-contaminant mixture determine bond strength	
Reacts - no adverse effect on adhesive cure	Adhesive-contaminant reaction product determines bond strength		
Reacts by accelerating or retarding adhesive cure	Properties of modified adhesive affect bond strength		

Forest Products Journal

Thermal Inactivation. Heating the wood surface to temperatures above 150 to 200°C definitely has been shown to cause poor adhesion with phenolic adhesives and, possibly, urea adhesives. Early research (60) verified that overheated wood was less wettable and tended to absorb less water from the adhesive. Whether this is caused by extractives, pyrolysis, oxidation of hydroxyl groups, or other chemical reactions at the wood surface has been the subject of much debate. Further, the mechanisms of thermal or "surface" inactivation may vary from species to species.

Chow and coworkers have studied the thermal inactivation of white spruce veneer in detail (61, 62, 63). They were not able to correlate bond quality to the concentration of fatty acids on the veneer surface. They found instead that surface inactivation was a function of time, temperature, and atmosphere. Even extractive-free veneer could be inactivated; the presence of extractives did accelerate the inactivation. The infrared spectra of the veneer surface showed increased carbonyl absorption (1,730 cm^{-1}) when veneer was heated in air to temperatures that caused inactivation. This led to the belief that oxidative carboxylation of hydroxyl groups caused the inactivation by eliminating hydroxyl groups that otherwise bonded covalently with the phenolic adhesive. To verify this concept further, the authors showed that treatment of the wet veneer surface with ozone even at room temperature would inactivate it to phenolic adhesives. In addition, spraying the veneer, before drying, with boron reducing agents prevented the inactivation (64). Although these experiments would seem to argue convincingly for oxidative carboxylation as the cause of inactivation, Chow and coworkers also found that heating wood in nitrogen caused inactivation, but at a much reduced rate. Thus, all of the initially proposed mechanisms for surface inactivation appear valid, to some degree, with oxidation of the surface possibly the most important.

Enhancing Adhesion

The fact that covalent bonding can be an important, and possibly necessary, contribution to water-proof adhesive bonds to wood has convinced many scientists to study methods of enhancing adhesion by increasing the probability of covalent bonding between wood and adhesive, or directly between wood particles. This subject is still in its infancy with solid wood, although pulp and textile fiber scientists have produced an enormous volume of literature from which wood scientists can draw.

One technique being used is to bond solid wood with monomeric materials that are highly reactive with hydroxyl groups such as those in the wood fiber wall. For example, Schoring *et al*. (65) synthesized particleboards using as a binder hexamethylene diamine and polyvinyl chloride at 140°C. They postulated that the polyfunctional amine covalently linked the wood components, especially lignin, to the polyvinyl chloride.

Collett (66) furthered the concepts of Schoring *et al*. (65) by oxidizing the wood surface to provide carboxyl groups for diamines and dialcohols to react with. The difunctional monomer was proposed as the bridge between wood particles through amide or ester linkages. Aqueous nitric acid and vapor mixtures of nitric oxide and oxygen (which form nitric acid *in situ*) were used with nearly equal effectiveness to oxidize the wood surface. Surprisingly good properties were obtained when oxidized particles were consolidated without the difunctional monomer. Hexamethylenediamine was most effective as the coupling agent, resulting in particleboards of specific gravity 0.8 to 0.9 with mechanical properties about 80 percent of those obtained for boards bonded with phenol-formaldehyde. Boards bonded with oxidized diamine, however, were superior to the phenolic-bonded boards in resistance to swelling in both boiling and cold water. But about 10 percent diamine, based on dry furnish, was consumed in the process. Other disadvantages of the process would be handling the nitric acid with its associated fumes plus the exothermic nature of the bonding reaction that could become unmanagable on a commercial scale.

Pohlman (67) used dibasic acids with catalysts to consolidate wood particleboards. He obtained composites that were nearly equivalent to boards bonded with phenol-formaldehyde by using the difunctional maleic anhydride with strong acid activators such as HCl and $FeCl_3$. The resistance of these boards to swelling in hot water was his basis for arguing that the bonding in part resulted from covalent coupling of wood particles by ester linkages to the difunctional acid. The activators used were quite strong oxidants, however, so bonding may have been, in part, from oxidative coupling of phenolics in the wood particles. He did show that the difunctional acid was essential to obtain durable bonds, however. The residual acid in the boards may pose problems with long-term durability, and the slightly higher density required to match the properties of phenolic-bonded boards would handicap this process.

Oxidizing wood and cellulose surfaces by treatment with ozone has been studied by Goring and Suranyi (68) and Kim and Goring (69). In the former study, cellulose bonded to itself with no binder, and in the latter wood veneers bonded to films of polyethylene and polystyrene by heat and pressure alone. Exposing the wood or cellulose surfaces to ozone substantially improved the dry bond strength. Treatment of the synthetic polymer films with ozone was especially effective at improving bond strength, whether or not the wood had been treated. Evidence was offered to substantiate oxidation of the surfaces (increased IR absorption at 1,720 cm^{-1}), but no estimate could be made as to whether the enhanced bonding was only because of increased polarity and thus increased physical attraction, or because of covalent bonds. Water resistance of the bonds was not measured.

Oxidative coupling of lignocellulosic materials has received major emphasis as a technique for improved adhesion to wood. The underlying principles of oxidative coupling of phenolic compounds are well established (70) and have been practiced to some extent for many years in the manufacture of wet-process hardboard by using ferric sulfate, aluminum sulfate, and sulfuric acid to enhance board properties. Allan *et al.* (71) measured the uptake of bromophenols by α-cellulose, masonite fiber, and kraft pulp in the presence of strong oxidants (potassium ferricyanide, ferric chloride, and a combination of potassium persulfate and ferrous sulfate). They proposed that the bromophenol covalently coupled to lignin-containing fibers by a free radical mechanism. Covalent bonding and the mechanism were inferred from the inability to remove the bromophenol from lignified fiber-oxidant mixture by washing with both aqueous acid and base, compared to the easy washing of the bromophenol from the fiber when oxidant was omitted from the reaction.

Recent interest has been shown in adapting phenolic oxidative coupling to other wood products. Stofko (72) formed discs of wood flour, microcrystalline cellulose, and lignin after treatments to enhance covalent coupling between particles. He obtained composites resistant to hot water with consolidated lignin and wood particles that had been treated with either acids or acids in conjunction with oxidants (H_2O_2 + $ZrCl_4$ or NaClQ + $FeSO_4$ + H_2O_2). In fact, the lignin-powdered, brown-rotted wood bonded best with heat only (300°F). Stofko proposed that oxidative coupling of phenolics was a major contributor to the bond strengths observed, although acid-catalyzed condensation could have occurred as well. Cellulosic materials bonded only under strongly acid conditions, which favored hydrolysis and reversion of carbohydrates as well as possible oxidative coupling reactions.

Johns and Nguyen (73) used peroxyacids to stimulate bonding between laminations of solid wood. Their system consisted of reacting hydrogen peroxide and an organic acid in the presence of catalytic quantities of a mineral acid, to form the peroxyacid. Acetic acid was used with both HCl and H_3PO_4. The peroxyacid mixture was sprayed on wood surfaces and the reaction completed at 150°C in a hot press. The better bond strengths were 50 to 80 percent of those obtained for phenol-formaldehyde bonding, whether tested wet or dry, and, in general, those conditions that favored greater peroxyacid concentrations enhanced bond strength. Specifically, higher concentrations of H_2O_2 and H_3PO_4 gave better properties. The explosive nature of selected treating liquids and extreme variability of bond strength were serious disadvantages to this system.

These pioneering studies in enhancing adhesion to wood, presumably by covalent chemical bonds, are yet a long way from commercialization. Costs of reagents, hazards from some reagents, and variability in performance are immediate deterrents to their implementation. The lack of gap-filling capability may be an inherent weakness in some of these systems. But the opportunity exists to duplicate by new technology the bonding that holds wood fibers together in nature. I hope such pioneering research will continue.

Frontiers

This status report suggests that wood adhesion is still more art than science, in spite of the real progress indicated with each major topic. Selected key techniques need to be developed and implemented to further this science. I call these needs the real frontiers in the science of wood adhesion.

The first frontier is the chemical composition of wood surfaces. Various spectroscopic techniques have been used to describe vaguely the average composition of these surfaces, but we need techniques to quantify specifically the functional groups available for bonding both before and after "adhesion-promoting" treatments. Only then can we cease speculating about bonding mechanisms and about better treatments to enhance bonding. Metallurgical scientists are far ahead of us in this area.

The second frontier is a nondestructive technique to monitor the development of an adhesive bond in place. Neither the theoretical approach of surface free energy nor the empirical approach of bond destruction are as desirable as being able to measure adhesive and cohesive strength as they are developing in the glueline. By constantly observing the science of other materials, we may obtain ideas and techniques that will let us move these and other frontiers in wood adhesion.

Literature Cited

1. Marian, J. E., and Stumbo, D. A., Holzforschung (1962) 16(5), 134-148.
2. Marian, J. E., and Stumbo, D. A., Holzforschung (1962) 16(6), 168-180.
3. Jurecic, A., Tappi (1966) 49(7), 306-310.
4. Halligan, A. F., Forest Prod. J. (1969) 19(1), 44-51.
5. Patton, T. C., Tappi (1970) 53(3), 421-429.
6. Collett, B. M., Wood Sci. and Tech. (1972) 6, 1-42.
7. Brown, H. P., Panshin, A. J., and Forsaith, C. C., "Textbook of Wood Technology," Vol. II. McGraw Hill Book Co., New York. (1952) pp. 185-227.
8. Salomon, G., in R. Houwink and G. Salomon, "Adhesion and Adhesives," Vol. 1. Elsevier Publishing Co., Amsterdam. (1970) pp. 1-128.
9. Horioka, K., Proceedings, IUFRO, Div. 5, Sept. 22-Oct. 12, Republic of South Africa, (1973) 2, 503-527.
10. Tarkow, H., Feist, W. C., and Southerland, C. F., Forest Prod. J. (1966) 16(10), 61-65.
11. Collett, B. M., Wood and Fiber (1970) 2(2), 113-133.
12. Smith, L. A., and Côté, W. A., J. Paint Tech. 44(564), 71.
13. Nearn, W. T., Wood Sci. (1974) 6(3), 285-293.
14. Schneider, M. H., J. Paint Tech. (1972) 44(564), 108-110.
15. Furuno, T., Nagadomi, W., and Goto, T., Mokuzai Gakkaishi (1975) 21(3), 144-150.
16. Tarkow, H., and Southerland, C., Forest Prod. J. (1964) 14(4), 184-186.
17. Okuro, A., Bull. Govt. Forest Expt. Sta., Meguro, Japan, (1970) 230, 143-154.
18. Mizumachi, H., and Fujino, M., Holzforschung (1972) 26(5), 164-169.
19. Mizumachi, H., and Kamidohzono, M., Holzforschung (1975) 29 (6), 229-231.
20. Troughton, G. E., Wood Sci. (1969) 1(3), 172-176.
21. Troughton, G. E., and Chow, S., J. Inst. Wood Sci. (1968) 21, 29-34.
22. Ramiah, M. V., and Troughton, G. E., Wood Sci. (1970) 3(2), 120-125.
23. Chow, S., Wood Sci. (1969) 1(4), 215-221.
24. Mizumachi, H., and Morita, H., Wood Sci. (1975) 7(3), 256-260.

25. Allan, G. G., and Neogi, A. N., J. Adhesion (1971) 3(1), 13-18.
26. O'Brien, R. N., and Hartman, K., J. Polymer Sci., Part C, (1971) 34, 293-301.
27. Iylengar, Y., and Erickson, D. E., J. Appl. Polymer Sci. (1967) 11, 2311-2324.
28. Chen, C. M., Mokuzai Gakkashi (1972) 18(9), 451-456.
29. Jordan, D. E., and Wellons, J. D., Wood Sci. (1977), in press.
30. Jordan, D. E., M.S. thesis, Oregon State University, Corvallis, 1974.
31. Luner, P., and Sandell, M., J. Polymer Sci., Part C, (1969) 28, 115-142.
32. Lee, S. B., and Luner, P., Tappi (1972) 55(1), 116-121.
33. Hse, C. Y., Forest Prod. J. (1972) 22(1), 51-56.
34. Smith, L. A., and Côté, W. A., Wood and Fiber (1971) 3(1), 56-57.
35. Parham, R. A., Forest Prod. J. (1975) 25(12), 19-25.
36. Fengel, D., and Kumar, R. N., Holzforschung (1970) 24(6), 177-181.
37. Koran, A., and Vasishth, R. C., Wood and Fiber (1972) 3(4), 202-209.
38. Borgin, K., Proceedings, IUFRO, Div. 5, Sept. 22-Oct. 12, Republic of South Africa, (1973) 2, 51-57.
39. Hare, D. P., and Kutscha, N. P., Wood Sci. (1974) 6(3), 294-304.
40. Chow, S., and Hancock, W. V., Forest Prod. J. (1969) 19(4), 21-29.
41. Chow, S., Holzforschung (1973) 27(2), 64-68.
42. Steiner, P. R., Wood Sci. (1974) 7(2), 99-102.
43. Chow, S., J. Polymer Sci. (1974) 18, 2785-2796.
44. Chow, S., and Troughton, G. E., Forest Prod. J. (1975) 25(8), 54-57.
45. Northcott, P. L., Kreibich, R. E., and Currier, R. A., Forest Prod. J. (1968) 18(5), 58-65.
46. Nack, L. E., and Smith, F. W., J. Paint Tech. (1974) 46(592), 47-50.
47. Marck, T. C., J. Paint Tech. (1974) 46(592), 51-56.
48. Gillespie, R. L., and River, G. H., Forest Prod. J. (1975) 25 (7), 26-32.
49. Kreibich, R. E., and Freeman, H. G., Forest Prod. J. (1968) 18(12), 24-26.
50. Wilkie, G., and Wellons, J. D., Forest Prod. J. (1977), in press.
51. Carroll, M. N., Forest Prod. J. (1974) 24(4), 24-30.
52. Carroll, M. N., and Warren, W. G., Forest Prod. J. (1976) 26 (3), 40-43.
53. Hse, C. Y., Forest Prod. J. (1972) 22(9), 104-108.
54. Hse, C. Y., Holzforschung (1972) 26(2), 82-85.
55. Chow, S., Steiner, P. R., and Troughton, G. E., Wood Sci. (1975) 8(1), 343-349.

56. Chen, C. M., Forest Prod. J. (1970) 20(1), 36-41.
57. Hergt, H. F. A., and Christensen, G. N., Holzforschung (1972) 26(1), 26-31.
58. Plomley, K. F., Hillis, W. E., and Hirst, K., Holzforschung (1976) 30(1), 14-19.
59. Wellons, J. D., Krahmer, R. L., Raymond, R., and Sleet, G., Forest Prod. J. (1977), in press.
60. Northcott, P. L., Hancock, W. V., and Colbeck, H. G. M., Forest Prod. J. (1962) 12(10), 1-9.
61. Chow, S., Wood Sci. and Tech. (1971) 5, 27-39.
62. Troughton, G. E., and Chow. S., Wood Sci. (1971) 3(3), 129-133.
63. Chow, S., and Mukai, H. N., Wood Sci. (1972) 4(4), 202-208.
64. Chow, S., Forest Prod. J. (1975) 25(5), 41-47.
65. Schoring, P., Roffael, E., and Stegmann, G., Holz als Roh-und Werkstoff (1972) 30(7), 253-258.
66. Collett, B. M., Ph.D. thesis, University of California, Berkeley, 1973.
67. Pohlman, A. A., M.S. thesis, University of California, Berkeley, 1974.
68. Goring, D. A. I., and Suranyi, G., Pulp and Paper Mag. Canada (1969) 70(10), 102-110.
69. Kim, C. Y., and Goring, D. A. I., Pulp and Paper Mag. Canada (1971) 72(11), 93-96.
70. Taylor, W. I., and Battersby, A. R., "Oxidative Coupling of Phenols," Marcel Dekker Inc., New York, 1967.
71. Allan, G. G., Mauranen, P., Neogi, A. N., and Peet, C. E., Tappi (1971) 54(2), 206-211.
72. Stofko, J., Ph.D. thesis, University of California, Berkeley, 1972.
73. Johns, W. E., and Nguyen, N. T., Forest Prod. J. (1977), in press.

11

Bonding of Lignocellulosic Surfaces by Oxidative Treatment and Monomeric or Simple Polymeric Crosslinking Agents

D. L. BRINK, B. M. COLLETT, A. A. POHLMAN,
A. F. WONG, and J. PHILIPPOU

University of California Forest Products Laboratory,
1301 So. 46th St., Richmond, Calif. 94804

The work summarized in this paper was initiated on the basis of a unique difference observed between pulps prepared by the nitric acid and alkaline processes and a premise involving the phenomena concerned with the bonding of lignocellulosic surfaces. In bonding such surfaces it was assumed a predominant role could be played by graft copolymerization; i.e., primary bonding forces, whether such bonding concerned veneers to form plywood, particulates to form composite board, or fibers to form hardboard, paperboard, or paper. By increasing the amount of covalent bonding that would take place between lignocellulosic surfaces and crosslinking agents the role played by secondary type bonds; e.g., van der Waal's forces and dipole-dipole interactions, would be diminished. Accordingly, the properties of the bonded products, largely attributed to secondary types of bonds in conventional bonding systems, could be substantially modified. The objective of our work over a period of years has been to devise a system wherein covalent chemical bonds between the lignocellulosic surfaces and crosslinking agents play a significant or dominant role. One important consequence of such a system should be to substantially increase dimensional stability of products formed.

The contribution made by a given type of bonding in a given system cannot be precisely defined but can be qualitatively described, can be demonstrated, and with study can be quantitatively estimated. For example, recognized properties of wood fiber in paper compared to wood particles in composite products provide a basis for qualitatively comparing differences in bonding systems. Pulp fibers, properly prepared, have a relatively large external surface, are relatively flexible and, deposited randomly from water in a mat, conform readily to produce a structure having properties on drying that are attributed, in large part, to the occurrence of hydrogen bonding. Based on this phenomenon, a spectrum of remarkable products can be made including paper, paperboard and hardboard which exhibit a myriad of properties. One property not exhibited without additional treatment is dimensional stability. Wood flakes, or larger pieces of wood, on the other hand, are relatively rigid forms of lignocellulosic material. The surface area of these which can be brought into sufficiently intimate contact to form hydrogen bonds is so small as to be insignificant in developing adhesion; i.e., no adhesion results on drying such surfaces brought into contact under the best of conditions. In order to bond such surfaces an agent is required that will bridge the gap and adhere to adjacent surfaces - the conventional function of an adhesive. The great variety of properties that are imparted to composite wood products using different adhesives attests not only to differences in properties of the resins used but, in all probability, to the extent that various mechanisms of bonding participate in the adhesion process. For example, the particulate system bonded by urea-formaldehyde resins displays good physical properties until exposed to water. Then, because of the adhesive used, bonding is rapidly destroyed. If, however, a phenol-formaldehyde resin is used bonding is substantially more resistant to water because of the properties of the adhesive including, perhaps, covalent linkages developed between the lignocellulosic substrate and the adhesive.

The unique property of nitric acid pulps compared to alkaline pulps, noted above, involved the great difficulty incurred in rewetting and redispersion of once-dried nitric acid pulps to give suspensions of individual fibers (1). This phenomenon was again observed in pulps prepared in a study of the aqueous nitric acid system under pressure (2) and has been observed consistently in all subsequent work. Based on the observations in the work cited and the knowledge that such pulps are oxidized, it was hypothesized that functional groups or reactive moieties were formed in the lignocellulose which, on drying, formed covalent linkages with other groups in the lignocellulose. In effect, the resistance of the nitric acid pulps to redispersion in water was a demonstration of a self-bonding system involving bonds considerably stronger than those normally developed in paper. This system provided one of the requirements for the bonding system envisaged; i.e., a lignocellulosic surface modified with respect to functional groups that can undergo further reaction. A second requirement of the system envisaged was that any two surfaces to be bonded must approach within a distance that would permit formation of covalent bonds or, alternatively, a crosslinking agent must be provided to bridge the distance between two surfaces. A third requirement was that a sufficient number of such bonds be formed to confer characteristic properties to the bonded surfaces.

It was assumed that in bonding lignocellulosic surfaces only fiber, and fiber only to a limited extent, could fulfill the second requirement given above without the incorporation of a crosslinking agent. Thus, a crosslinking agent was assumed to be a requirement of the system and necessarily introduced its own set of requirements which had to be realized to develop the bonding system envisaged. One fundamental requirement of a crosslinking agent was that it should be difunctional or polyfunctional and that two or more functions should be capable of forming covalent bonds. A second requirement was that the crosslinking agent should have dimensions such that it would bridge the gap between lignocellulosic surfaces. Then the functional groups produced in the modified lignocellulose and those of the crosslinking agent could be induced to form a system involving covalent bonding; i.e., lignocellulosic particulate : crosslinking agent : lignocellulosic particulate. Of the various types of chemical linkages or physical forces that can be utilized, covalent linkages should produce the strongest and most durable bonded systems.

The hypothesis for the bonding system described above was conceived in work carried out using nitrogen dioxide as a bleaching agent. In that work and the subsequent investigation of the use of nitric oxide - oxygen ($NO-O_2$) and hydrogen peroxide as bleaching agents (3) gas phase as well as liquid phase reactions of lignocellulose were employed. The objective in the use of a gas phase reaction was to effect a greater selectivity of reaction than might be possible in a liquid phase. Reasons for the greater selectivity included the possibility of greater uniformity of reagent distribution throughout the cell wall of the particulate material being treated. Then, depending on reaction rates, the probability of reaction with the lignin moiety rather than polysaccharide moieties could be enhanced. Based on the background developed in pulping with nitric acid in the liquid phase and work that had been planned for using $NO-O_2$ in the gas phase it was assumed that the same reaction system could be used to selectively modify lignocellulose in different ways depending upon the selection of conditions used. Thus, it appeared possible that lignocellulose could be cleaved under one set of conditions so that a good grade of pulp could be produced. Under a different set of conditions it was visualized that the lignocellulose could be modified with the formation of new functional groups that could then be caused to undergo polymerization by one of several possible mechanisms; e.g., as outlined by Billmeyer (4). Accordingly, equipment was designed, fabricated, and has been used to study the effects of reaction parameters on the properties of lignocellulosic substrates using both the liquid and the gas phase reactions. Some of the results obtained in studies of parameters of the reaction were reported in considerable detail (5). In this report the general oxidative mechanism so lucidly set forth by Levitt (6) was invoked to explain reactions taking place under acidic conditions. Thus, control of parameters using the same reagents; i.e., $NO-O_2$, NO_2, HNO_3(conc.), and reaction system favorably directed one or more of the several types of lignin reactions that could occur to give the desired results. These reactions include electrophilic substitutions, electrophilic displacement of ring substituents, cleavage of alkyl-aryl ether linkages, and oxidation (7). Conditions we have used in the oxidative reactions have been tailored specifically to enable selective delignification of various substrates (8,9,10). Similarly, in parallel studies (11, 12,13,14) we have used conditions to modify lignocellulose without delignification that provide a modified substrate which can then be copolymerized for use in the manufacture of composite wood products. Also, the oxidative mechanism applied to the $NO-O_2$ reaction is applicable to equivalent oxidizing agents under acidic conditions as discussed by Levitt; e.g., chlorine, chlorine dioxide, hydrogen peroxide, peracetic acid and ozone (7,15). Moreover, products similar to those produced by the nitric oxide systems are produced by various oxidizing agents used in an alkaline medium (15).

It is the purpose of this presentation to summarize the results of the studies carried out specifically to develop the bonding system described with more complete information given or to be given elsewhere.

Experimental

Reaction systems designed were based on the use of rotary vessels with capabilities for controlling temperature and recording temperature and pressure. Three equivalent systems were fabricated having capacities of: .5 - 5 liters (round bottomed flasks); 24 liters (a round bottomed flask) and 185 liters (a 6.5 ft^3 stainless steel rotary digester). Temperatures within the vessels were regulated by partial immersion in controlled (to 100°C) water baths. Aqueous nitric acid was introduced as either a fine stream or spray into the vessels; both nitric oxide and oxygen flow rate and pressure drop were independently monitored into the systems.

Equipment used in forming particle board included Carver hydraulic presses provided with electrically heated platens that were used to form 4" x 4" boards in frames, an Eimco hydraulic press provided with electrically heated platens that was used to produce 10" x 10" boards in frames or to stops and a 24" x 24" hydraulic press with Dowtherm heated platens used to prepare boards 14" x 18".

In the studies being summarized a variety of experimental conditions has been used. Some of these are given herein but for complete details reference is made to the more detailed reports.

Discussion and Results

Polyamine Crosslinking Agents.

Collett (11) carried out an extensive series of aqueous phase nitric acid, and gas phase ($NO - O_2$) reactions to extend the background previously developed in pulping studies relative to the major effects of reaction parameters. In this study the specific object was to establish conditions for preparing an activated substrate that could be used for particleboard formation. Using these activated substrates the procedures and conditions for adding crosslinking agents and catalysts and for adjusting moisture contents were established. As the final step in board formation the formulated substrates assembled as mats were pressed according to cycles developed during the studies. The range of parameters investigated is presented in Table I.

TABLE I
Parameters Studied

Reaction conditions		
Reagent	$HNO_3(aq)+O_2(g)$	$NO(g)+O_2(g)$
Mode of addition	Liquid and gas	Gas
Usage, %[a]	10-42	5-15
Concentration, %[b]	10-55	---
Particulate, form	Flakes; chips	Flakes
Temperature, °C	20-90	20-90
Time, hr	1-24	1-24
Moisture content, %[a]	---	10-156
Pretreatment		
Water washing	Unwashed, washed	Unwashed
Refining[c]	Chips only	none
Crosslinking agent		
Type		
None	Control	Control
Diamines	Hexamethylene (HDA)	HDA
	Ethylene (EDA)	---
	Phenylene (PDA)	PDA
Glycols	Ethylene (EG)	EG
	1,6-Hexanediol (HG)	HG
Usage, %[a]	10	10
Application	Mist	Mist
Moisture content, %[a]		
To press, %[a]	15-108	2.0-48
From press, %[d]	1.4-28	0.3-3.7

[a]Oven-dried wood basis
[b]Basis: weight of HNO_3+H_2O in wood and acid
[c]8-in. Bauer refiner, single pass, breaker plates, water flush
[d]Basis: oven-dried board weight

In activating the substrate one of two procedures was used in introducing the reagent into the reaction vessel as the particulate wood was being tumbled. One procedure involved simple atomization of aqueous nitric acid into the reaction vessel. In the second procedure, calculated volumes of nitric oxide and oxygen were monitored into the reaction vessel. These gases react very rapidly to form nitrogen dioxide which, in turn, reacts with water present in the wood to form nitric acid and a complex array of oxidation and reduction products. Even though the moisture content of the particulate wood is adjusted when using this procedure so that a theoretical acid concentration would be equivalent to that calculated on the addition of aqueous nitric acid, substantial differences would be expected in the amounts of reaction products formed. This is a consequence of the great differences in concentration of reactant species. Such differences should also be a function of temperature of the system and rates at which reactants are distributed throughout the substrate.

Important parameters were varied over relatively wide ranges producing particle boards with acceptable properties; i.e., properties similar to those obtained using a phenol-formaldehyde resin as the control. Conditions used in both the liquid and gas phase reaction systems gave results shown by the examples in Table II.

In order to achieve results comparable to phenol-formaldehyde bonded particle boards which had densities in the range of .71 to .79 g/cc, densities of boards crosslinked with HDA were varied from .77 to .88 g/cc. Thus, using the same weight of flakes, boards prepared using the activated substrates exhibited similar properties but slightly higher densities.

Moisture content of the formulated substrates proved to be an important parameter. During pressing, water vapor tended to become trapped in the mat causing internal blisters to form parallel to the faces of the boards. As moisture content was increased the entrapped vapor caused blow-outs of the mat. Moisture contents of 15-23% were found to be satisfactory when substrates were prepared using aqueous nitric acid. With the gaseous system, however, the problem of internal blister formation appeared to be a function of both moisture content and NO usage. Thus, at 15% NO usage the problem was observed at a moisture content of 14.8%. At 12% and 10% NO usages the problem was not apparent at the highest moisture contents used; i.e., 13.3% and 16.9%, respectively.

Properties of the HDA boards prepared from the activated substrate may be compared to the phenol-formaldehyde (PF) boards prepared using 6% resin solids. Comparison may also be made to the property requirements from commercial standard CS 236-66, type 2 medium density (below 0.8 g/cc), Class 1 particleboard fabricated using durable and highly moisture resistant and heat resistant binders suitable for interior and certain exterior applications. Specifications of this class of particleboard, generally prepared from PF resins, are MOR, minimum = 2,500 psi; MOE, minimum = 450,000 psi; IB, minimum = 60 psi; and linear expansion, maximum = 0.25%. Properties of all examples given in Table II substantially surpass the comercial standard and also surpass the properties of the PF control boards with respect to linear expansion, thickness swelling, and absorption in the one-hour boil test. Internal bonds, both initial and after the one-hour boil test, and water immersion results of the HDA experimental boards were slightly inferior to results of these tests for the PF control boards. Properties of urea-formaldehyde control boards were similar to the PF control boards excepting in the one-hour boil test where complete failure of all bonding took place in about 15 minutes.

TABLE II
Typical Conditions[a] and Results

Reaction conditions	HDA Experimental Boards						PF[b]
Reagent	HNO_3(aq)+O_2(g)		NO(g) + O_2(g)				
Mode of addition	liquid mist		gas				
Usage, %[c]	30		15		12	10	--
Concentration, %[d]	30						
Temp., °C step 1	90		80	20	20	20	--
step 2	--		90	--	90	90	--
Time, hr step 1	1.0		4	24	3.4	3.5	--
step 2[e]	--		.25	--	0.5	1.1	--
Moisture content							
to press, %[c]	23.1	14.8	18.2	17.9	9.6	15.7	9.6
from press, %[f]	1.9	3.0	1.2	1.9	1.5	2.1	3.1
Properties[g]							
Thickness, in.	.337	.340	.351	.350	.346	.343	.358
Density, g/cc	.88	.81	.84	.85	.80	.84	.72
Internal bond, psi	142	104	124	132	123	101	175
One hour boil							
Swell, %[h]	19.0	18.6	15.6	17.4	17.1	17.6	26.0
Absorption, %[i]	42.0	46.2	45.8	49.2	45.3	52.7	93.6
I.B., psi	52	48	61	71	68	41	152
I.B., retained, %[j]	39	47	49	54	56	41	86.8
Water immersion							
2 hour							
Swell, %[h]	5.3	9.2	7.8	11.2	12.8	11.6	5.4
Absorption, %[i]	29.7	45.4	34.1	43.2	55.5	43.1	11.4
24 hour							
Swell, %[h]	12.2	16.0	11.9	15.5	14.3	14.5	11.0
Absorption, %[i]	41.6	59.9	49.0	50.2	60.8	52.7	36.4
Linear expansion							
7 day, %[h] L.E.	.116	.075	.137	.087	.160	.095	.184
Swell, %[h]	9.6	8.3	2.9	2.6	5.1	4.0	10.2
MOR, psi	3663	--	--	--	3657	--	4021
MOE, psi x 10^{-3}	1061	--	--	--	1005		505

[a]Constant conditions for examples in Table are: unwashed flakes, no refining; crosslinking agent, hexamethylenediamine, 10%
[b]Control boards; phenol-formaldehyde, 6%
[c]Basis, oven-dried flakes
[d]Basis, weight of aqueous HNO_3 plus water in flakes
[e]Increasing temperature from step 1 to step 2 required about 1 hr.
[f]Basis, oven-dried board weight
[g]ASTM Test Methods: D-1037-72, Pt. 16, with exceptions of certain sample sizes and the nonstandard 1-hour boil
[h]Basis, dimension at conditioned humidity
[i]Basis, weight at conditioned humidity
[j]Basis, initial I.B.

Surface smoothness or its inverse property, grain raising, was not quantified. Qualitatively this property was changed only slightly in the instance of the HDA experimental boards after the one-hour boil test; whereas, that of the PF control boards had substantially deteriorated.

Based upon the properties of the HDA experimental boards it was demonstrated that a high degree of dimensional stability of the modified substrate has been effected. It also appears that water absorption of the product, almost the same in the one-hour boil test as in the 24-hour immersion test, was largely due to filling of the exterior capillary spaces; i.e., voids in the boards, and not to absorption of water by the modified lignocellulosic structure. Accordingly, it is suggested this type of absorption might be substantially reduced by an appropriate post-press sizing treatment.

In the addition of reagents two distinctly different procedures were used. In one the reagents were added to the reaction system as it was held isothermally at either 20 or 80°C. In the other, reagents were added at 20°C in step 1. The temperature of the system was then increased over a period of about one hour to 90°C. Finally, the system was held at 90°C in step 2 until the evolution of gases subsided. When material prepared at 20°C was heated, a strongly exothermic reaction always occurred in a temperature range of 54 to 72°C. At this stage external heating was discontinued until the exothermic reaction had largely subsided. Then heating was continued until a bath temperature of 90°C was reached.

Although a wide range of acid concentrations and usages were studied in a preliminary investigation only those listed in Table II were used in activating flakes for board formation. Work was not undertaken to minimize the use of nitric acid once the validity of this bonding system had been demonstrated. Instead, emphasis was placed on application of the gaseous mode of reagent addition in view of the greatly improved uniformity of either surface treatment or surface treatment plus in-depth penetration that would be possible.

In the application of the gaseous mode of addition, both of the procedures described above gave good results as shown by the examples given in Table II. Moisture content of the flakes was an important parameter since it controlled concentration of reactive species in the aqueous phase.

Total water contents in the systems for the reactions given in Table II (O.D. flake basis) were varied over a wide range; i.e., 70%, 10%, and 21% for the aqueous nitric acid system, the gas system at 15% NO usage, and the gas systems at 10% and 12% NO usages, respectively. Similarly, usage of NO was varied over a wide range; 5 to 15% (O.D. flake basis). Concerning the critical reaction parameter, NO usage at 15% appeared to be too high because of the blistering phenomena described. Equivalent results, with no blistering, were obtained at 10 and 12% usage. From a practical point of view usage should be minimized.

Finally, in this discussion, it should be noted that several crosslinking agents were screened for possible use. Emphasis was placed on the use of HDA since it appeared to be most promising. Poorer results with ethylenediamine were attributed to its being too short to "bridge the gap" between surfaces. Phenylenediamine gave encouraging results but probably suffered from the same limitation as EDA in addition to its being a bulky molecule that would have steric limitations. Also, under the conditions being used difunctional alcohols and acids did not give as promising results.

Clearly, satisfactory conditions were found for effecting excellent bonding. However, there were a large number of variables involved and each required considerably more study in order to optimize results.

In addition to the strongly exothermic reaction observed in heating flakes treated at 20°C with either the aqueous or gaseous reagents, highly exothermic reactions were observed during pressing. Flakes treated with aqueous nitric acid under the conditions given in Table II and then formulated with phenylenediamine could not be dried below a moisture content of about 20% (O.D. basis) without combusting spontaneously. When this mixture was pressed at moisture contents above 20% it was necessary to use a low platen temperature and a very slow rate of heating in order to avoid a blowout. When the activated substrate was washed with water before formulating with phenylenediamine, it could be dried to any water content and pressed at high temperature without suffering an exothermic reaction causing blowouts. A simple but effective means for following this reaction consisted of implanting a thermocouple in the center of a mat before pressing. The occurrence of an exothermic reaction was recognized when the temperature of the mat exceeded that of the press platens. Results obtained using this technique are summarized in Table III.

These results, though not precise, showed two separate exothermic reaction zones. The first was highly dependent upon moisture content to the press, initial platen temperature, and rate of heating. The second exotherm occurred, at a temperature of about 190°C, over a wide range of conditions. The shortest reaction times of the second exotherm were realized when: the final reaction temperature in activating flakes was 90°C; the moisture content of the formulated substrate before pressing was below about 21%; HDA was used as the crosslinking agent; and a high initial platen temperature was used. When control boards were made neither phenol-formaldehyde nor urea-formaldehyde gave the exothermic reactions. Water washing used to control the exothermic reaction invariably had a detrimental effect on the properties of the boards prepared. This was attributed to the removal of an active component in bonding. In this connection the amount of exudate, that formed after addition of HDA or PDA when boards exhibiting good properties were made, was visibly reduced. The exudate was hardened on pressing and became insoluble in boiling water. Thus, it is involved in forming the bonded product.

In the study with Collett it was shown that boards having excellent dimensional stability and other physical properties similar to phenol-formaldehyde bonded particleboard were produced using the concept outlined. Since the product was produced using diamines it was characterized as an amide-type board though specific types of bonding were not elucidated. Reactions utilizing the gas phase mode of reagent addition were most readily carried out attaining maximum uniformity of treatment. Final activation temperature of 90°C and moisture contents of the formulated substrates below 20% gave boards exhibiting the excellent properties noted in the shortest pressing times used.

TABLE III

Exothermic Reactions -- Typical Conditions

Activation reaction							$NO(g)+$
Reagent	$HNO_3(aq)+O_2(g)$						$O_2(g)$
Usage, %[a], HNO_3	30						--
NO	--						15
O_2	excess						8-14[b]
Temperature, °C							20-90
Step 1 - Step 2	20	--	90	--	90	--	(80-90)
Occurrence of exotherm							
Flakes	54 - 58						
Chips	68 - 72						
Substrate treatment[c]	U	W	U	W	U		U
Crosslinking agent[d]	PDA		PDA		HDA		HDA
Moisture to press, %[a]	--	--	23.2	44.0	14.8		15-21
Press reactions							
Maximum							
Time, min.							
First	75	34	5	4	6	2	1 - 3
Second	90	59	26	26	26	19	9 - 14
Mat temp., °C							
First	108	109	171	170	172	122	141-149
Second	173	190	242	238	191	190	188
Platen temp., °C							
First	66	94	164	167	104	120	135-141
Second	102	162	218	216	161	164	163-166
Initial platen temp., °C	77	68	163	166	88	121	141-143

[a]Basis, oven-dried flakes
[b]Stoichiometric amount to form HNO_3 based on NO usage = 12.0%[a] O_2
[c]U = unwashed; W = washed after reaction
[d]PDA = phenylenediamine; HDA = hexamethylenediamine

Polybasic Acid Crosslinking Agents; Acid-Peroxide Activators.

Based on encouraging cursory results obtained with glycols, a study was undertaken with Pohlman (12) to effect bonding using crosslinking agents that could, among other possibilities, form ester linkage; e.g., di- or polyfunctional acids or alcohols. After consideration of a number of possible crosslinking agents and activation procedures, attention in this work was soon focused on acid activation of lignocellulosic materials using dibasic acids. Maleic anhydride was selected for comparative purposes. A number of hydro, oxy, and organic acids and acid salts were investigated as acid activators. Hydrogen peroxide and various organic peroxides were also used as activators in combination with the acids.

The usual procedure selected for preparing formulated material for pressing involved spraying a solution containing the crosslinking agent, acid activator, and, when used, peroxide activator on the particulate. The latter, generally wood flakes, were tumbled in a spray chamber during application of the solution. Some of the results are summarized in Table IV.

Based on dimensional stabilization, the results obtained were in the order listed with hydrochloric acid being most effective and acetic acid least effective. Internal bond appears to be directly related to acid strength. Activities of the strong acids in 1.0 molal solutions at 25°C or pK's of weaker acids are given under Acid, column 1, of Table IV. It will be noted the strength of internal bonds decreased with acid used in the order $HClO_4 > H_2SO_4 > HCl > H_3PO_4 > HNO_3 >$ oxalic $>$ acetic. With two exceptions the acids fall in the order of decreasing acid strength. The first exception is that of nitric acid. As previously shown nitric acid is reduced rapidly giving products of substantially lower acidity in the temperature range of 55°C to 100°C which is considerably below the temperature required for satisfactory board formation; i.e., 135°C to 150°C. Thus, the amount of nitric acid available for activation is substantially reduced before the activation reaction takes place. The second exception is that of oxalic acid which should be equal to or slightly better than phosphoric acid based strictly on acid strength. It is possible that the oxalic acid concentration is also effectively reduced during the course of the activation reaction by participating itself in the crosslinking reaction or perhaps acting as an oxidizing agent under the conditions of pressing with conversion to a less effective acid.

The good performance of ferric chloride was attributed to the relative ease with which the first chlorine will form hydrochloric acid by hydrolysis under the conditions used. Similarly, the poorer performance of ferric nitrate is attributed to the formation of nitric acid by hydrolysis and the reaction of the latter to form products of substantially lower acidity. The excellence of hydrogen chloride in increasing dimensional stability could be due to its greater mobility and increased probability of reaction throughout the cell wall compared to the other acids used.

In Table V the effects of strong mineral acids in combination with salts and benzoyl peroxide as activators are shown using 6%

Table IV
Effects of Acid Activator on Particleboard[a] Properties

Activator			Board Properties			
Acid	Equivalents[b]	Moles[b]	Density g/cc	Internal Bond, psi	1 Hour Boil Swell., %	1 Hour Boil Absorp., %
HCl	0.011		0.88	80	18	35
Activity	0.025		0.95	98	11	27
Coeff.= 0.809	0.036		0.91	140	6	16
at 1.0 molal	0.050		0.92	183	6	16
	0.074			blew in press		
$HClO_4$	0.003		0.79	68	104	91
Activity	0.005		0.81	84	48	62
Coeff.= 0.823	0.010		0.89	133	30	44
at 1.0 molal	0.020		0.78	123	13	31
H_2SO_4	0.006	(.003)	0.83	53	112	94
pK - 2nd H	0.010	(.005)	0.77	73	77	88
= 1.92	0.018	(.009)	0.81	108	39	58
	0.036	(.018)	0.84	130	20	37
H_3PO_4	0.012	(.004)	0.86	75	F	F
pK - 1st, 2nd	0.027	(.009)	0.82	70	99	91
3rd=2.1, 7.2	0.054	(.018)	0.82	95	79	74
12.7, resp.	0.110	(.037)	0.87	125	60	57
HNO_3	0.008		0.79	35	F	F
Activity	0.016		0.81	40	F	F
Coeff.=0.724	0.032		0.86	48	110	100
at 1 molal	0.064		0.89	68	86	78
COOH		0.013	0.82	25	F	F
COOH		0.042	0.79	43	119	110
pK - 1st, 2nd=		0.087	0.89	83	72	74
1.23, 4.19, resp.		0.169	0.89	95	61	67
HCOOH	0.022		0.80	20	F	F
pK = 4.75	0.043		0.86	35	F	F
	0.065		0.85	30	F	F
	0.087		0.88	38	F	F
$FeCl_3$		0.018	0.85	35	F	F
		0.038	0.88	85	33	56
		0.057	0.88	93	17	34
		0.118	0.96	160	3	12
$Fe(NO_3)_3$		0.031	0.80	15	F	F
		0.061	0.86	43	F	F
		0.092	0.88	63	F	F
		0.186	0.87	105	83	80

[a]Crosslinking agent=maleic anhydride, 6%[b]; mixed flakes: Douglas-true firs, ponderosa-Jeffery-sugar pines; pressed-150°C, 15 min.
[b]Usage per 100 g flakes (OD)

maleic acid as the crosslinking agent. Good results were obtained when 0.11 moles of nitric acid, 0.08 moles of ammonium chloride, and 1.0% of benzoyl peroxide were used together as activators.

TABLE V
Effects of Acid-Salt-Benzoyl Peroxide Activators on Properties of Particleboards[a] Using 6% Maleic Anhydride as Crosslinking Agent

Activator[b]			Board Properties			
Acid	Salt	Benzoyl Peroxide	Density	Internal Bond	1 Hour Boil	
Moles[c]	Moles[c]	g[c]	g/cc	psi	Swell.,%	Absorp.,%
HNO_3	NH_4Cl					
0.03	0.0	1.0	0.85	28	F	F
0.05	0.08	"	0.88	63	19	40
0.08	"	"	0.90	73	9	20
0.11	"	"	0.87	130	9	27
0.16	0.0	"	0.90	20	77	79
"	0.03	"	0.92	74	11	43
"	0.05	"	0.82	80	12	41
"	0.08	"	"	104	3	29
"	0.11	"		blew in press		
"	0.15	"		" " "		
	$CaCl_2$					
"	0.04	1.0	0.91	134	7	8
	KCl					
"	0.08	1.0	0.94	88	11	22
HCl						
0.013	0.0	1.0	0.94	75	34	61
0.025	"	"	0.87	88	2	25
0.05	"	0.0	0.93	88	7	30
"	"	0.05	0.89	91	5	30
"	"	0.10	0.91	119	5	25
"	"	1.0	0.91	122	5	22
H_2SO_4						
0.036	0.0	1.0	0.85	101	6	29

[a]Flakes-mixed species (see[a] Table IV); pressed 165°C for 15 minutes total time including initial temp. cycle
[b]Added to solution of 6% maleic anhydride in acetone and sprayed on flakes
[c]Usage per 100 g flakes (OD)

Results obtained using 0.05 moles of hydrochloric acid and 0.1 to 1.0% of benzoyl peroxide were similar. When nitric acid was doubled and 0.04 moles of calcium chloride were substituted for the ammonium chloride the best board properties were obtained. Sulfuric acid, 0.036 moles and 1% benzoyl peroxide gave slightly inferior results but at a significantly lower density. Hence, the results outlined can be considered as roughly equivalent.

The press temperature used in the experiments summarized in Table V, 165°C, was significantly above the optimum established by the experiments summarized in Table VI. It may be anticipated that properties of boards, made at the higher temperature, would have been improved by use of a press temperature more nearly at the optimum setting.

TABLE VI
Effects of Pressing Conditions on Particleboard Properties[a]

Platen Temp.[b] Initial Cycle A B C °C	Time, min D	at C	E	MC[c] %	Board Properties Dens. g/cc	Int. Bond psi	1 Hour Boil Swell. %	Abs. %
				Platen Temperature				
135-125-135	5	10	15		0.99	163	9	23
150-140-150	5	10	15		0.95	169	6	23
175-160-175	9	6	15		0.96	94	11	29
185-167-185	12	3	15		0.92	94	6	27
				Moisture Content				
150-140-150	5	10	15	6.5[d]	0.96	124	4	26
150-138-150	7	10	17	11.5	1.00	105	7	24
150-138-150	6	10	16	16.5	0.99	35	5	29
150-139-150	6	10	16	21.5	0.89	32	6	29
				Press Time				
150-140-150	5	1	6	SMC[d]	0.87	18	9	24
150-140-150	6	4	10	"	0.85	105	8	33
150-140-150[e]	5[e]	10	15[e]	"	0.86	130	7	31
150-140-150	5	15	20	"	0.87	80	7	28
				Density				
STtC[e]				SMC[d]	0.66	40	11	47
"				"	0.83	72	10	43
"				"	0.86	88	9	33
"				"	0.88	90	7	29
"				"	0.91	120	8	30
"				"	0.92	140	9	23

[a]Flakes-mixed species (see[a] Table IV); usage: 0.05 moles HCl; 0.1g succinic peroxide; 6 g maleic anhydride per 100 g flakes (OD)
[b]Initial platen temp. = A, minimum platen temp. depression = B, final platen temp. = C. time for initial cycle, A B C = D, total press time = E
[c]MC = moisture content, % OD flake basis
[d]SMC = standard moisture content adopted = 5 to 7%
[e]STtC = standard temp.-time cycle adopted = 150-140-150; D = 5 min; E = 15 min

The series of experiments given in Table VI were carried out to establish best press conditions using conditions for formulation which had given good results in previous work; i.e., 0.05 moles HCl, 0.1% succinic peroxide, 6% maleic anhydride (all on the oven-dried flake basis). First, it was shown that the optimum press temperature was approximately 150°C. The initial cycle given was typical for the temperature depression and recovery sustained in heating the mat to the final platen temperature. Then a moisture content of the flakes, which determined the final moisture of the mat before pressing, was shown to give best results when adjusted to about 6.5%, the lowest used. A range of about 5-7% was adopted as a standard (SMC) for subsequent work. Using the preferred initial platen temperature of 180°C and moisture content of about 6.5%, it was then shown that a 5 minute period used to bring the press back to the final temperature followed by a 10 minute period at the final temperature gave best board properties. These conditions were adopted as the standard temperature-time cycle (STtC) in subsequent work. Finally, it was shown as density was increased from 0.83 to 0.92 that internal bond doubled, absorption was decreased by about 50%, and swelling was decreased slightly.

An extensive study, summarized in Table VII, was carried out which established the effect of hydrochloric acid-succinic anhydride as combined activators on properties of boards made using maleic anhydride as the crosslinking agent. Data in Table VII is presented in two parts. In Part A experiments are summarized in which either one or two of the three additives were used in the formulations; in Part B all three additives were used.

In general, internal bonds above 100 psi, swelling of less than about 10%, and absorption of less than about 50% were obtained when using 0.025 moles of hydrochloric acid or more and 6% of maleic anhydride or more at densities of 0.91 g/cc (Part A) and 0.89 g/cc (Part B) and higher. Most exceptions to this generalization could be accounted for by the press temperature used. Also, when swelling and water absorption percentages were low (5-10% and 14-23%, respectively) a high degree of crosslinking was indicated by density (above 0.91 g/cc) and higher values of internal bonds. In those instances where internal bond was unexpectedly low, considering other results under similar conditions, it appears that incipient blister formation may have occurred; e.g., the two results given using 0.04 moles of hydrochloric acid (Part A) illustrate such a situation.

Data in Table VII can best be interpreted by arranging results in increasing order with respect to hydrochloric acid usage, then maleic anhydride usage, and finally board density. When this is done it is observed at 0.025 moles of hydrochloric acid usage and lower that internal bonds were higher and swelling and absorption percentages were lower for a given density when 0.1% succinic peroxide was used than in its absence. Thus, the same internal bonds and dimensional stabilities were obtained but the addition of the peroxide to the activator lowered the density at which the values could be obtained, at least for the level of peroxide used in these experiments. At hydrochloric acid usages of 0.035 moles and higher this effect was no longer apparent.

TABLE VII
Effects of Hydrochloric Acid and Succinic Peroxide Activators and Maleic Anhydride on Particleboard[a] Properties
Part A; Results Using Either 1 or 2 Formulating Agents

Activator		Cross-linking Agent[b]	Board Properties			
HCl	Succinic Peroxide[b]		Density	Internal Bond	1 Hour Boil	
					Swell.	Absorp.
Moles[b]	g	g	g/cc	psi	%	%
0.025	0.0	0	0.83	19[c]	31	61
0.038	"	"	0.82	31[c]	10	38
0.05	"	"	0.83	35[c]	13	45
0.075	"	"		Blew in Press		
0.0	0.1	"	0.82	9[d]	F[e]	F
"	0.0	10	0.82	38[d]	F	F
"	0.1	10	0.83	49[d]	F	F
0.025	"	0	0.87	25[d]	23	50
0.05	"	0	0.85	38[c]	9	43
0.013	0.0	6	0.88	80[d]	18	35
0.025	"	4	0.91	65[d]	18	34
"	"	6	0.95	98[d]	11	27
"	"	"	0.95	115[d]	8	20
"	"	8	0.92	130[d]	8	23
"	"	"	0.93	100[d]	13	37
"	"	"	0.94	143[d]	8	19
"	"	10	0.95	113[d]	9	18
"	"	"	0.96	125[d]	5	16
"	"	12	0.96	160[d]	6	14
0.038	"	6	0.91	140[d]	6	16
0.05	"	"	0.93	88[c]	7	30
"	"	"	0.92	183[d]	6	16

(for footnotes see Part B)

Clearly, properties were directly correlated with usage of hydrochloric acid and maleic anhydride. With hydrochloric acid alone internal bond was poor but dimensional stability was improved appreciably as the activator usage was increased. When the acid activator and maleic anhydride were used together, board properties were improved markedly as hydrochloric usage was increased and, especially, as maleic anhydride usage was increased. All boards were pressed using a frame (3/8 in. thick) and, except in the density study (Table VI), using a constant weight of flakes (on the OD basis). The densities of boards produced using activa-

TABLE VII (continued)
Effects of Hydrochloric Acid and Succinic Peroxide Activators and Maleic Anhydride on Particleboard Properties
Part B; Results Using 3 Formulating Agents

Activator		Cross-linking Agent	Board Properties			
HCl	Succinic Peroxide		Density	Internal Bond	1 Hour Boil	
					Swell.	Absorp.
Moles[b]	g[b]	g[b]	g/cc	psi	%	%
0.013	0.1	6	0.92	43[d]	43	61
"	"	"	0.91	86[d]	39	62
"	"	8	0.94	70[d]	27	30
"	"	10	0.85	52[d]	41	38
0.025	"	4	0.88	73[d]	16	36
"	"	6	0.87	80[d]	19	44
"	"	"	0.92	110[d]	9	20
"	"	"	0.94	144[d]	9	26
"	"	8	0.94	130[d]	8	15
"	"	10	0.95	173[d]	9	19
0.038	"	6	0.91	113[d]	7	26
"	"	6	0.97	124[d]	10	24
0.05	"	1	0.85	65[c]	8	39
"	"	2	0.84	72[c]	12	41
"	"	4	0.87	69[c]	9	40
"	"	6	0.86	130[d]	7	31
"	"	"	0.89	125[c]	8	18
"	"	"	0.89	133[d]	6	23
"	"	"	0.91	120[d]	8	30
"	"	"	0.95	169[d]	6	23
"	"	"	0.96	124[d]	4	26
"	"	8	0.89	114[c]	5	32
"	"	"	0.93	182[c]	8	23
"	"	10	1.00	153[c]	5	15
"	"	12	1.00	167[c]	5	19
"	"	14	1.00	220[c]	3	14

[a]Flakes, mixed species (see [a] Table IV)
[b]Usage per 100 g flakes (OD)
[c]Pressed at 165 C, 15 min total time
[d]Pressed using STtC (see [e] Table VI)
[e]F = complete board failure in test

tor only or peroxide and maleic anhydride only were 0.82-0.83 g/cc. When the combined activator was used this increased to 0.85-0.87 g/cc. With hydrochloric acid and maleic anhydride densities increased from 0.88 to 0.96 g/cc with increasing usages. Finally when all three components were present (Table VII, Part B) the values ranged from 0.85 g/cc using 1% maleic anhydride to 1.00 g/cc using 10% maleic anhydride and above. The mass of the materials used to make these boards was held constant, except for the amount of additive. The increase in density noted might be attributed solely to the increase in additive except for an accompanying decrease in thickness amounting to as much as 10%. This decrease in thickness with increasing mass as additive is increased is a result of greater consolidation of the lignocellulose.

In this study a number of dibasic acids or their anhydrides were used. Results obtained using succinic anhydride and maleic acid, summarized in Table VIII, are particularly pertinent. When

TABLE VIII
Properties of Particleboards[a] Formed Using Succinic Anhydride and Maleic Acid

Activator		Cross-linking Agent	Board Properties[b]			
HCl Moles[c]	Succinyl Peroxide g[c]	g[c]	Density g/cc	Internal Bond psi	Swell. %	Absorp. %
		Succinic Anhydride				
0.025	0.1	6	0.92	133	12	33
0.038	"	"	0.92	103	10	31
0.05	"	4	0.93	86	8	23
"	"	6	0.93	141	9	26
"	"	8	0.93	129	8	26
		Maleic Acid				
0.0	0.0	10	0.86	46	F	F
0.25	"	3.5	0.91	53	12	28
"	"	7.0	0.91	103	10	26
"	"	9.3	0.95	160	6	14

[a]Flakes, mixed species (see [a], Table IV
[b]Pressed using STtC (see [c], Table VI)
[c]Usage, per 100 g flakes (OD)

succinic anhydride was used in place of maleic anhydride, other conditions being equal, very similar results were obtained. Also, results using maleic acid are quite similar to those obtained using either maleic anhydride or succinic anhydride. Thus, the double bond in maleic anhydride and the use of the anhydride rather than the acid appear to play minor, if any, roles in the bonding that has taken place.

The improvement in physical properties as a function of increasing crosslinking agent accompanying the demonstrated greater consolidation of the lignocellulose provides strong evidence for the participation of the crosslinking agent in covalent bonding. Additional independent support for this conclusion was provided by showing that the crosslinking agent could not be extracted from finely subdivided board products using a solvent extraction sequence. Also, accelerated aging tests showed good aging characteristics (12). Thus, all results obtained are consistent with those obtained with Collett (11) and of the bonding concept outlined previously.

Furfuryl Alcohol and Formaldehyde Crosslinking Agents. Although the boards produced using acid activation were excellent on the basis of criteria set forth previously, the use of mineral acid is undesirable on the basis of the acidic residue that would remain in the products. Also, practical problems would be incurred in pressing wood flakes that have been treated with hydrochloric acid unless a means is provided to scavenge the chloride ions. To overcome these problems, work was undertaken with Wong for further investigating the use of the $NO+O_2$ system in the activation of lignocellulose. One advantage of this system of activation is that no mineral acid residue should persist in the product; i.e., nitric acid or oxides of nitrogen at the state of oxidation of nitric oxide or higher will be reduced under conditions used in pressing. Also, the acid activator can be simply applied in a well designed and engineered system.

The procedures followed in applying $NO+O_2$ were the same as those previously described. Some of the results obtained in this study are summarized in Table IX, Parts A and B.

It will be noted that properties using depithed bagasse treated with $NO(5\%)+O_2$ as the activator and maleic anhydride as the crosslinking agent gave highly satisfactory results at a density of 0.69 g/cc in the absence of hydrochloric acid as an activator. By replacing maleic anhydride with furfuryl alcohol results were essentially equivalent. Upon use of maleic anhydride with hydrochloric acid activator only; i.e., in the absence of $NO+O_2$ treatment, dimensional stabilization was significantly decreased. In the absence of both activators results were unsatisfactory.

Using $NO(5\%)+O_2$ treated bagasse and 10% formaldehyde, only, a board was produced at a density of 0.50 having an internal bond of only 15 psi and a water absorption of about 100%. However, it was highly significant that in the 2-hour boil test this board expanded only 30%. Thus, as defined in our work, crosslinking had occurred.

When using furfuryl alcohol and wood flakes treated with $NO(5\%)+O_2$ internal bonds were significantly increased in products having densities (0.74 and 0.76) similar to those of the bagasse products just discussed. However, dimensional stabilization was inferior to results obtained with bagasse as well as those reported in Table VII.

TABLE IX

Furfuryl Alcohol and Formaldehyde Crosslinking Agents

Part A: Furfuryl Alcohol vs Maleic Anhydride as Crosslinking Agents

Activator $NO+O_2$ g[a]	Activator HCl Moles[a]	Cross-linking Agent g[a]	Board Properties: Density g/cc	Internal Bond psi	24-Hour Immersion Swell. %	24-Hour Immersion Absorp. %	2-Hour Boil Swell. %
			Depithed Bagasse				
			Furfuryl Alcohol				
0	0.011	10	0.72	57	57	-	90
5	0	10	0.72	83	13	-	27
5	0	10	0.79	85	18	-	28
			Maleic Anhydride				
0	0	10	0.59	40	60	-	108
0	0.011	10	0.72	80	34	-	46
5	0	10	0.69	91	11	-	19
			Flakes, Mixed Species[b]				
			Furfuryl Alcohol				
5	0	10	0.74	102	29	71	42
5	0	10	0.76	149	21	62	38
5	0	20	0.81	233	12	14	23
5	0.011	10	0.78	154	18	55	26
			Part B: Formaldehyde as Crosslinking Agent				
5	0	10	0.5	15	12[c]	100[c]	30

[a]Usage per 100 g lignocellulose (OD)
[b]See [a] Table IV
[c]4 hours

By increasing furfuryl alcohol usage to 20%, an exceptionally high internal bond and good dimensional stabilization were realized. Also, an amount of hydrochloric acid lower than any usage reported in Table VII, used with 10% furfuryl alcohol, gave board properties approximating those of the phenol formaldehyde control listed in Table II.

The work summarized in Table IX, Part A demonstrated that furfuryl alcohol was an excellent crosslinking agent for use in the bonding system described. Moreover, the cursory result using formaldehyde had demonstrated some bonding properties even though it is capable of spanning only a small distance between surfaces.

In a continuation of our work Phillipou sought to optimize the system using furfuryl alcohol and formaldehyde as crosslinking agents. In Table IX, Part C, results obtained using mixtures of furfuryl alcohol and formaldehyde are given.

Table IX
Furfuryl Alcohol and Formaldehyde Crosslinking Agents
Part C: Furfuryl Alcohol--Formaldehyde Mixtures

Activator		Particulate Form[b]	Crosslinking Agent		Board Properties			
$NO+O_2$	MA[a]		FA[a]	Form[a]	Density	Internal Bond	2-Hour Boil	
							Swell.	Absorp.
%c	%c		%c	%c	g/cc	psi	%	%
0	0	Fl	10	5	0.75	0	-	-
2	0.5	Fl	10	0	0.74	18	-	-
3	0.5	Fl	10	0	0.75	40	-	-
5	0.5	Fl	10	0	0.73	39	-	-
2	0.1	Fl	10	2.2	-	70	-	-
3	0.25	Fl	10	1.9	0.73	70	14	70
3	0.5	Fl	10	1.9	0.78	85	6	26
3	0.1	Sp	10	2.2	0.75	125	9	26
3	0.5	Sp	10	2.2	0.75	68	12	38
3[d]	0.25	Fl	10	2.2	0.78	69	6	35
5	0.4	Fl	10	2.2	0.75	101	-	-
5	0.4	Fl	10	2.2	0.75	85	-	-

[a]MA = Maleic anhydride; FA = Furfuryl alcohol; Form = Formaldehyde
[b]Particulate Form Fl = flakes; Sp = splinters
[c]Basis 100 g chips (OD)
[d]14-inch x 18-inch board; MOR = 3949 psi; MOE = 1,137,000 psi

Boards exhibiting excellent dimensional stability and internal bonds were produced using 3 and 5% NO. Once again it appears that splinter-type particles rather than flakes give highest internal bonds. Densities of all boards reported ranged from 0.73 to 0.78. Molar ratios of furfuryl alcohol to formaldehyde used were approximately 3:2.

Results, very similar to those given in Table IX, Part C, were obtained using a furfuryl alcohol-formaldehyde resin prepared (13) having a mole ratio of furfuryl alcohol to formaldehyde of approximately 1.5. A summary of some of the results obtained in testing boards prepared from this resin is given in Table IX, Part D.

As in Part C, excellent properties were exhibited by boards produced using 2 to 5 % of NO and 0.1 to 0.5 % of maleic acid to activate the lignocellulose. The amount of crosslinking agent was decreased to 7.5% based on the OD lignocellulose. The density of these products varied from 0.70 to 0.77 g/cc and conditions were found in which internal bonds were well above 100 psi. Results of tests on the large boards (14-inch x 18-inch) demonstrated that the modulus of rupture of these boards was equivalent to and the modulus of elasticity was greater by a factor of 2 compared to the same physical properties of the phenol-formaldehyde control boards (see Table II).

Table IX
Furfuryl Alcohol and Formaldehyde Crosslinking Agents
Part D: Furfuryl Alcohol--Formaldehyde Resin

Activator		Particulate Form[a]	FA-Form Resin[b]	Board Properties			
$NO+O_2$	Maleic Acid			Density	Internal Bond	2-Hour Boil	
						Swell.	Absorp.
%[c]	%[c]		%[c]	g/cc	psi	%	%
0	0.1	Fl	7.5	0.79	7.5	F	F
0	0.5	Fl	7.5	0.74	18	-	-
0	0.1	Sp	7.5	0.72	31	F	F
0	0.1	Sp	7.5	0.76	50	F	F
2	0.1	Fl	7.5	0.72	105	15	38
2[d]	0.1	Fl	7.5	0.73	114	8	60
3	0.1	Fl	7.5	0.70	67	9	35
3	0.3	Fl	7.5	0.71	43	-	-
3	0.3	Fl	7.5	0.74	60	-	-
3	0.2	Fl	7.5	0.75	78	12	60
3	0.3	Fl	7.5	0.76	58	-	-
3	0.3	Fl	11.3	0.76	93	-	-
3	0.3	Fl	7.5	0.77	93	8	26
3	0.1	Sp	7.5	0.72	159	13	38
5	0.1	Fl	7.5	0.72	78	20	74
5	0.2	Fl	7.5	0.72	83	17	73
5[e]	0.1	Fl	7.5	0.73	97	11	51
5	0.1	Fl	7.5	0.74	155	11	33

[a]Fl = flakes, Sp = splinters
[b]Furfuryl alcohol--formaldehyde resin
[c]Usage per 100 g particulate (OD)
[d]14-inch x 18-inch board; MOR = 4,227 psi; MOE = 1,109,000 psi
[e]14-inch x 18-inch board; MOR = 3,750 psi; MOE = 997,000 psi

It will be noted that dimensional stabilities of products in Table IX, Part D, are slightly lower than for products made using mixtures of furfuryl alcohol and formaldehyde. This result is due to the great facility with which furfuryl alcohol penetrates into the wood structure; whereas, diffusion of the resin into the surface is expected to be substantially slower. The greater depth of penetration of the furfuryl alcohol is readily seen by the extent of chip discoloration visible on examination of cut faces of the particleboards.

The dimensional stabilities realized in the products listed in Table IX are striking. After the 2-hour boil, grain raising is just perceptible. To the extent that these products have been evaluated, it has been shown that products are made having specifications suitable for exterior exposure.

Conclusions

Particleboards have been prepared by activation of lignocellulose followed by crosslinking using difunctional or polyfunctional monomers and polymers. Excellent physical properties including internal bond, MOR, MOE, and dimensional stability have been exhibited by the products made. These properties demonstrated that extensive bonding has taken place. A rationale is presented to support the hypothesis that covalent bonding must play a dominant role in the process. The variety of crosslinking agents and types of functional groups studied; i.e., diamines, dibasic acids, saturated and unsaturated anhydrides, furfuryl alcohol, furfuryl alcohol with formaldehyde as mixtures and uncured resins, provide evidence that a general mechanism has been discovered for bonding lignocellulosic materials.

This mechanism has been described in the introduction of this paper and is supported by the four studies described.

A method is outlined for activating lignocellulose so that it can then be crosslinked using a variety of agents. This method has been described in various publications cited and also in a patent (14) assigned to the University of California.

Literature Cited

(1) Brink, D.L., Merriman, M.M., Schwerdtfeger, E.J., Tappi (1964) 47 (4) 244-248.
(2) Mauch, R., "Off-Gas Analysis and Reaction Rate Study in Nitric Acid Pulping", M.S. Thesis, University of California, Berkeley, 1964.
(3) Martin, D., "Use of Nitrogen Oxides in Pulp Bleaching Sequences", M.S. Thesis, University of California, Berkeley, 1972.
(4) Billmeyer, F.W., "Textbook of Polymer Science", 255-376, 2nd Ed., Wiley-Interscience, N.Y., 1971.
(5) Brink, D.L., Merriman, M.M., Collett, B.M., and Lin, S.Y., "Reaction of Lignocellulose with Oxides of Nitrogen", Abstracts of Papers, Division of Cellulose, Wood and Fiber

Chemistry, 12, 167th ACS National Meeting. Los Angeles, Ca., Mar. 31-Apr. 5, 1974.
(6) Levitt, L.S., J. Organic Chem., (1955) 20 , 1297-1310.
(7) Dence, C.W., in "Lignins - Occurrence, Formation. Structure and Reactions", Ed. by Sarkanen, K V. and Ludwig, C.H., 373-432, Wiley-Interscience, N.Y., 1971
(8) Brink, D.L., LaCoste, M. (unpublished data).
(9) Lin, S.Y., "Two Stage Pulping with Nitric Oxide-Oxygen: Sodium Hydroxide", Ph.D. Thesis, University of California, Berkeley, 1975.
(10) Wang, L.T., "Water as a Parameter in Two Stage Pulping Using Oxides of Nitrogen:Sodium Hydroxide", M.S. Thesis, University of California, Berkeley, 1976.
(11) Collett, B.M., "Oxidative Mechanisms for Polymerization of Lignocellulosic Materials:Nitric Acid and Nitrogen Oxides", Ph.D. Thesis, University of California, Berkeley, 1973.
(12) Pohlman, A.A., "Solid Phase Polymerization of Lignocellulose and Dibasic Acids Using Acid Activation", M.S. Thesis, University of California, Berkeley, 1974.
(13) Technical Bulletin No. 131-A, The Quaker Oats Co.
(14) Brink, D.L., U.S. Patent No. 3,900,334, August 19, 1975.

12

Chemical Changes in Wood Associated with Wood Fiberboard Manufacture

HOWARD A. SPALT

Department of Forestry,[1] Southern Illinois University, Carbondale, Ill. 62901

Over twenty billion square feet (1/8-inch basis) of wood fiberboard are produced annually. Most of this footage goes into house and furniture construction as sheathing, acoustical tile, siding, paneling, underlayment, dustproofing, drawer bottoms, case backs, and doors. The industry had its beginning in 1914 with the development of insulation board by Muench (1). Mason's work on thermomechanical pulping (2, 3) and hot-press densification (4, 5) expanded the range of density and properties attainable with felted wood fibers. The classification of these products is given in Table I.

Table I
Classification of Wood Fiberboards

Class	Density, lbs./ft.^{3}	Densified?
Softboard (Insulation)		
Semirigid	1.2-10	No
Rigid	10-30	No
Hardboard		
Medium Density	30-50	Yes
High Density	50-75	Yes
Special Densified	84-90	Yes

Manipulation of raw material and process enables fiberboards of diverse properties to be

[1]The research reported herein was conducted under the direction of the author when he was Manager of the Basic Research Section, Research Center, Masonite Corporation, St. Charles, Illinois.

produced. Low-density boards with heat-insulating or sound-absorbing qualities also possess sufficient strength for sheathing and dropped-ceiling applications. High-density boards possess machinability, embossability, and printability to take high-fidelity decorative designs in color and texture and to wear well as interior paneling, door skins, or furniture components. Medium-density boards combine a suitable level of machinability and embossability with high weatherability and coatability built in to perform well as exterior siding. Exterior applications require high durability of substrate and coating but are less demanding of fidelity in color and texture designs than are interior products.

Raw materials for fiberboard manufacture come from several sources including whole wood, sawdust and mill residues, waste paper, agricultural wastes, and plant tissue other than woody stems. Additives usually enable attaining the desired properties but whole wood reduced to virgin fiberboard pulp often can be converted to board of impressive properties with little or no additives. High lignin content of virgin fiberboard pulps is usually cited as the source of this superiority.

Processes used to manufacture softboards and hardboards are basically similar and readily divisible into a furnish-preparation phase and a board-conversion phase. Hardboard processes differ in that board conversion uses pressure to densify the sheet whereas softboard processes do not. Both processes subject the felted sheets to high temperatures in the board-conversion phase. Hardboard processes use more severe conditions and effect more extensive physical and chemical changes; consequently, these processes offer more insight into chemical changes associated with board manufacture. The results reported in this paper are derived from studies of hardboard but they should apply equally to softboards and softboard processes.

Furnish Preparation Processes

Most virgin wood pulp used in fiberboard manufacture today is generated by thermomechanical pulping developed by William Mason in 1925 (2, 3). Few thermomechanical pulping processes existing today use the original Mason process now that more streamlined operations have evolved but all stem from Mason's development.

Three major furnish preparation processes (Figure 1) are now in use today. These are the original Masonite wet-form process, the wet form process using pressurized refining, and the dry form process.

The Masonite Wet-Form Process. Masonite Corporation was founded to exploit Mason's patents and still practices the original furnish-preparation process shown schematically in Figure 1. Cooking is accomplished by charging chips into a vertical cylinder outfitted with a slotted plate at the bottom that supports the chip stack. Below the plate is an hydraulically-operated valve that provides passage to the atmosphere.

Once filled with chips, the cylinder is sealed and pressurized with saturated steam at pressures up to 1000 psig. Usually pressures below 800 psig are used and the program of steam pressurization varies. Steam pressure may be raised steadily until discharge to the atmosphere at target pressure. Or steam pressure may be admitted at a prescribed rate until a target pressure is attained, held for a prescribed time, and discharged. Another version uses controlled rate of steam pressurization to target pressure, hold at pressure up to 90 seconds, and raise rapidly to a higher pressure (shooting pressure) and discharge.

Regardless which cooking program is used, the chips are permeated by the saturated steam and develop high internal steam pressures. At the instant of valve opening, the contents are forced by the high pressure differential through the slotted plates which have apertures too narrow for the chips. Extensive size reduction occurs by shredding followed by defibration effected by the sudden decompression. The result of this shooting of chips is gun fiber which is a mixture of fiber, fiber bundles, and hard shives. After cooking and shooting, the mixture is slurried in water and refined to reduce the coarse fraction.

Following the first (primary) refining, the stock is washed on countercurrent, brown-stock washers to remove solubles generated by the cooking. Secondary refining reduces the stock to target freeness and after additives are blended, the furnish is ready for the forming machine. Additives commonly used are acid-insoluble phenolic resins and petrolatum waxes. Machinability benefits greatly even from low resin add levels and water repellency increases sensitively with wax addition.

Wet Form Process Using Pressurized Refining. Thermomechanical pulping has progressed rapidly since the advent of the pressurized refiner. Advances in the knowledge of cell wall structure and composition enables understanding the workings of the versatile thermomechanical pulping process. The research of Fergus *et al.* (6) reveals lignin to be concentrated in the compound middle lamella and holocellulose to be concentrated in the secondary wall. Lignin can be softened at temperatures between 130C and 190C, the lower temperatures suffice at high moisture content (7).

Saturated steam above 130C will soften the lignin-rich middle lamella as it penetrates wood chips. The pressurized refiner applies shear to wood chips at the same time that the middle lamella is softened by saturated steam. Very fine, anatomically whole fiber is produced with low mechanical energy consumption. Furthermore, low cooking steam pressures minimize solubles formation and weight yields of pulp are high. Koran's (8) scanning electron photomicrographs of thermomechanical hardboard fiber reveal cell corner ridges on the fiber surface which confirms that the zone of separation is the middle lamella. By extracting the lignin from the fiber surface with acid chlorite, Koran uncovered the random microfibrillar structure of the primary wall which confirms that the entire cell wall structure is intact.

At saturated steam temperatures below 130C, the middle lamella does not soften. Koran's (8) scanning electron photomicrographs of pressurized refiner pulp produced at temperatures slightly below 130C reveal the S_2 layers of the cell wall. Simply by lowering the cooking and refining steam pressure, the zone of failure moves from the middle lamella to the region between the S_1 and S_2 layers. Consequently, the lignin-rich portions of the wood can be separated as fines from the secondary cell wall to produce holocellulose-rich refiner groundwood.

The evidence supports the hypothesis that thermomechanical pulping exploits the thermoplastic character of lignin to produce lignin-rich fiberboard fiber or holocellulose-rich refiner groundwood fiber for papermaking.

Once the fiber is generated, washing and blending operations complete the preparation of furnish for the forming machine.

Dry Form Process. The advent of the pressurized refiner led eventually to dry-formed wood fiberboards because high-consistency pulps can be produced. Green wood chips contain equal parts water and dry wood matter and can be pressure refined in saturated steam with little change in moisture content. Fiber from the refiner at fifty percent consistency can be made fluffy and readily suspendable in air. This enables vapor-phase dewatering in hot-air driers to produce a fine, dry fiber that can be formed in air and dry pressed. Without liquid-phase dewatering, solubles formed by the steam cooking of chips remain in the fiber furnish going to the board machine.

Blending of additives such as phenolic-resin binder and wax is usually accomplished at the point where the fiber is discharged from refiner pressure to the atmosphere. This location in the blow line provides turbulent mixing to effect good additive distribution but also has the advantage of occurring before the blowers or the windage refiners used to fluff the fiber. Clumps formed by poor additive dispersion can be broken up by fluffing before hot-air drying.

Chemical Changes in Wood Effected by Furnish Preparation Processes

Wood, and the fiber obtained from it, is a complex arrangements of polymers and macromolecules that are exceedingly difficult to analyze chemically. But it is shown in Figure 1 that wet processes produce two streams, pulp from cooked fiber for board production and a liquor containing water solubles extracted from the stock. The liquor is easier to analyze and its composition may afford insight into the chemical changes effected by furnish preparation.

The rate of hot water-solubles formation as a function of cooking steam pressure is shown in Figure 2. Hardboard operations in the past have used steam pressures up to 1000 psig at which as much as fifteen percent of the cell wall can be solubilized in sixty to ninety-second cooks. Extraction of these solubles by the washers produces liquors of pH 3.8-4.5 at solids of 3.5-4.0 percent. Lower pH occurs at higher steam pressures. The sudden increase in hot-water solubles between 300 and 400 psig steam pressure is believed caused by cleavage of acetyl groups on the hemicelluloses, which gives rise to acetic-acid-catalyzed hydrolysis of cell wall components.

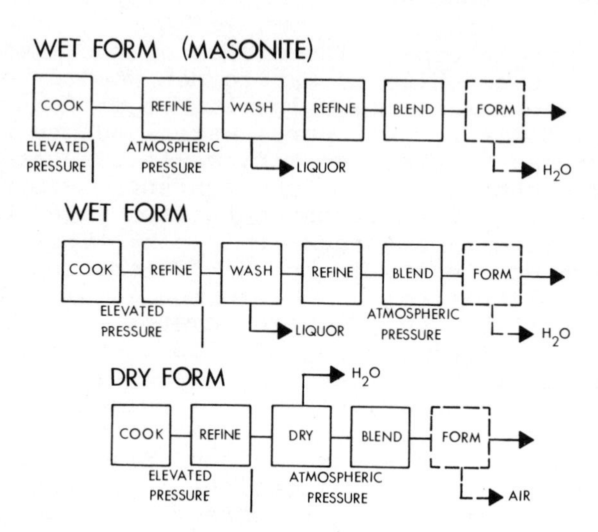

Figure 1. Process schematics for the typical furnish preparation (defibering) processes used in wood fiber-board manufacture

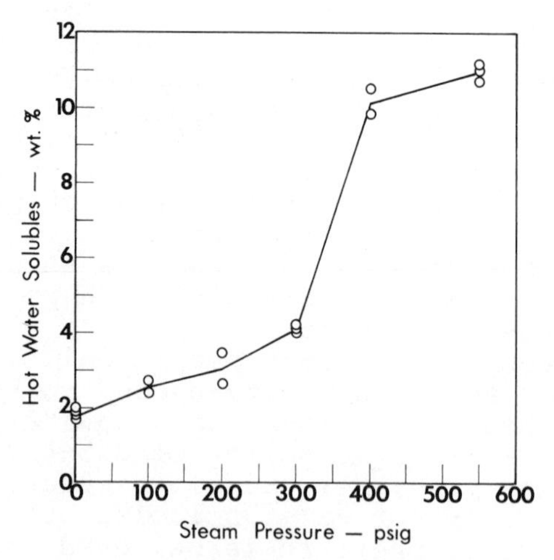

Figure 2. Hot-water extractability of wood chips as a function of saturated steam pressure used in the preheat segment of the digesting operation

When the extract is passed over an anion-exchange resin, approximately 40-45 percent of the solids are retained on the resin and the remainder passes through. Infrared absorption spectra of the two fractions are presented in Figure 3. The neutral fraction shows absorption of infrared at 3 microns (hydroxyl), 5.9 microns (aldehydic carbonyl), and at 9-10 microns (C-O-C and C-O-H), with little absorption at 3.4 microns (C-H stretch). This spectrum resembles that of the common simple sugars when scanned as a melt between NaCl optics.

The acidic fraction, which adsorbs to the resin and must be eluted by acidic extraction, exhibits a different infrared absorption spectrum in the 6-7 micron range. Sharp bands at 6.3 and 6.6 microns indicate absorption assigned to distortion of the aromatic nuclei occurring in lignin breakdown products.

Considering the ligno-holocellulosic composition of wood, it is not surprising that the liquor is divisible into a carbohydrate (neutral) fraction and an aromatic (acidic) fraction, the former of holocellulosic origin and the latter of lignin origin.

Careful acid hydrolysis of the whole extract from the washers produces a three-fold increase in reducing power (Figure 4). When calculated as the reducing power of glucose, the extract solids as discharged from the washers run 20 percent reducing sugar on total solids. Acid hydrolysis at 100C raises the reducing sugar to 60-65 percent of solids. If this increase is attributable exclusively to reducing sugar formation, the oligosaccharides in the extract have an average degree of polymerization of three anhydride units. This conclusion agrees with earlier estimates of 450-550 grams/mole obtained from measures of boiling-point elevation. A tarry precipitate equivalent to 15 percent of liquor solids forms on acid hydrolysis, also. Crude analyses of this material indicate a furfural condensation product.

Gas-liquid chromatography of the hydrolyzate after hydrogenation with borohydride and acetylation with acetic anhydride according to the procedure of Sawardeker, Sloneker, and Jeanes (9) produces chromatograms with six distinct peaks. Five of the peaks occur at the same retention times as the alditol acetates of the five sugars commonly found in wood. These are the pentoses L-arabinose and D-xylose and the hexoses D-mannose, D-galactose, and D-glucose. The sixth sugar, which elutes first

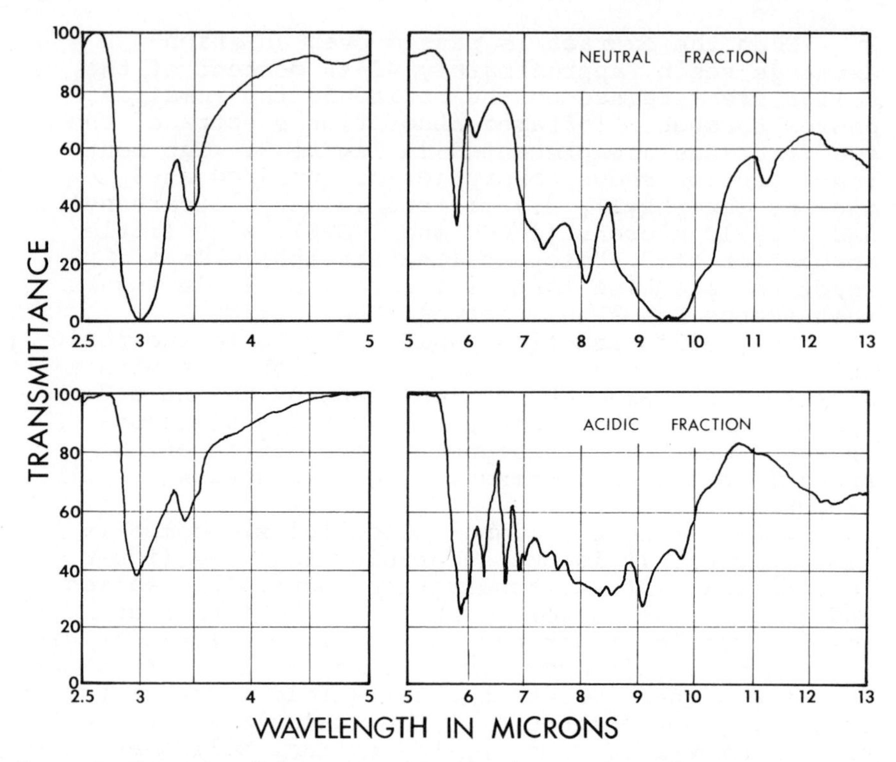

Figure 3. Top: infrared absorption spectrum of the neutral furnish extract fraction that passes through an anion-adsorption resin. Bottom: infrared absorption spectrum of the acid furnish extract fraction that adsorbs to the anion-adsorption resin.

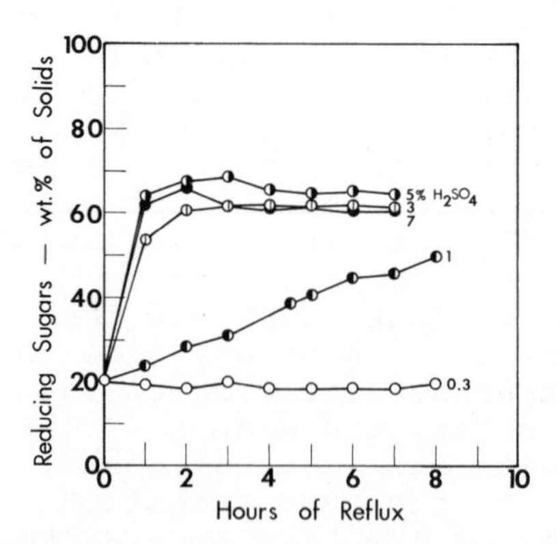

Figure 4. Reducing sugar content (Somogyi–Nelson Technique) of extract solids as a function of reflux time at several acid concentrations

from the column, has the same retention time as L-rhamnose. The presence of 5-methyl furfural in the hydrolyzate is strong evidence of a 6-deoxy hexose in the extract and only L-rhamnose fits all the evidence.

The relative proportions of sugars in the hydrolyzates from hardwood liquors and from softwood liquors are reported in Table II. The mannose: galactose:glucose ratio in softwood liquors is the

Table II
Distribution of Simple Sugars in Hydrolyzates of Mill Liquors

Sugar	Sugar, Percent of all Sugars	
	Hardwood	Softwood
Rhamnose	2	2
Arabinose	5	5
Xylose	77	25
Mannose	7	40
Galactose	4	13
Glucose	5	15
Total	100	100

same 3:1:1 ratio for softwood galactoglucomannans reported by Timell (10). Mannose and glucose are present in roughly equal quantities in extracts from all hardwood cooks which reflects the mannose: glucose ratio of hardwood glucomannans between 1:1 to 2:1 (10). The evidence reveals that the thermomechanical pulping process attacks only the hemicellulose polysaccharides and does not affect the cellulose fraction to any degree.

The noncarbohydrate fraction can indeed be traced to lignin source. Gas-liquid chromatography of ether extracts of the washer liquors indicates the presence of guaiacol syringaldehyde, vanillin, vanillic acid, syringic acid, and hydroxybenzoic acid.

When worked up sequentially with sodium bisulfite, sodium bicarbonate, and sodium hydroxide, the noncarbohydrate portion divides into fractions rich in aldehyde groups, carboxyl groups, and free phenolic hydroxyl groups, all exhibiting strong aromatic bands in infrared spectra (Figure 5). A fourth fraction remains after extraction by the other three reagents which shows some hydroxyl and aldehydic functionality but mainly aliphatic hydrocarbon structure. This neutral fraction is derived from the propane portion of the phenyl-propane building block of lignin.

All of the evidence supports the hypothesis that cooking wood chips with saturated steam above 130C causes the lignin-rich middle lamella to fail when subjected to the shearing action of the refiner. Coincident with thermoplastic separation of fibers is the hydrolytic degradation of hemicellulose and lignin forming water-soluble oligosaccharides and aromatic and aliphatic lignin fragments.

Board Conversion Processes

Conversion of fiberboard furnish to board begins with the felting operation. Fibers suspended in a liquid or gaseous fluid are deposited on a moving wire and concentrated into an interfelted mat by the removal of the suspending fluid. Application of pressure between rolls or platens in the cold-press operation further reduces interfiber void space and expresses the forming fluid.

The conversion processes are shown schematically in Figure 6. Softboards or insulation boards are cold-pressed and hot-air dried to specific gravities of 0.2 to 0.5. Air temperatures of 200-250C are used to evaporate residual water.

Hardboard processes require hot pressing to further densify the sheet above specific gravity 0.5. Three different board-conversion processes are used:

1. Wet/Wet - wet formed mats at 50 percent consistency are pressed between a hot plate and a screen to form a smooth-one-side board.
2. Wet/Dry - wet formed mats are hot-air dried to anhydrous condition and hot pressed between two smooth plates to form a smooth-two-sides board.
3. Dry/Dry - dry formed mats of six percent moisture content are pressed between two smooth plates to form a smooth-two-sides board.

Although forming and cold press operations densify the interfelted sheet, little chemical change is effected at the temperatures used which are well below the boiling point of water. Major changes in board properties and some changes in

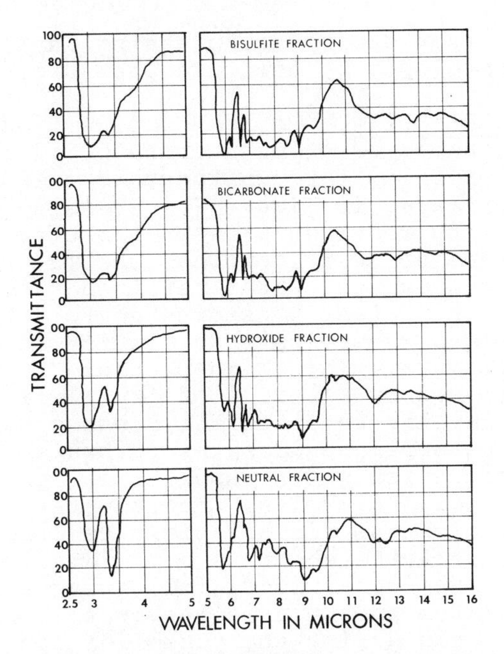

Figure 5. Infrared absorption spectra of the acidic furnish extract fraction fractionated further by sequential extraction into aldehydic (bisulfite), hydroxyl (bicarbonate), phenolic (hydroxide), and aliphatic (neutral residual) fractions

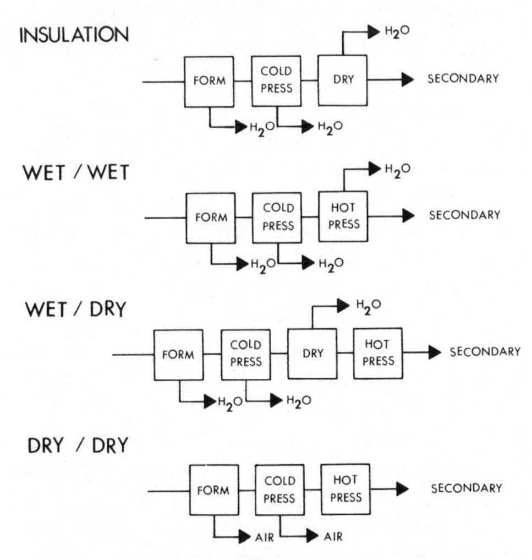

Figure 6. Process schematics for the conversion of furnish to board by wet- and dry-press operations

composition occur during hot pressing as higher densities are developed by heat and pressure.

Wet/Wet Pressing. Representative press cycles, conceptual schematics of mass transfer in the sheet, and density profiles through the thickness of the sheet are portrayed in Figure 7. Pressing of wet mats starts with a steady pressure rise to 400 psi platen pressure so as to compress the mat to minimum void volume and express water retained from cold pressing. Platen steam pressures up to 400 psig heat the mat and reduce water viscosity and raise its vapor pressure. This high-pressure inversion cycle is followed by a period of low platen pressure intended to dry the sheet to anhydrous condition. Low platen pressure provides the pathways in the sheet and screen for water vapor to escape. During inversion and drying cycles, the temperature of the sheet at midthickness rises rapidly to 100C and remains there until the sheet is dry. When the core temperature rises above 100C, indicating anhydrous condition, platen pressure is increased to a target pressure and held during the consolidation cycle when sheet properties are developed. Generally, board core temperatures rise to 190-200C before the cycle terminates with platen decompression. Density is high at the smooth surface and declines through the sheet to a minimum in the screen texture. Because low-molecular-weight, water-soluble fractions are removed earlier in washing and forming operations, wet pressing causes little loss in sheet dry weight.

Wet/Dry Pressing. Wet/dry board conversion enables smooth-two-side board to be produced by very short press cycles. Pressing anhydrous sheets eliminates the time-consuming dewatering aspects of wet pressing but higher platen steam pressures (450 psig) and press pressures (up to 500 psi) are needed to soften and densify the dry sheets. The press cycle shown in Figure 7 has an inversion cycle to harden the surfaces by densification as they are heated and to densify the core so as to promote rapid heat transfer to the center. As heat penetrates and softens the sheet, the inversion pressure is lowered to holding pressure with occasional breathing cycles to release water vapor and gases generated by the high platen temperatures. Without breathing cycles, the internal pressures will usually blister the tight surfaces. Because both surfaces are pressed directly against hot plates with high inversion pressures, two dense surfaces are produced with the core density varied by the holding pressure. The smooth-two sides board with controlled core density is particularly well suited for tileboard use.

Dry/Dry Pressing. Dry/dry pressing also enables short press cycles because dewatering is avoided but, dry-formed mats have no cohesive properties and must be compressed slowly to allow air to escape without mat rupture. Slow platen closure plus high plate temperatures at platen steam pressures up to 550 psig cause premature heating of the mat surfaces before the inversion pressure peak is reached. This leads to precure, so to speak, of the surfaces with low surface density and properties.

After the inversion cycle in dry-mat pressing, a consolidation cycle is used to develop density and properties through the sheet. Heat transfer to the core is augmented by mass transfer of water vapor and low-molecular-weight material generated by cooking and retained in the furnish. Mat moisture content is not allowed to go below six percent because of the hazard of dust explosions. This moisture and retained solubles cause high volatiles content which poses problems pressing between smooth plates at such high platen temperatures. One or more breathing cycles are included in the press cycle to enable release of the gases.

Heating of the surfaces before densification pressures are reached and mass transfers to the core cause the density distribution through the sheet shown in Figure 7. Mass transfer to the core not only adds mass to the core but transfers heat and acts as sorbed material to plasticize core fibers effecting greater density there. Soft surfaces are detrimental to machining and coating operations. Dense cores increase difficulty in on-site working properties and nailability.

Wet-Strength Properties of Hot Pressed Boards

Lignocellulosic fibers, when interfelted and densified under heat and pressure, develop impressive wet strength. Evidence for this claim is shown in Figures 8 and 9. A wet/dry tileboard grade of 1.0 gms/cc. density soaked in near boiling (97C) water for up to thirteen days still exhibits measurable bending strength when tested wet. When tested after drying to 6-8 percent moisture content, these boards

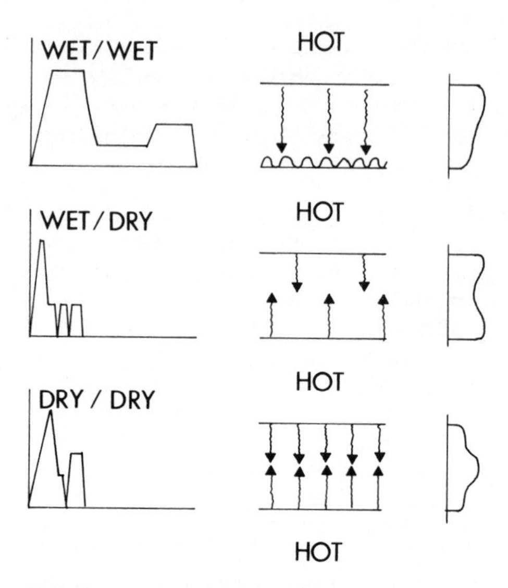

Figure 7. Conceptualizations of pressing pressure/time schedules (left), volatiles movement in the board (center), and density profiles through the board thickness (right) for wet- and dry-pressed boards

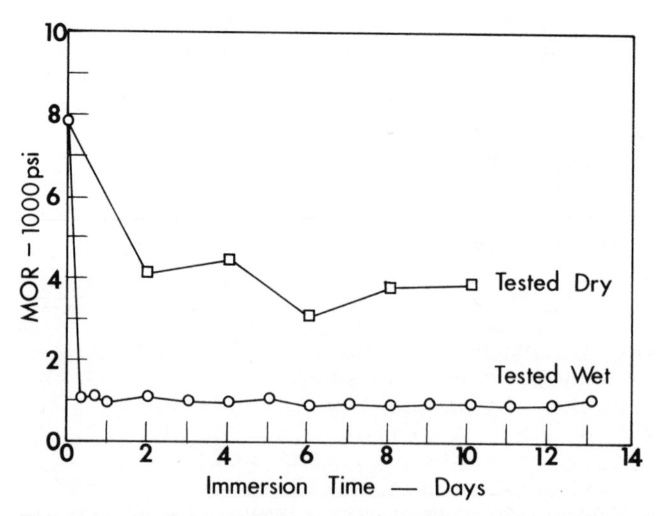

Figure 8. Modulus of rupture of wet/dry boards soaked in 97°C water and tested in the soaked and redried condition

retain half of their original bending strength despite having lost about forty percent of their out-of-press density due to thickness swelling in soaking (Figure 9). Much of the loss in bending strength can be attributed to the density decrease as shown in Figure 10. There is a close linear relation between the modulus of elasticity and modulus of rupture in bending. The curves in Figure 10 show that reducing specific gravity from 1.0 to 0.6-0.8 will cut bending strength in half. Not only does the interfiber bond resist breakdown in near-boiling water, it enables most of the original strength of the board to be recovered upon drying. If it were not for permanent thickness swelling, lignocellulosic fiberboards would show the same wet/dry reversibility of strength that whole wood exhibits.

Mechanism of Wet-Strength Properties

Two mechanisms are advanced to explain the water resistance of the hardboard lignocellulosic bond. One is a polycondensation of cell wall substances to form a water-insoluble bonding substance. Runkel and Wilke (11) provide one of the most lucid descriptions of this mechanism. They propose that the water-insoluble, resinous substance is developed in two phases. The first phase is hydrolytic and leads to the formation of carbohydrate and lignin breakdown products. Upon further degradation in acidic media, the pentoses from hydrolysis of hemicellulose are converted to furfural and other aldehydic functionalities and substituted phenols are obtained from lignin. These are argued to form phenol-aldehyde condensation products analogous to phenol-formaldehyde resins in reactions that occur at high temperatures in the hot press.

The other mechanism draws heavily from the thermoplastic model of macromolecules which is a logical extension of Goring's argument on the thermoplasticity of lignin. Goring (7) concludes in his studies of the thermal softening of lignin that lignin will undergo glassy transitions at temperatures between 130-190C, depending upon moisture content. This mechanism views lignin as a thermoplastic adhesive which develops tack when heated above 130C, enabling lignin-rich surfaces to be fused. In this mechanism, lignin is analogous to a hot-melt adhesive.

Neither mechanism is supported by rigorous evidence in the literature. A few authors point to

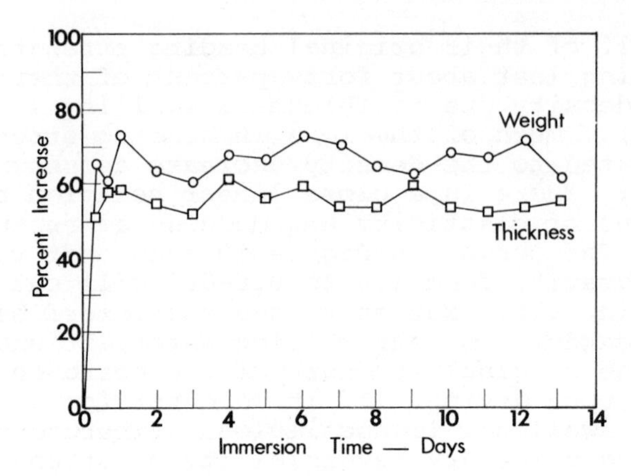

Figure 9. Increases in weight and thickness of wet/dry boards soaked in 97°C water

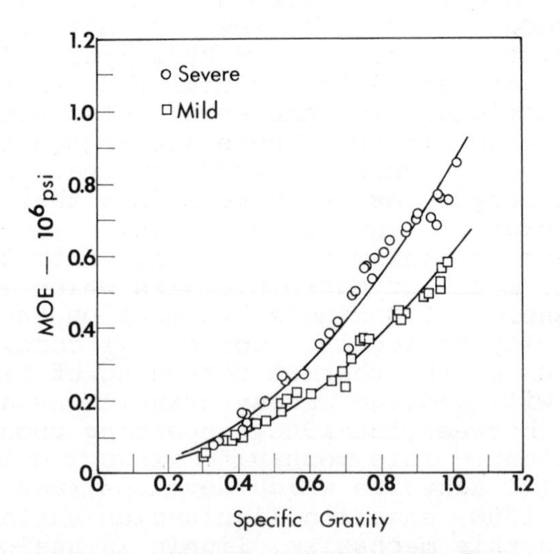

Figure 10. Modulus of elasticity in bending as a function of specific gravity for wet/wet boards made from furnish prepared from a high steam-pressure cook (severe) and a mild cook

analytical and experimental evidence to support a vaguely-described, molecular-weight-building condensation reactions. Naveau (12) reports hydrolysis of carbohydrates and condensation of hemicelluloses during hot pressing of particleboard. Potutkin and associates (13) contend that lignin enters into condensation reactions during the hot pressing of particleboard despite increases in hot-water extractability of the board. In another paper, Potutkin and associates (14) report changes in the lignin isolated from hot-pressed particleboards compared to that from raw birchwood from which the boards were made. Lignin isolated from boards pressed at 180C showed increased infrared absorption at 800 cm^{-1} and 1300-1500 cm^{-1}, which is interpreted to indicate condensation and demethoxylation, respectively. An appearance of a new band at 1370 cm^{-1} is assigned to aryl-alkyl ether bonds formed by reactions of lignin with other wood breakdown products to yield pseudolignin.

Chemical Changes in Wood Effected by Board Conversion

Studies conducted at Masonite Corporation and discussed earlier in this paper reveal that defibration is indeed hydrolytic and produces furfural and mononuclear aromatics. Clearly, the elements of the polycondensation mechanism are available as described by Runkel and Wilke (11). Boehm (15), Klauditz and Stegmann (16), and Gupta *et al.* (17) go as far as to claim that the high lignin contents of Asplund and Masonite pulps are due to condensation of lignin at high temperatures.

Although impressive, the evidence to support the hydrolysis-condensation sequence of reactions is subject to other interpretation when the response of wood to heating is viewed more broadly. Topf (18) found that upon heating to 180C, lignin breaks down and volatilizes more readily than does cellulose. Kollmann and Topf (19) report that lignin-rich materials are more susceptible to pyrolysis and autoignition. Alekseev and Reznikov (20) observe that above 100C acidified wood undergoes primarily degradation. Merritt and White (21) report that pyrolysis begins below 180C and becomes exothermic at 240-260C with the principal condensible vapors produced being acetic acid, furfural, and methanol. Katzen *et al.* (22) identifies the acetyl groups on hemicellulose as the source of acetic acid and methoxy groups on lignin as the source of methanol. Furfural comes from the degradation of pentoses.

At one time, cross-linking of cellulose chains through ether linkages was thought to be the cause of the reduced hygroscopicity and dimensional change of heat-stabilized wood (23). Later studies (24) revealed that heat-stabilized wood swells less in water but more in other liquids such as pyridine, morpholine, or 18 percent aqueous NaOH. This discredits the cross-linking mechanism for whole wood and casts suspicion on the polycondensation mechanism generally. In fact, effective heat treatments of wood are known to be attended with substantial dry-weight losses which is greatest when wood is heated in a closed system, in the presence of air, and under steaming conditions.

Hot Pressing of Felted Mats. An experiment intended to identify which of the two mechanisms is operative was designed and executed. The approach taken was based on the argument that each mechanism requires certain cell wall components be present and that removal of all or most of the component will be detrimental to wet-strength development. The phenol-furfural condensation mechanism requires hemicellulose as a source of furfural. The thermoplasticity of lignin mechanism requires lignin. Consequently, three pulps were prepared containing:

1. all cell wall components from whole wood.
2. cellulose and lignin with hemicellulose removed.
3. cellulose and hemicellulose with lignin removed.

Analyses of these pulps are given in Table III. It is acknowledged that the selective extractions

Table III
Analyses of Specially-Prepared Pulps

Treatment	Yield	Percent, by Weight		
		Cellulose	Hemicellulose	Lignin
None	100.00	56.03	24.10	19.52
10% NaOH (2 hrs)	81.28	51.09	5.38	16.64
10% NaOH (4 hrs)	80.71	50.35	4.81	16.11
Acidic $NaClO_2$ (twice)	85.70	50.80	24.90	4.40
Acidic $NaClO_2$ (thrice)	75.00	45.60	29.40	2.20

are not exhaustive. Because preparative methods that selectively remove hemicelluloses or lignin also remove waxes and other low-energy materials that can affect interfiber bonding, part of the whole furnish was extracted with petroleum ether to remove these low-energy components. Wet/dry and dry/dry boards were prepared from these furnishes. Changes in apparent cellulose and lignin caused by hot pressing wet-formed boards at 425F are indicated in Table IV.

Table IV
Change in Cellulose and Lignin in Hot Pressing

Furnish	Cellulose		Lignin	
	Before	After	Before	After
Whole, unextracted	55.8	50.1	17.2	20.0
Whole (petroleum ether extracted)	56.5	50.5	19.5	20.9
HC-free (petroleum ether extracted)	47.9	42.3	15.0	15.8
Lignin free (petroleum ether extracted)	46.4	--	3.1	4.7

There is only a small increase in apparent lignin caused by pressing. Rough measures of dry weight losses in pressing indicate only 1-2 percent loss when wet-formed boards were pressed compared to 10 percent dry weight loss when dry-formed boards were pressed. Some of this weight loss with dry-formed boards is moisture adsorbed by the fluffy mat but the magnitude of weight loss reveals low-molecular-weight fragments are desorbed, also.

Dry and wet strengths of boards made from the special furnishes are shown in Figures 11-13. Wet strength applies to boards tested wet after 24-hour immersion in 97C water. Removal of low-energy materials by petroleum-ether extraction improves bond strength and static and impact bending properties. Removing hemicelluloses, the source of furfural for polycondensation with lignin breakdown products, effects further improvements in wet and dry properties. Removing lignin, however, has a consistently detrimental effect on strength.

Additional evidence against the polycondensation mechanism is given in Figure 14, for two commercially produced hardboards, one produced with high cooking pressure and the other with low cooking pressure. The severely-steamed furnish exhibits much lower hygroscopicity than the mildly-steamed furnish after conversion into board. But immersion in liquid ammonia followed by desorption of ammonia and adsorption of water causes the two boards to have almost the same sorption capacity. This recovery of

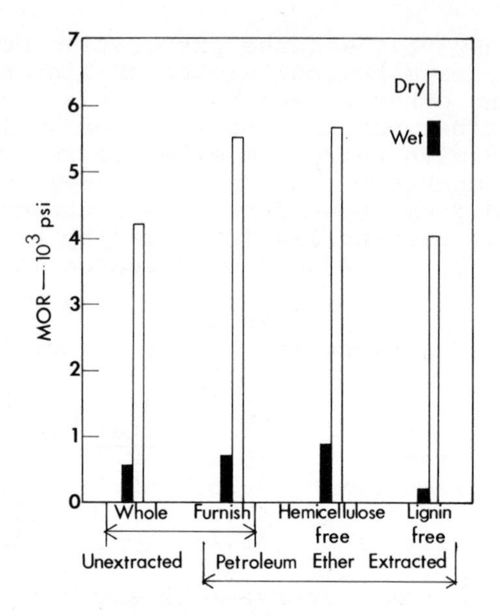

Figure 11. Modulus of rupture for boards made of specially prepared furnishes (see text) *and tested in the wet and redried condition after 24-hr immersion in 97°C water*

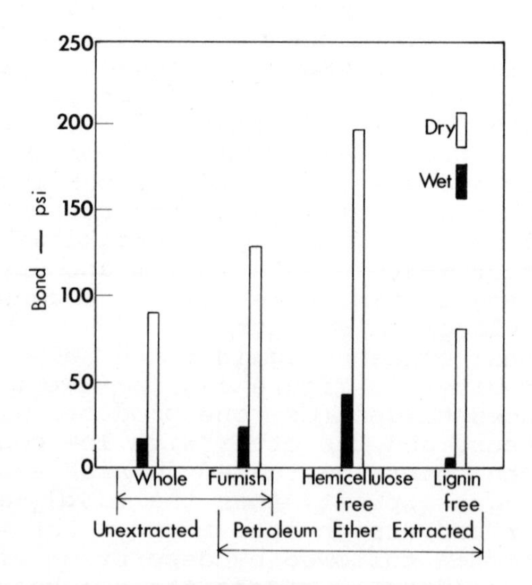

Figure 12.—Internal bond strength of boards described in Figure 11

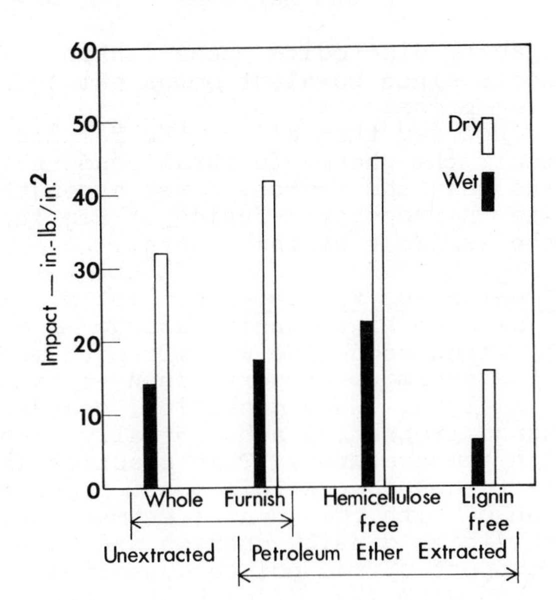

Figure 13. Impact strength of boards described in Figure 11

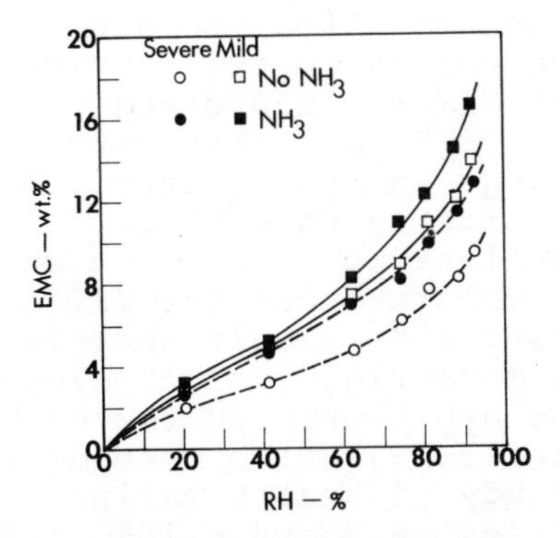

Figure 14. Adsorption isotherms at 32°C for boards with or without liquid ammonia treatment before water sorption. Severe and mild refer to cooking steam pressures.

sorption capacity discredits cross-linking of cell wall components since covalent bonds should not fail under swelling pressure alone.

It is concluded that all of the results presented above discredit the phenol-furfural condensation mechanism and that the source of wet strength in hardboards is thermoplastic-fusion of lignin concentrated on the surfaces of the fibers.

Heat Treating of Fiberboards. Hot pressing of low-density mats to high-density boards is a poor operation by which to decide wet-strength mechanisms because of the enormous property changes caused by densification alone. Once densified, boards are known to change properties substantially when oven baked at high temperatures. This practice is used quite widely to develop certain board properties. Property changes with this heat treatment occur without significant density changes and have been offered as support of the polycondensation mechanism.

Klinga and Back (25) offer heat treating results in support of the polycondensation mechanism. They note that wet-pressed hardboard exhibits progressive and permanent increases in thickness when subjected to alternating high and low humidities. Oven heat treating the board between humidity cycles arrested the progressive changes and gave essentially reproducible dimensional values at each humidity equilibrium. But heat treating produced a permanent decrease in board thickness.

Klinga and Back (25) ascribe this behavior in terms of a cross-linking of carbohydrate chains through hemi-acetal bonds. These bonds are claimed by Klinga and Back to offer opposition to re-expansion of the boards to their prepress thickness. But the range of dimensional change after heat treatment is as large or larger than before baking, revealing no change in the moisture content differential between humidity equilibria. Cross-linking forms rigid, covalent bonds between chains which must reduce the moisture content differential.

Cross-linking also should increase strength properties as a more rigid, three-dimension polymer is formed. This hypothesis is tested by the data given in Figures 15-19, which were derived from a comprehensive study of heat treating. The boards had been oven dried 16 hours @ 105C before heat treating so as to be anhydrous going into the baking ovens. It is shown in Figure 15 that heat treating these 3/8-inch, wet-pressed boards of

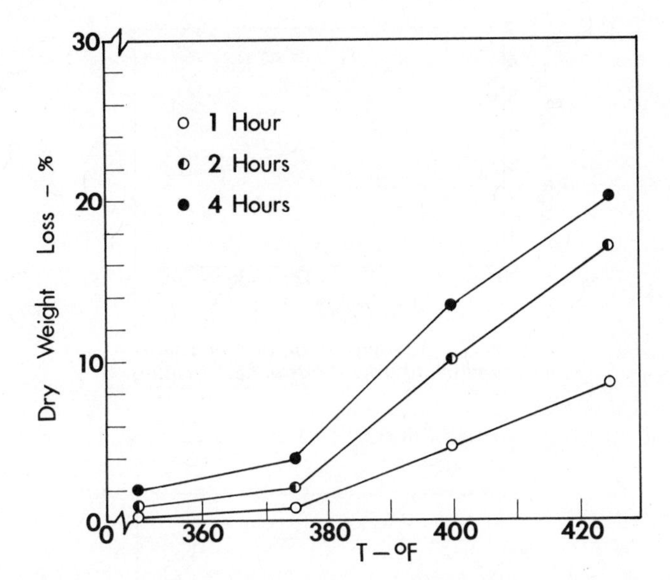

Figure 15. Dry weight loss of anhydrous boards produced by heat treating at several baking temperatures and times

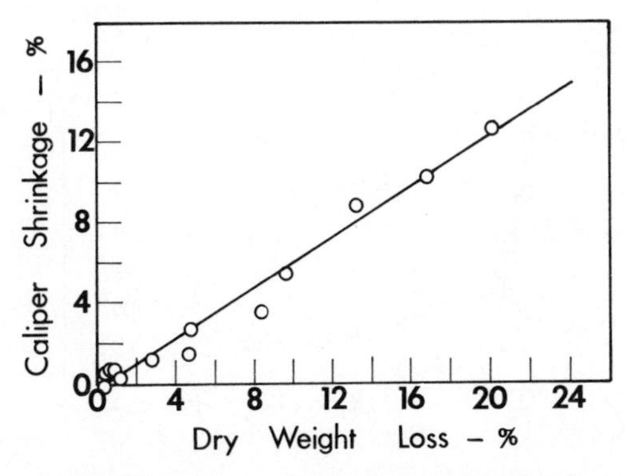

Figure 16. Caliper (thickness) shrinkage of boards caused by weight loss in heat treating

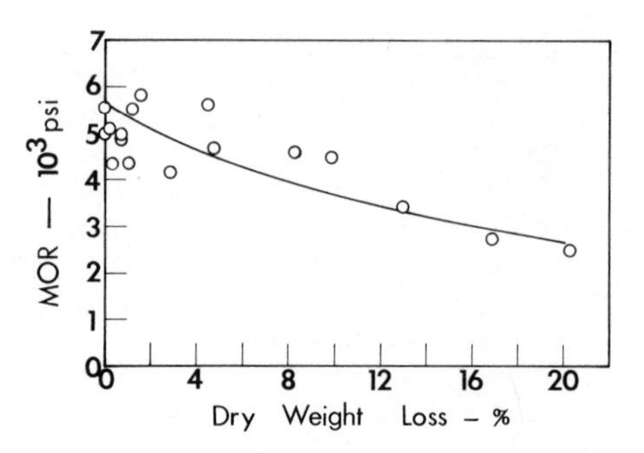

Figure 17. Modulus of rupture of boards as a function of weight loss in heat treating

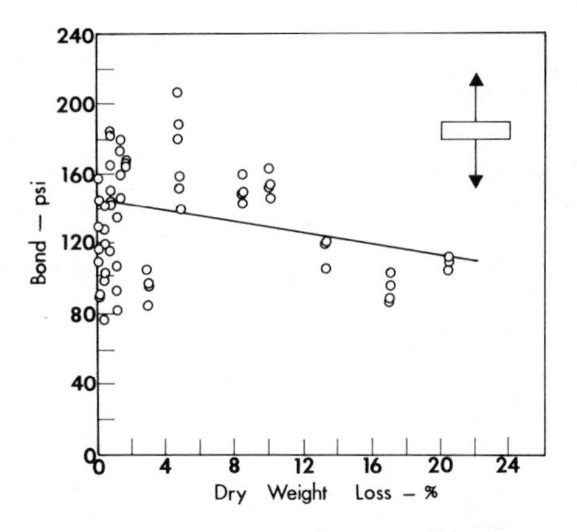

Figure 18. Bond strength of boards as a function of weight loss in heat treating

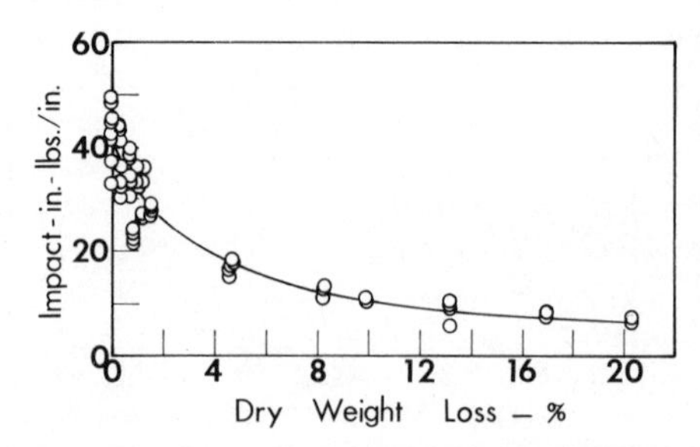

Figure 19. Impact strength of boards as a function of weight loss in heat treating

southern pine furnish is attended by sizable dry weight losses. Board thickness also decreases with baking directly proportional to weight loss (Figure 16). Similar results are seen when adsorbed water is driven from the board by oven drying. These results show that cell wall substances are gassified by heat treating just as water is gassified to cause weight loss and shrinkage in thickness. The board response to heat treating appears to be typical of sorption and reveals no extraordinary changes that cross-linking or polycondensation would produce.

This tentative conclusion is further supported by the changes in strength with weight loss caused by heat treating (Figures 17-19). Bending strength shows little change with small weight losses but decreases progressively at large weight losses. Cross-linking or polycondensation should produce sizable bending strength increases. Even internal bond strength (Figure 18) shows no improvement with heat treating and, in fact, suffers at large weight losses caused by heat treatment. Impact strength (Figure 19) undergoes rapid and extensive decline with weight loss as does whole wood (24).

Besides changes in weight, dimensions, and strength, heat treating wood fiberboards also increases water repellency and reduces springback (irreversible increases in board thickness). These important changes are effected through the mechanism of gassification of volatiles trapped in the board. Sauer and Haygreen (26) provide the evidence for this explanation. They report that when hardboard is subjected to long-term bending loads, creep is greater when the material undergoes a change in moisture content during loading than when moisture content is maintained constant. This they attribute to the freedom of movement of cell wall molecules undergoing spatial rearrangement as water is added or subtracted under bending stress. Extending their explanation to wood fiberboards undergoing desorption of water and cell wall materials during hot pressing and heat treating provides a view of cell wall structural molecules rearranging to relieve compression stresses generated by hot pressing. As a result of this plastic relaxation of compression stresses, well heat-treated boards are freed of the residual compression stresses that cause the board to swell irreversibly in thickness when moisture is adsorbed.

How water repellency is enhanced by this scheme is demonstrated by crude hot-pressing and heat-treating studies conducted at Masonite. A press and oven outfitted so that gases evolving could be collected, condensed, and analyzed reveals that the materials being desorbed from the board during hot pressing and heat treating are methanol, acetic acid, furfural, and aromatic particulates (smoke). A substantial portion of the vapor phase is comprised of petrolatum wax added to the furnish to enhance weatherability. The wax vapor tends to condense on all surfaces because of its marginal volatility and it is concluded that cell wall surfaces within the board also condense wax. Heat treating in the press or oven serves to redistribute the wax into a monomolecular film on all fiber surfaces with a coincident increase in water repellency. Both springback decreases and repellency increases are attributable to the desorption of materials by the board conversion operations conducted at high temperatures.

Considering the volatiles that evolve from hot pressing and heat-treating operations, it is concluded that the major chemical changes occurring in this stage are pyrolytic. Methanol, acetic acid, furfural, and ligneous tars are the common volatiles produced by the slow pyrolysis of wood practiced in destructive-distillation processes. The temperatures used in the board conversion operations approach pyrolysis temperatures of wood and the evidence indicates that pyrolysis is indeed active in board conversion.

Summary and Conclusions

All of the chemical evidence that can be marshalled indicates that wood fiberboard manufacture exploits the thermoplastic properties of lignin. Defibering is effected by the thermal softening of lignin in the middle lamella at saturated steam pressures above 130C. Interfelted fiber mats are consolidated with or without densification pressure by the thermoplastic fusion of lignin-rich fiber surfaces at high board conversion temperatures.

The major chemical changes in wood caused by fiberboard manufacture are secondary side reactions which are both beneficial and detrimental to the final properties achieved. Defibering is accomplished by hydrolytic breakdown of lignin and hemicelluloses under wet acidic conditions combined with high process temperatures. Board conversion and consolidation is attended by pyrolytic reactions which further breakdown and gassify low-molecular-weight products generated by defibration hydrolysis.

Literature Cited

1. Muench, C. S. (1920). U.S. Patent 1,339-254.
2. Mason, W. H. Paper Trade Journal (1927) 84(8): 131.
3. Mason, W. H. (1926). U.S. Patent 1,578,609.
4. Mason, W. H. (1928). U.S. Patent 1,655,618.
5. Mason, W. H. (1938). U.S. Patent 2,120,137.
6. Fergus, B. J. et al. Wood Science and Technology (1969) 3(2):117.
7. Goring, D. A. I. Pulp and Paper Magazine of Canada (1963) 64(12):1517.
8. Koran, Z. Wood and Fiber (1970) 2(3):247.
9. Sawardeker, J. S., Sloneker, J. H., Jeanes, A. Analytical Chemistry (1965) 37(12):1602.
10. Timell, T. E. Cellular Ultrastructure of Woody Plants (1965). Syracuse University Press.
11. Runkel, R. O. H. and Wilke, K. D. Holz als Roh- und Werkstoff (1951) 9:260.
12. Naveau, H. P. Chem. Abstr. (1968) 70:21079t.
13. Potutkin, G. F. et al. Chem. Abstr. (1968) 70:69401u.
14. Potutkin, G. F. et al. Chem. Abstr. (1969) 71:40445w.
15. Boehm, R. M. Paper Trade Journal (1944) 118 (13):35.
16. Klauditz, W. and Stegmann, G. Holzforschung (1951) 5(1):68.
17. Gupta, P. R. et al. Pulp and Paper Magazine of Canada 63(1):T21.
18. Topf, P. Chem Abstr. (1971) 76:15875 g.
19. Kollmann, F. F. P. and Topf, P. Journal of Fire and Flammability (1971) 2:231.
20. Alekseev, A. D. and Reznikov, V. M. Chem. Abstr. (1970) 77:63618t.
21. Merritt, R. W. and White, A. A. Industrial Engineering Chemistry (1943) 35(3):297.
22. Katzen, R. et al. Industrial Engineering Chemistry (1943) 35(3):302.
23. Stamm, A. J. et al. Industrial Engineering Chemistry (1946) 38(6):630.
24. Seborg, R. M. et al. Journal Forest Products Research Society (1953) 3(9):59.
25. Klinga, L. O. and Back, E. L. Forest Products Journal (1964) 14(9):425.
26. Sauer, D. J. and Haygreen, J. G. Forest Products Journal (1968) 18(10):57.

13

Review of Particleboard Manufacture and Processing

M. W. KELLY

Department of Wood and Paper Science, North Carolina State University, Raleigh, N.C. 27607

The particleboard industry has grown rapidly since its beginning in Europe during the Second World War. The original development was in response to the demand for an inexpensive panel product which could be produced from low-quality wood unsuitable for plywood. In the mid 1950's the particleboard industry was established in the United States not in response to a demand for the product but in response to the availability of cheap residues generated by planer mills and sawmills.

Particleboard is produced in large, capital intensive plants with highly automated equipment. Many equipment variations exist from plant to plant, but all plants have to adhere to similar processing steps during manufacture. These steps are particle preparation, particle drying, adhesive-particle blending, mat formation, hot pressing, and finishing. Only manufacture and processing of platen-pressed particleboard will be discussed. A relatively small quantity of particleboard is produced annually by an extrusion process whereby the adhesive treated particles are forced between heated dies which polymerizes the adhesive and forms a continuous particleboard ribbon, which is then cut into desired lengths. The production of extruded particleboard has been steadily decreasing and, due to the limited mechanical properties, the use of this material is basically restricted to that of corestock for furniture.

Definition

Matformed particleboard is an engineered panel product of machined particles bonded together with an adhesive under controlled heat and pressure. A basic difference between particleboard and medium density fiberboard is in the degree of disintegration of the wood macrostructure. The component particles of fiberboard are either individual wood fibers or fiber bundles; interparticle bonding is accomplished by an added adhesive system. Hardboard is also composed of fibers and fiber bundles but interparticle bonding is due to the self-bonding in the high density panel, not to an external adhesive. The gross macrostructure of the wood is retained in the component particles of particleboard and an external adhesive is required for interparticle bonding. A wide range of particle sizes, from sawdust granules to flakes three inches long is commonly used, although rarely is the entire size distribution included in a single panel.

Raw Materials

The common raw materials of particleboard are wood, adhesive, and wax emulsion. High quality particleboard at the optimum production rate demands continuous monitoring of the wood material to determine when adjustments should be made in the process or the adhesive. Monitoring of the adhesive and wax emulsion quality is a critical, but often ignored, factor in particleboard manufacture.

Wood. The wood content on a dry basis for most particleboard is between 90 and 95 percent. Any particle configuration can theoretically be used although certain physical properties will be adversely influenced if adequate particle uniformity is not observed. Also, physical properties can be engineered into the panel by using different particle sizes or configurations in the core and surfaces. For example, long particles at the surfaces significantly increase the bending strength of the panel but they also result in a rough, difficult to finish surface. Hence, if finishing characteristics are more critical than bending strength for a particular application, smaller particles, which result in a smoother surface, are used on the surface. Normally, the particle size and configuration, as well as the distribution of the various sizes through the panel thickness, is adjusted to optimize the desired properties with a minimum effect on the remaining properties.

Many wood species, both hardwoods and softwoods, are used for particleboard; however, the density of the particleboard should be higher than the density of the raw material to efficiently utilize the adhesive system. The compression of the particles, which is required for consolidation into the finished product, enhances the particle-particle contact, producing more interparticle adhesive bonds as well as reducing the total void volume in the panel. With wood of density higher than the finished particleboard, the compression of the particles is lower and the resultant reduced interparticle contact and higher void volume adversely influence the physical and mechanical properties of the particleboard.

The acidity of the wood should also be monitored to allow adjustments in the adhesive system to maintain the same polymerization rate. The adhesives are pH sensitive and excessive fluctuations in the wood pH may retard or speed the polymerization process.

Adhesive. Urea-formaldehyde water-based dispersions are the most widely used particleboard binders. The low-cost, rapid curing, and colorless properties of urea-formaldehyde adhesives make them the adhesive of choice for most interior particleboard. These adhesives have been continuously improved by the resin manufacturers, resulting in reduced press times without detrimental effects on their storage life or handling characteristics.

The urea-formaldehyde polymer is formed by a multi-step reaction process between urea and formaldehyde. The initial phase is a methylolation of the urea under slightly alkaline conditions with a formaldehyde-urea (F/U) molar ratio of 2.0:1 to 2.4:1. Condensation of the methylolureas from the methylolation reaction is at atmospheric reflux with a pH of 4 to 6. This condensation polymerization continues to a pre-determined viscosity, at which time the pH is adjusted with a suitable base to 7.3 to 8.0. The adhesive is then concentrated to a total solids content of 50 to 60 percent by vacuum distillation. Additional urea is then normally added to produce a final F/U molar ratio of 1.6:1 to 1.8:1.

The final polymerization of the urea-formaldehyde adhesive occurs in the hot press and is one of the critical steps in particleboard manufacture. If the adhesive cures at any point in the manufacturing sequence other than in the hot press with the mat compressed to the desired thickness, an unsatisfactory product will be produced. Since the polymerization of urea-formaldehyde adhesives is much faster under acidic conditions and at elevated temperatures, optimum curing rates (minimum press times) are attained if the adhesive pH is 3 to 5. However, urea-formaldehyde adhesives are supplied at slightly alkaline levels to retard polymerization during transit and storage but this increased stability lengthens the press time. Many wood species, such as the oaks and southern pines, are acidic and contribute to a rapid pH decrease in the adhesive. Also, resin suppliers attempt to limit the buffering capacity of their adhesives by using a volatile base to adjust the final pH. When the adhesive and wood are exposed to the elevated temperatures in the hot press the volatile base is rapidly evaporated, the pH drops, and polymerization follows rapidly. Adjustments such as these and a better understanding of polymers and polymerization mechanisms have resulted in continuous reductions in the press times for urea-formaldehyde bonded particleboard. Since the pressing operation normally controls the production capacity of a particleboard plant, even small reductions in the press time result in increased production levels.

The disadvantages of the urea-formaldehyde adhesives lie in their lack of durability and in their characteristic pungent formaldehyde odor. For particleboard applications subject to high temperature and moisture exposure, phenol-formaldehyde adhesives are required, since the urea-formaldehyde polymer is hydrolyzable and hydrolysis is enhanced with moisture and heat.

Formaldehyde evolution, at the hot press and in applications where adequate ventilation is lacking, result in relatively high levels of free formaldehyde fumes with the urea-formaldehyde resins.

Phenol-formaldehyde adhesives are the only other adhesive system used in significant quantity in particleboard production. The increased durability of this class over that of the ureas results in phenolics as the adhesive of choice for exterior particleboard. However, phenolic adhesives are only used where the additional durability is required since they are more expensive and require longer curing times.

Phenol-formaldehyde adhesives are produced by a condensation polymerization reaction between phenol and formaldehyde. The phenolics used for exterior particleboard are made at a formaldehyde/phenol ratio greater than 1.0; i.e., they are classified as resoles and additional formaldehyde is not required to complete the curing reaction to a highly cross-linked network structure. Many characteristics can be incorporated into the adhesives by changes in the F/P ratio, condensation pH, and condensation time. The reactive solids content is normally between 40 and 50 percent since the stability and viscosity are adversely affected at higher solids.

Wax Emulsions. The final component in most particleboard is a sizing agent to reduce the absorption of liquid water. This is normally a paraffin wax emulsion which is supplied to the particleboard manufacturers at approximately a 50 percent wax solids in water. Less than 1 percent wax solids based on the ovendry wood weight is used in most particleboard; levels above 1 percent tend to interfere with interparticle bonding while levels below 0.75 percent do not offer maximum water resistance.

The above three components--wood, adhesive, and wax--are the only ingredients in most particleboards. Only limited quantities of fire retardant and preservative treated particleboard is presently produced. However, with increased flammability requirements and applications of particleboard in locations subject to biological degradation, an increase in production of both preservative and fire retardant particleboard is expected.

Manufacturing

Particle Preparation. The initial step in particleboard production is reduction of the wood raw material into the desired configuration for the particular particleboard to be manufactured. The wood may be received from a number of sources and in a variety of forms. Roundwood, chips, planer shavings, plywood trim, and sawdust are the most common raw material forms and rarely does one particleboard plant use more than two or three of the above sources. Different processing steps are required to produce quality particles from each of the above sources; consequently, each plant is limited to the source which is compatible with their wood reduction system.

The various reduction systems can be classified into knife, hammer, and attrition units, each type producing a characteristic particle. A tramp metal detection system is included in all reduction steps to protect the equipment from serious damage. The shape and integrity of the component particles strongly influence the quality of the resultant particleboard; therefore, the optimum in particle preparation is achieved when the desired particle is obtained with no damage to the structure of the wood. Wood failure within the particle will result in a particleboard of lower strength than one formed from intact particles.

Chippers and flakers are the most widely used knife reduction systems. Chippers produce coarse particles from roundwood, slabs, plywood trim, and other residues from the primary wood industry. Chippers are the initial reduction step in particle preparation and further size reduction is necessary to produce a satisfactory particle. The chipping operation may be located at the particleboard plant or the chips may be delivered to the plant from an in-woods chipping operation, a chip-n-saw mill, or from another primary wood industry. Screening after the chipper removes all fines and oversize chips before they enter the secondary reduction step. Oversize chips are recycled to the chipper and fines are normally sent to the boiler for fuel.

Flakers are used for initial reduction of roundwood and for further reduction of chips. Flakers designed for primary reduction of roundwood have knives projecting from a rotating drum or disc with the axis of rotation parallel to the log. The length and thickness of the resulting flake are controlled by scoring knives and the cutting knife projection, respectively. Normally, all roundwood will have the bark removed before the flaking operation.

Flakers used for secondary reduction of chips have an entirely different design. The knives in these flakers are mounted on a rotating ring which rotates against an impeller ring. The chips enter the machine in the center and are thrown to the perimeter and held across the projecting knives by both the centrifugal force and the impeller ring; the flakes are then produced by the rotating knives.

Seldom is particleboard produced from particles generated only by knives, most flakes will be further reduced by either a hammermill or an attrition unit. Hammermills reduce planer shavings, chips, plywood waste, and trimmings by mechanically breaking and tearing the wood. Rotating hammers beat the material against breaker plates until sufficient size reduction has occurred and particles can exit the machine through a screen. Hammermills are used for additional size reduction of flakes and planer shavings. Since the wood will cleave readily along the grain, hammermills are used to reduce the flake width without significantly reducing the flake length or thickness.

Attrition units are commonly found in plants producing

particleboard in which smooth surfaces are required. Flakes, planer shavings, and sawdust are reduced in attrition mills to small particles which will mat well and form a smooth surface. Attrition mills are either single or double rotating discs which mechanically grind the material into small particles.

Many particle preparation processes exist, the one used in a given particleboard plant will depend upon the raw material source and the type of particleboard produced. The particle quality from each process is a function of the wood moisture content, degree of maintenance of the equipment, and the form of the raw material. Good particle quality does not guarantee a quality particleboard but high quality particleboard cannot be produced from low quality particles.

Particle Drying. All particles used in particleboard manufacture are dried to a uniformly low moisture content before the adhesive is applied. There are a number of dryers used by the industry, details of which will not be presented here. The particles are quickly dried to a moisture content of 3 to 6 percent (based on wood ovendry weight) with commercial dryers. The particles are exposed to the high temperatures of combustion gases from oil, gas, coal, or wood as they are rapidly moved through a closed chamber. The evaporation of water and the short dwell time within the chamber minimizes the potential fire hazard. Continuous monitoring of the incoming particle moisture content is required to allow corrective action in the dwell time to prevent insufficient or excessive drying. Dwell time in the chamber and fuel consumption adjustments are the most common methods of correcting for changes in the incoming particle moisture content. Rapid fluctuations in the moisture content of the wet particles entering the drier should be avoided.

Blending. Addition of wax, adhesive, and other additives to the dry wood particles is called blending and is normally done by spraying the aqueous adhesive system and wax emulsion onto the particles as they are moved through a blender. The adhesive level is based on the ovendry weight of the particles; no attempt is made to monitor the total particle surface area. Consequently, smaller particles with a larger area to weight ratio have substantially more adhesive on a weight basis if both surfaces have equal adhesive coverage per unit of area. Particleboard quality is strongly dependent upon interparticle bonding and, as particle size decreases, more interparticle bonding per unit weight is required to produce the same density particleboard. Consequently, there is a need for higher resin levels on smaller particles when resin content is measured on a weight basis. However, as will be shown in the mat formation section, most small particles are placed at the surfaces for improved surface quality and smoothness in the final board, and better consolidation and more efficient use of the adhesive occurs at the surface;

therefore, less adhesive on an area basis is required for the small surface particles. Passing the small particles rapidly through the blender limits their time of contact with the adhesive spray and prevents excessive adhesive pickup.

Paraffin wax emulsion and other additives are also added to the particles during blending.

The blending operation is an important step in the production of quality particleboard--uneven distribution of the adhesive will result in regions of low interparticle bonding and weak particleboard. Strict monitoring of both the adhesive and particle streams delivered to the blender is required for optimum blending.

Mat Formation. The process by which the blended particles are deposited in a continuous ribbon on a moving belt is called felting or mat formation. Significant equipment advances in the forming machines have resulted in much more uniform mats with much less density variation across the board than was common with earlier generations of felters. The forming process is entirely automated, the particles fall from the felter as a curtain, forming a continuous ribbon of particles on the moving belt. Usually, more than one felter is required to build the desired mat thickness; multiple felters allow more uniform mat formation since less material is deposited by an individual felter. Also, multiple felters are required for layered particleboards in which larger particles are used in the core and smaller particles at the surfaces. Accurate and uniform felting is an extremely critical step in the production of particleboard. Density variations in a poorly formed mat cannot be eliminated and these will be present in the finished particleboard. The movement of the belt and the curtain of chips have to be finely adjusted to insure proper chip deposition to produce the target board density after compaction and resin hardening in the press. Changes in the wood species, particle size, and particle moisture content have to be accounted for by adjustments in the felting operation. The advantages of maintaining raw material uniformity to assist in this process are obvious.

Three classes of particleboard are commonly recognized, based on particle size distribution in the thickness direction. These are: 1) homogeneous - all particle sizes are distributed equally; 2) layered - large particles in the core and small particles at the surfaces; and 3) graduated - large particles in the core with progressively smaller particles from the core to the surfaces. The surface particles in the layered and graduated mats normally have a higher adhesive content (on a weight basis) than do the larger core particles. The small particles produce a smoother, more continuous surface which is easier to finish than are the rougher surfaces formed by larger particles.

Homogeneous mats are formed by depositing a mixture of particles on the moving belt without size segregation. Multiple formers are normally used with each former depositing a portion of the total mat thickness. Layered particleboard is produced by using small particles in the formers depositing the surfaces and larger particles in the core formers. Duplicate blending and transport systems are normally used for the surface and core particles in layered particleboard allowing control of the adhesive level in each layer but also requiring higher initial capital investment.

Graduated particleboards are similar to layered particleboards since large particles are present in the core and small particles at the surface. However, particle separation based on size is done in the felting operation which eliminates the need for two conveying and blending systems. A minimum of two formers are required, each former depositing half the mat thickness. Particle separation is accomplished by subjecting the falling particle curtain to air or by throwing the particles with a mechanical device.

As the continuous ribbon is conveyed from the forming station it may or may not be consolidated by a cold press. This prepressing operation reduces the mat thickness and increases the mat density which improves the handling characteristics of the mat, but does not initiate adhesive polymerization. The continuous ribbon is also trimmed to width and cut into individual mats, the length of which is equal to the length of the hot press. The individual mats are placed in the press loader which serves as a temporary storage area for the mats prior to hot pressing.

Pressing Operation. The consolidation of the particle mat and polymerization of the adhesive to produce a particleboard panel is accomplished in a hot press. The mat is compressed and held at the desired thickness until the adhesive on the particle surfaces has polymerized and established adequate bridges between particles. The panel is then removed from the press, cooled, and sent to the finishing phase.

The pressing operation is extremely important and is highly dependent upon previous processing steps. If a poor mat has been delivered to the press, a poor particleboard panel will result. Particles with insufficient adhesive from a poorly functioning blender or a mat with excessive moisture cannot be tolerated if quality particleboard is to be produced. The press is the most expensive equipment in a particleboard plant and the output of a plant is controlled by the pressing operation. Consequently, it is imperative that the press function efficiently with as short a cycle as possible. Many physical and mechanical properties of particleboard are influenced by the pressing operation; therefore, a clear understanding of the pressing function is required.

Most particleboard plants have multiple-opening hot presses which produce one panel per opening per press cycle. The press loader also has storage area for the number of mats equal to the openings of the hot press. When the press loader is filled and the press opens, all mats in the press loader are simultaneously transferred to the hot press, and the finished panels from the previous press cycle are removed to the press unloader. It is imperative that the forming line be operating at the proper speed to produce sufficient mats to have the press loader filled when the press opens. The press controls the plant production capacity and it should be operating continuously; it should not be held open waiting for additional mats. Consequently, most plants are designed with variable speed forming lines which can be synchronized with the press cycle.

Multiple-opening hot presses presently used in the particleboard industry are simultaneously closing; i.e., all openings close together at the same rate subjecting all mats to the same press cycle. In the earlier presses, which closed from the bottom, the mat in the lowest opening was subjected to a significantly longer press cycle than the mat in the top opening. Consequently, all the particleboard panels produced in the same pressing cycle did not have the same properties; the properties were influenced by the particular press opening in the hot press. Simultaneously closing hot presses have eliminated this source of variation, since all mats are subjected to the same press cycle.

Mechanical stops placed on two edges of each press opening are often used for thickness control. As the press is closed platens compress the mat until contact is made with these stops, at which point compression of the mat ceases and the particleboard thickness is equal to the thickness of the stops.

The mat surfaces are rapidly heated to the temperature of the platens as the mat is compressed. The water in the particles at the surface is vaporized and migrates into the cooler portion of the mat, i.e., toward the core. Condensation of this steam releases heat which increases the mat temperature quicker than could be accomplished by conduction through wood. However, the press is compacting the mat to target board thickness before the mat is completely heated. The compressive strength of wood is much lower at elevated temperatures and, since the mat is compressed when only the surface region is heated, compressive failure of the wood within the hot surface region occurs. The mat is compressed to thickness before the core is heated; consequently, there is a vertical density gradient in the thickness direction of hot-platen-pressed particleboard. High density surfaces and low density cores are produced with the average particleboard density falling between these two extremes. The low density core resulting from this vertical density gradient reduces the screw holding strength, shear resistance, and tensile strength of this region. Various vertical density gradients can be obtained for the same average board density by adjusting the rate at which the press is closed. However, long press closing times are to be avoided since the adhesive on the surface particles may harden before adequate interparticle contact is obtained. This condition is commonly referred to as surface precure.

The moisture migrating to the mat core also presents

difficulties in platen-pressed particleboard. The moisture evaporates from both surfaces and progressively migrates to the core as the temperature increases from the surfaces to the core. However, the temperature of the core eventually exceeds 100°C, turning the water into steam. The water in the core, in the form of steam, has to escape from the board during the pressing operation. The press time has to be sufficiently long to allow the steam to escape or the panel will delaminate when the hot press is opened and this steam rapidly expands. Also, the water interferes with the condensation polymerization reaction of the curing resin, limiting the curing rate and lengthening the press time. Therefore, the moisture in the particle mat assists in heat transfer to the core but restricts the adhesive cure and is a potential source for delamination at the panel midplane. The mat moisture content at which these two effects can be balanced will vary for each particleboard plant, depending on particle size and species. A technique commonly used is to have a non-uniform moisture distribution in the particle mat. A high surface moisture content, to assist in heat transfer, and a low core moisture content is widely used to minimize the press time.

Finishing

The type and extent of the finishing process for particleboard is determined by the product grade--floor underlayment and mobile home decking are simply squared and sanded to thickness while the industrial grade used in furniture applications is subjected to much more elaborate procedures. Painted and simulated grain surfaces can be formed directly on the particleboard while veneers, vinyl, and other surfacing materials are bonded to the particleboard by adhesives. Regardless of the finishing method the goal is to produce an attractive and functional surface with the required durability at a minimum cost. The finishing processes discussed here are those commonly used for the industrial grade particleboard.

Edge Finishing. Most industrial particleboard is produced with large particles at the core and smaller surface particles for surface smoothness. This graduated or layered construction, together with the platen-pressed method of manufacture, produces a particleboard with a low density porous core and higher density surfaces. Consequently, a non-uniform panel edge is present which does not machine or accept paints and finishes uniformly. Various techniques are available to mask these edges which are to be exposed in the completed furniture, cabinet, or shelf. The methods commonly used for edge finishing are tapes, lumber strips, T-mouldings, or V-grooving.

All edge finishing, except the lumber banding, is done after the panel surface has been finished. The wood or vinyl tapes are glued to the edges with PVA or hot melt adhesives. Plastic

T-mouldings are used to give a machined edge effect not possible with the flat tapes; a projection on the back is inserted into a machined groove in the edge of the particleboard.

Lumber banding consists of gluing lumber strips, 1/2 to 2 inches in width, on the particleboard edges. These strips are normally used in applications where the particleboard is to be covered with wood veneers. The solid wood strip can be machined to decorative edges and, with the veneer surfaces, the panel is fully as functional and attractive as a solid wood panel, but at a lower cost. The lumber bands are normally bonded to the particleboard with polyvinyl acetate or urea-formaldehyde adhesives, cured rapidly by either contact or high frequency heating.

A relatively recent development, in which vinyl covered particleboard is self-edged, is by V-grooving. V-grooving is accomplished by machining V-shaped grooves through the particleboard substrate to, but not into, the vinyl film. Adhesive is then applied to these grooves and, using the vinyl film as a hinge, the particleboard is folded back on itself. This edge finishing method results in continuous vinyl film at the corners and edges which are normally the prime locations for film and edge tape delamination.

Surface Finishing. All particleboard surfaces have rough, irregular surfaces; the degree of roughness and irregularity is a function of the surface particle size. As the particle size decreases and the density of the particleboard increases, smoother surfaces are obtained on which less work is required to obtain satisfactory finishes. The degree of surface smoothness required is dictated by the particular finishing method employed. Grain printing requires smoother surfaces than does veneering and more surface preparation is required prior to printing than is necessary for veneering.

The simplest construction for veneered particleboard is a face and back veneer glued to the particleboard core. It is imperative that veneers of equal thickness, grain direction, and dimensional stability be used to insure a balanced panel. A balanced panel reduces the potential for bowing and warp on subsequent exposure to changes in ambient relative humidity.

Particleboard which has been lumber banded and conditioned is sanded to insure uniform thickness of the particleboard and edges. Conditioning after edgebanding is imperative to allow equalization of the water from the edgebanding adhesive throughout the assembly. Premature sanding and veneering creates the possibility of subsequent dimensional changes which will produce a panel with a distinct border from the lumber band telegraphing through the surface veneer. This border cannot be removed and will always be evident in the panel. Sanding after proper conditioning insures an equal thickness for both the particleboard and lumber edges.

Water-based adhesives, usually urea-formaldehyde, are used in the veneering operation. The water in these adhesives can result in excessive surface particle swelling which will "telegraph" through the veneer, being especially evident with high-gloss finishes. Addition of a cross band veneer between the particleboard and surface veneer will prevent most problems of telegraphing. The cross band veneer is normally thicker and lower quality than the surface veneer and is placed with the longitudinal grain direction at a right angle to the grain direction of the surface veneer. However, if a cross band is used below the surface veneer, an equal cross band has to be used between the back veneer and the particleboard to retain a balanced construction. The resulting 5-ply construction is much more stable than the 3-ply to changes in ambient relative humidity.

Other surfaces are also commonly bonded to particleboard substrates; these include vinyl overlays, high density overlays, and low density overlays. Low density overlays are melamine impregnated paper which bond to the substrate with the melamine formaldehyde adhesive present in the overlay. High density overlays are highly durable and resistant sheets of phenolic resin impregnated paper with a top sheet impregnated with a melamine-formaldehyde resin. Contact adhesives are commonly used to bond high density overlays to particleboard cores for applications requiring high durability such as countertops.

Vinyl overlays are thin sheets of polyvinyl chloride, often with a simulated grain pattern, which are glued to the particleboard to obtain an inexpensive finish of relatively low durability. Particleboard panels containing a 3-dimensional design can be vacuum laminated with vinyl films, provided sharp corners are not present in the design. The thermoplastic vinyl film is heated to the softening temperature and, as the air is withdrawn from between the film and the machined panel, atmospheric pressure from above forces the film to conform to the contours of the panel.

All of the above finishing techniques are characterized by addition of a separate surface layer or film to the panel by adhesive bonding. Particleboard is also finished by applying liquid finishes directly to the surfaces and initiating a physical or chemical reaction of the finish to form the desired durability and appearance. Painting and grain printing are common examples of this method. Particleboard panels precut to the required final dimensions are commonly painted or printed on an automated finishing line. The painted or printed panels are then assembled and the final topcoats applied. Print lines are more sophisticated and technologically advanced than paint lines and will be described in detail.

Grain Printing on Flat Panels. Grain printing of particleboard surfaces is rapidly expanding, primarily due to refinements and advances in the techniques and equipment for printing. With multi-colored printing and rapid line speeds this method of finishing particleboard closely simulates wood but with costs below those of other finishing methods. Printed particleboard is widely used in vertical applications for cabinets, casegoods, and other applications in which highly durable surfaces are not required. Extremely close control on both the panel smoothness and the print rolls is necessary to maintain quality of the printed panel. Most simulated grain printing is done with high volume, automated finishing lines on precut flat panels prior to assembly of the finished item.

A wide range of equipment and techniques are used to print particleboard with simulated grain pattern. However, the process essentially consists of sanding and filling the particleboard to obtain a smooth surface, followed by applying a basecoat for the background color and, finally, the grain pattern. The final sealer finish is often applied to the assembled item on a production finishing line in the furniture plant. One of the previously mentioned edge-finishing procedures is also required with printed panels for applications with exposed edges. The reverse surface should be coated with a material possessing similar permeability as the top surface to maintain a balanced construction and minimize bowing and warping difficulties. Adequate equipment maintenance and careful panel preparation is mandatory for production of quality printed panels.

Accurate sanding of the particleboard is essential for surface smoothness and uniform thickness. Sanding to thickness with 50-100 grit sandpaper is followed by finer grit paper to obtain the required smoothness. The total sanding operation is normally done with multi-head sanders, with progressively higher grit paper and a final smoothing bar.

The sanded panel is brushed and vacuumed to remove surface dust and debris. Minute wood particles remaining on the surface through subsequent finishing steps will result in defective filling and printing operations.

The next step is the filling operation in which the small depressions and voids between the surface particles are filled with a high solids, high viscosity coating. Due to the inherent structure of particleboard small interparticle voids and depressions will always be present at the surface, regardless of the surface particle size or the sanding technique. The filler is normally applied by reverse roll coater which forces the filler into these voids and produces the extremely smooth surface required for printing. Fillers with a wide range of chemical properties are available; the one chosen by a given producer is determined by the curing equipment in the plant. UV curable polyester fillers are widely used due to the speed at which they cure but a UV radiation source is required. UV curable filler can be hardened in 10-15 seconds, thereby significantly increasing finishing line speed as well as shortening the overall length of the line. Vinyl, polyurethane, and urea-alkyd filler systems cure by heat or high air velocity and require substantially longer curing times. A second filling after a light sanding of the first fill coat results in a much smoother surface and produces a better printed particleboard.

A basecoat is applied to the panel after the filling operation. This is normally a pigmented lacquer or vinyl based material whose function is to hide the surface and provide a uniform color on which the grain will be printed. The basecoat is commonly applied by a curtain coater or more commonly with a roller coater. Obviously, skips in the basecoat, whether the result of deformed application rolls or panel unevenness, cannot be tolerated. Surface depressions not completely filled during the filling operation will not be basecoated and will appear as small white dots on the panel.

Basecoats are normally cured in heated ovens with high air velocity or with infrared heaters; excessive panel temperatures should be avoided to prevent drying of the particleboard. A light scuff sanding of the basecoat is used to remove high spots, followed by a brush cleaning to remove the dust and debris.

The grain pattern is then printed on the panel with one to three printers in tandem; each printer has a different color which allows better grain pattern simulation. Lines equipped with three printers can produce a four-tone pattern since the basecoat is normally a different color than the printers.

The inks used for graining are drying inks which must be compatible with the fill, basecoats, and other finishing materials used in the process. Three rolls are used by each printer to transfer the ink from the ink tray to the panel. The application roll transfers ink from the tray to the print roll which has been etched with the desired grain pattern. A doctor blade removes excess ink from the etch roll before the rubber covered transfer rolls remove the pattern from the etch roll and transfers it to the panel. Ink not completely transferred from the transfer roll to the panel has to be cleaned with another doctor blade to insure continuously sharp grain patterns. If more than one printer is used, they have to be synchronized to insure proper grain patterns result. Each etched drum has a portion of the total grain pattern and they must revolve in sequence to produce the desired simulated grain.

The edge finishing and seal coat may be applied on the finishing line but is more commonly done on the furniture assembly and finishing lines. Careful handling of the printed panel is required to prevent chipping at the edges and corners during furniture assembly. With proper care during assembly and shipping, simulated grain printed particleboard results in attractive, inexpensive furniture for the mass market.

Conclusion

The particleboard industry has rapidly evolved, in the relatively short period of its existence, from small, low capacity, highly labor intensive plants to high volume, highly automated facilities. The product from these modern plants also bears little resemblance to the initial particleboard with its limited physical and mechanical properties which restricted its application to cores for decorative veneers. Many researchers have made significant contributions in equipment, process developments, and expanded applications which resulted in rapid growth of this industry. Continued investments in particleboard research will expand applications and improve the processing technology to more efficiently utilize our renewable wood resource.

Literature Cited

Additional information pertaining to particleboard manufacture and processing is available in the books listed below as well as in numerous technical and trade journals.

1. Mitlin, L. (ed.). "Particleboard Manufacture and Application." Pressmedia Ltd., Sevenoaks, Kent, U.K., 1968.
2. Maloney, T. and A. L. Mottet, Particleboard. In "Modern Materials." (B. W. Gonser, ed.), pp. 1-38. Academic Press, New York, 1970.
3. Moslemi, A. A. "Particleboard" Vol 1 and 2. Southern Illinois University Press, Carbondale, 1974.
4. Kollmann, F. P., E. W. Kuenzi, and A. J. Stamm. "Principles of Wood Science and Technology, Vol. 2. Wood Based Materials." pp. 312-550. Springer-Verlag, New York, 1975.

14

Bark Extracts as Bonding Agent for Particleboard

ARTHUR B. ANDERSON

Forest Products Laboratory, University of California, Richmond, Calif. 94804

Until recently the particleboard industry has enjoyed a favorable environment for growth through ample supplies of relatively low-cost bonding agents, including urea-formaldehyde and phenol-formaldehyde resins. In the manufacture of water-resistant exterior-type particleboard, phenolic and phenol-resorcinol modified resins are employed. The particleboard industry is facing, at present, a shortage and competing market for phenol and high cost polyhydroxy phenol, such as resorcinol (3). It becomes apparent that it would be desirable to find another low-cost source for phenolics, preferably from a renewable resource.

One of the ubiquitous by-products of the forest products industry at plant site is bark residues. While some bark is used as fuel and in agriculture, in particular soil applications, considerable amounts of bark remain unused (6,7). A recent survey in Oregon indicated that the lumber and plywood industries generated annually a total of 3.0 million dry tons of bark. Nearly half (46 percent) of the bark producedwas used as fuel, 10 percent for other purposes--mainly for soil application, and 44 percent was not used (6). Bark was utilized least of all types of residues. And bark disposition is becoming more of a problem due to the increased restrictions on the incineration of bark residues.

Chemical processing of bark is limited and the principal chemical products produced commercially from barks are based on the barks' phenolic content (1,12). Barks generally are richer than wood in quantity and complexity of extractive components, the most important being a) the monomeric polyphenols or flavonoid compounds, and b) the polymeric phenolics, such as tannin, phlobaphenes and phenolic acids.

Use of the phenolic components of bark extracts in preparing adhesive components used in plywood and particleboard manufacture has been proposed from time to time (8,15). Such preparations are based on the reaction of bark phenolic components with an aldehyde, usually formaldehyde.

The chemical reaction of western hemlock bark extract with

formaldehyde has been proposed as a bonding agent for plywood (13, 18). Mangrove tannin-formaldehyde resin has been investigated as a strong water resistant adhesive for plywood (4). Wattle tannin is being used in Australia as a waterproof adhesive in the manufacture of plywood and particleboard (19,20). *Pinus radiata* and ponderosa pine bark extracts have also been investigated as possible bonding agents for particleboard (2,9).

This paper is a report on preliminary studies evaluating the suitability of bark extracts from four West Coast softwood species as bonding agents for particleboard. The barks investigated include white fir (*Abies concolor*), ponderosa pine (*Pinus ponderosa*), Douglas fir (*Pseudotsuga menziesii*) and western hemlock (*Tsuga heterophylla*).

Material and Preparation

The bark was air dried and then put through a hammermill using a 1/16-inch screen. The entire product was used in the preparation of extractives.

Wood particles used for the outer layers were comprised of that fraction of Pallmann milled particles which passed a 10-mesh screen and were retained on a 16-mesh screen with random lengths to 1/2-inch. Wood flakes which remained on a 10-mesh screen and between 0.008 and 0.012-inch thick, with random lengths to 3/4-inch and random width to 1/4-inch were used as core in the three-layer board.

The wax used contained 50 percent solids and the source of formaldehyde was powdered paraformaldehyde.

Bark Extracts

Each of the barks used in the present investigation was analyzed for extractive content and the results are summarized in Table I. Table II shows the tannin and reactive polyphenols formaldehyde-hydrochloric acid contents (5).

On the basis of data from previously reported experiments relating to use of aqueous sodium carbonate as an extractant and the addition of sulphites to increase extract stability, pulverized bark (1500 gms oven-dry basis) was extracted at from 70° to 80°C, after addition of 2 percent sodium carbonate (o.d. bark basis) (2,9). The bark slurry was held at the extraction temperature for 30 minutes, then filtered and washed at the same temperature by stirring the filter cake for 15 minutes. The wash liquor, from which bark had been removed by filtration was then used in extracting the second lot of bark (1500 gms) and so on.

A mixture of 0.25 percent sodium bisulphite and 0.25 percent sodium sulphite (o.d. bark) was added to the combined filtered extracts. The extracts were concentrated under reduced pressure at 35° to 55°C in a circulating vacuum evaporator. The extract was concentrated to 36 to 47 percent solids and stored in a refrigerator for future use.

The extract yields and percent reactive polyphenolics are summarized in Table III.

Three-Layer Particleboard

Weighed amounts of prepared wood particles were tumbled in a laboratory blender and a wax emulsion (1 percent solids oven-dry basis) was added from a spray gun, after which concentrated bark extract (8 percent solids oven-dry basis) was sprayed onto the tumbling mixture. Powdered paraformaldehyde (1 percent oven-dry basis) was added slowly to the stirred mixture.

Boards of 3/8- and 3/4-inch in thickness were prepared, using a 10-1/2 square inch wooden deckle box. Weighed amounts of bark extract treated wood particles were hand-felted into the forming frame to form the face, and this was followed by hand-felting the core furnish and subsequently hand-felting the particles for the third layer. The coarse core comprised of about 50 percent of the total furnish. The mat was prepressed at 250 psi for 1 minute. After removing the frame, the compressed mat was put between aluminum cauls and transferred to the hot press. The 3/8-inch board was pressed for 3 minutes at 180°C platen temperature, including 30 seconds for closing to stops. The 3/4-inch boards were pressed for 8 minutes at 180°C. The density of the boards was varied by adding various quantities of core material, while keeping the weight of the outer layers constant.

Testing

After the boards were removed from the hot press they were conditioned at room temperature for several days before testing. Three 2 x 10-inch strips were cut from each board for determining breaking load. A 0.24-inch-per-minute loading rate and a 9-inch span were used. Thickness of each strip was measured at the point where the load was to be applied, and after the strip had been broken each half was cut into two 2 x 2-inch samples. These samples were used to obtain data necessary for determination of oven-dry density, water absorption and thickness swelling, internal bond, and a 2-hour boiling-in-water thickness swelling test. A 3 x 6-inch sample was used for the linear expansion test; in this instance the sample was conditioned at 50 percent R.H. measured and then exposed for 3 weeks at 90 percent R.H. and then measured again.

Results and Discussion

As indicated in Table I, the hot-water soluble extractive content of the various barks varied from 12.9 to 14.7 percent. The hot-water soluble contains tannin, which is normally deter-

Table I. Percentage Extractives (oven-dry basis)

Bark	Solvent: Ether	Solvent: Hot Water	Solvent: Ethanol
White Fir	8.4	12.9	19.0
Ponderosa Pine	5.5	14.0	20.5
Douglas-fir	8.0	13.9	21.6
Western Hemlock	4.9	14.7	20.4

Table II. Percentage Tannin and Polyphenols (oven-dry basis)

Bark	Soluble Solids	Tannin (hide-powder)	Non-Tannin	Form-HCl Phenolics	% Form-HCl Phenolics of SS
White Fir	12.7	7.7	5.0	10.5	82.6
Ponderosa Pine	15.3	7.5	7.8	10.2	66.6
Douglas-fir	14.1	8.0	6.1	10.0	70.9
Western Hemlock	14.7	8.1	6.6	10.7	72.7

Table III. Yield of Extract and Percent Reactive Phenolics

Bark	Yield (o.d. bark) %	Soluble Solids %	Form-HCl Reactive Phenolics %
White fir	17.5	42.6	79.4
Ponderosa Pine	16.1	45.0	69.2
Douglas-fir	17.9	43.3	69.7
Western Hemlock	17.6	43.0	78.3

mined by the ALCA hide-powder method (5). Tannin is a polyphenolic polymer and since some phenolics do not react with hide-powder, a more relevant measurement would be the reaction in which condensed tannin and other phenolics undergo the Stiasny formaldehyde-hydrochloric acid condensation (5). Table II shows tannin content according to the hide-powder and Stiasny's method. The reactive phenolics varied from 66.6 to 82.6 percent of the total soluble solids. Table III gives the average yields of extract solids obtained from each of the barks on extraction with dilute sodium carbonate as extractant. The total yields varied from 320 to 360 pounds per ton of dry bark. The amount of reactive phenolics in each of the bark extracts varied from 69.2 to 79.4 percent based on total soluble solids.

Table IV gives the average physical property values of the three-layer bark extract bonded particleboards, involving each of the four coniferous bark extracts. The results reported are the averages of 9 boards, varying in density from 0.68-0.76 for each bark extract bonded particleboard. The properties of all boards meet medium density particleboard specifications in bending (i.e., 2500 psi) and internal bond (i.e., 60 psi). Additionally, these boards have low water absorption and thickness swelling values, together with relatively good stability in thickness swelling after the 2-hour boiling-in-water test. These properties indicate that the bark extract bonded boards may be classified as waterproof comparable to synthetic phenolic bonded particleboard (Type 2, Class 2 (phenolic-bonded) Density B (37-50 lbs/ft^3) by Commercial Standard CS236-66 Mat-formed Wood Particleboard). This suggests that the bark extract responds as a highly reactive polyhydroxy phenolic compound comparable to resorcinol since gelation time for each is immediate in the presence of formaldehyde-hydrochloric acid at room temperature (14).

Catechin is among the principal polyphenolic monomers in white fir and western hemlock barks (10,11,13). Quercitin occurs in ponderosa pine bark, while dihydroquercitin is found in Douglas fir bark (16,17). The structure of these compounds are as follows:

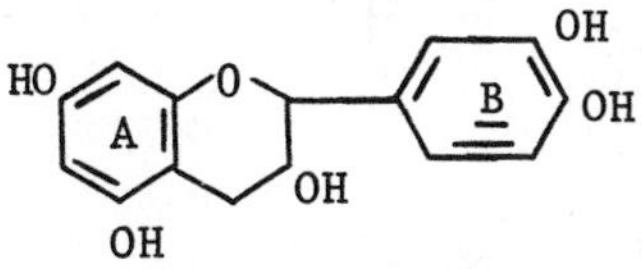

Catechin

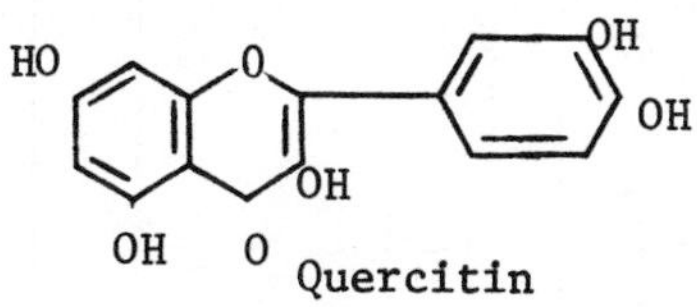

Quercitin

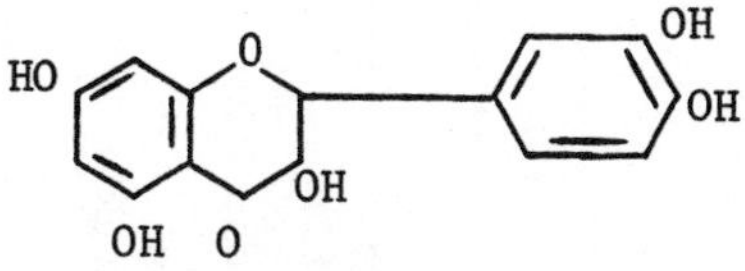

Dihydro-quercitin

Table IV. Three-Layer Bark Extract Bonded Particleboard

Bark Extracts	Density O.D. (gm/cc)	Modulus of Rupture (psi)	Internal Bond (psi)	24-hr Water Absorption (%)	Immersion Thickness Swelling (%)	2-hr Boil Test Thickness Swelling (%)	Linear Expansion (%)
3/8-Inch							
White Fir	0.71	2879	179	20.7	7.6	30.8	0.26
Ponderosa Pine	0.72	2572	165	24.5	8.7	35.8	0.24
Western Hemlock	0.69	2564	149	25.6	9.5	33.4	0.19
Douglas-fir	0.70	2584	138	27.0	9.8	36.8	0.26
3/4-Inch							
White Fir	0.72	3104	141	14.4	3.3	24.7	0.22
Ponderosa Pine	0.72	2804	132	15.5	4.1	28.6	0.21
Western Hemlock	0.72	3562	157	15.3	4.2	28.7	0.14
Douglas-fir	0.71	2804	116	18.7	5.0	33.7	0.21

As indicated by the structures of these molecules, the A ring contains resorcinol phenolic hydroxyls, while the B ring contains the catechol or adjacent phenolic hydroxy groups, both of which would be expected to be highly reactive in resin formation. This high reactivity would also hold for the condensed tannins present in the bark extract, since they are polymeric flavonoids (14).

Summary and Conclusion

The yield of bark extracts from four West Coast coniferous barks varied from 320 to 370 lbs of extract solids per ton of oven-dry bark. When a small amount of paraformaldehyde is added to wood particles which have been sprayed with bark extract and processed into board, formaldehyde released during the hot-press cycle reacts *in situ* with polyphenolic compounds present in the extract and forms a boil-proof bonding agent.

The bark extract bonded particleboards met specifications requiring the inherent durability provided by phenolic adhesives. These products are used for floor decking for modular homes, specialized furniture uses, home siding, garage door panels and more recently, as a wall and roof sheathing and single layer floor decking in conventional home construction. Thus, phenol and phenol-resorcinol modified resins can be replaced by a low-cost bark product. This use of bark would be a profitable outlet for bark residues and could lead to virtual independence of the wood particleboard industry from the petrochemical industry.

Literature Cited

1. Anderson, A. B. Econ. Bot. (1967) 21(1):24-27.
2. Anderson, A. B., A. Wong and K. T. Wu. For. Prod. J. (1974) 24(8):48-53.
3. Anon. For. Prod. J. (1974) 24(1):7.
4. Brandt, T. G. Tectona (1953) XLII p. 137-150.
5. Chang, Y. and R. L. Mitchell. TAPPI (1955) 38(5):315-320.
6. Corder, S. E., T. C. Scroggins, W. E. Meade and G. D. Everson. Wood and Bark Residues in Oregon (1972) Res. Paper 11, Oregon State University Forest Products Lab., Corvallis, 16 pp.
7. Dost, W. A. For. Prod. J. (1965) XV(10):450-452.
8. Hall, J. A. Utilization of Douglas-fir bark. (1971) Pac. N.W. Forest & Range Exp. Sta. For. Serv. USDA Portland, Oregon, pp. 84-85.
9. Hall, R. B., J. A. Leonard and G. A. Nicholls. For. Prod. J. (1960) 10(5):263-272.
10. Hergert, H. L. and E. F. Kurth. TAPPI (1953) 36(3):137-144.
11. Hergert, H. L. and E. F. Kurth. Jour. Org. Chem. (1953) 18(5):521-529.
12. Hergert, H. L. "Economic importance of flavonoid compounds

in Geisman, T. A. The Chemistry of Flavonoid Compounds." pp. 553-593, N. Y. MacMillan. (1962)

13. Herrick, F. W. and L. H. Bock. For. Prod. J. (1958) 8(10): 269-274.
14. Herrick, F. W. and R. J. Conca. For. Prod. J. (1960) 10(7): 361-368.
15. Hillis, W. F. "Wood Extractives." pp. 196-198, Academic Press, New York (1962).
16. Kurth, E. F. and J. K. Hubbard. Ind. Eng. Chem. (1951) 43, 896-900.
17. Kurth, E. F. TAPPI (1953) 36(7):119A-122A.
18. Maclean, H. and J. A. F. Gardner. Pulp and Paper Mag. of Canada (1952), pp. 111-114.
19. Plomely, K. F. CSIRO Div. of For. Prod. Tech. Paper No. 46, (1966) Melbourne, Australia, pp. 16-19.
20. Plomely, K. F. and A. Slashevski. CSIRO For. Prod. Newsletter, No. 363 (1969), Melbourne, Australia.

15

Composition Boards Containing Bark

RAYMOND A. CURRIER

Forest Research Laboratory, Oregon State University, Corvallis, Ore. 97331

Competition among various segments of the forest products industry for wood residues is becoming more intense each year. For example, some pulp mills now can use sawdust and shavings, which until recently were the major residue utilized for furnish in manufacturing wood particleboard. In addition, since the energy shortage in late 1973, large quantities of wood and bark residues have found increased markets as fuel. Efforts to use bark in composition boards predate the energy shortage, but they appear to be slated for renewed interest as the competition for clean woody residues accelerates. A review of past efforts in North America to utilize bark in composition boards is desirable to assess potential needs and the possibilities of producing salable composition boards wholly or partially from bark. The term composition boards is meant to include both fibrous and particle products. It includes what are commonly termed insulation board, hardboard, medium density fiberboard, and particleboard.

This review of composition boards containing bark will not include references on the use of bark or bark extracts in the role of bonding agents for composition boards. That subject has been covered in the paper by Dr. A. B. Anderson.

Amounts of Bark Available

An appropriate beginning is an attempt to answer the question, "How much bark might be available for composition board furnish?" Estimates of total bark available in the United States have been difficult to obtain, and the published estimates have shown considerable variation.

A recent estimate made by Ellis (1) is based upon four regional compilations in 1973. Ellis estimated 17 million tons (ovendry) of bark are produced annually, of which 7 million tons presently are unused (Table I). At least 1 million tons presently are unused in each region. The entire amount, of course, would not be available for composition boards as other potential uses, such as fuel and mulch, would siphon off part of the unused bark.

Table I. Estimated 1973 U.S. Production of Bark Residues and Amounts Presently Unused, in Millions of Tons, Ovendry Basis.

Region	Total production	Unused
Pacific Coast	7	2
Rocky Mountains	2	1
South	6	2-3
North	2	1-2
Total	17	7

A more specific tabulation of bark production and present uses for one state, Oregon, during 1972 was compiled by Schuldt and Howard (2). Their findings, shown in Table II, indicate the unused portion of bark is 22.5%, which is under the estimated unused amounts shown for any region listed in Table I. Use of bark for fuel in Oregon came to 61.5% of the total produced, and the amount no doubt has increased greatly since 1972.

Table II. Production and Disposition of Bark Residues in Oregon, 1972, in Tons, Dry-weight Basis.

Total bark produced	Used			Unused
	Pulp & board	Fuel	Miscellaneous	
3,556,103	40,470	2,188,155	529,131	798,347
% of total	1.1	61.5	14.9	22.5

From the above, we can conclude that sufficient bark still is available for use as composition board furnish. The amounts available, however, vary greatly from one geographical region to another. Potential volumes are large enough to be a source of furnish for production of composition boards.

Previous Bark Board Bibliographies and State-of-the-Art Review

An excellent comprehensive compilation of the literature on all phases of tree bark was prepared by Marian and Wissing and published over a span of 2 years (1956-1957) in 11 different issues of Svensk Papperstidning (3). Their bibliography lists the subject under 12 different major subheadings; bark composition boards may be found under the subheading "The Utilization of Bark Fibers."

Another excellent bibliography by Roth and coworkers (4) lists 1,339 references with a concise abstract for each. Since the original compilation was published in 1960, two supplements have appeared, I in 1968 (5) and II in 1973 (6). References to bark composition boards are listed under the subheading "Utilization."

The Chemical Utilization Division of the Forest Products Research Society published a "Review of Chemical Utilization" in 1960 (7). The authors, Pearl and Rowe, included a section titled "Bark," and its use in composition boards was reviewed. This review was followed up by another published 3 years later by Rowe (8).

Gregory and Root in 1961 (9) prepared what they termed a "statistical analysis" of the literature covering bark utilization and, in addition, reviewed examples of commercial and pilot plant operations. They found 52 references on use of bark in composition boards. The report concludes with sections covering "Limitations and Hurdles in Bark Utilization" and a discussion of "Future Opportunities."

Ross (10) in 1966 compiled references that have been published since Roth et al. (4) reported on the bark literature in 1960. Ross categorized the latest references under one of 12 headings; bark composition boards were included under the title, "Bark Fiber, Cork, and Dust Products, Boards, Panels, Adhesives, and Tiles."

A general state-of-the-art plus bibliography was published by Harkin and Rowe in 1969 (11). Included was a section labeled "Wood-base Materials," which covered insulation board, hardboard, fiberboard, and particleboard containing bark.

The same year (1969), Walters (12) prepared a report specifically reviewing the current status of bark used for board products. Information was given regarding potential maximum amounts of bark utilizable in various types of composition boards, based on research up to that time.

A short review by Currier and Lehmann (13) on use of bark in composition boards was contained in the proceedings from a conference in 1971 on "Converting Bark into Opportunities."

Hall also came out the same year (1971) with a comprehensive technical review and bibliography of the uses for Douglas-fir bark (14). Included is a state-of-the-art report under the subtitle "Board and Tile Manufacture."

Another review of utilizing a single bark species in composition boards was presented by Scroggins and Currier for western redcedar bark (15).

One of the few textbooks covering particleboard was published in 1974 with Moslemi as the author (16). An entire chapter covers the subject "Bark in Particleboard," and available literature is reviewed extensively.

The most recent state-of-the-art report was published last year under the sponsorship of the Bark and Residues Committee of the Forest Products Research Society. The author, Bhagwat (17), included a section on particleboard, hardboard, and molded products where bark was a constituent in the furnish.

Review of Efforts to Use Bark in Composition Boards

The first published results of efforts to utilize bark as furnish for composition boards in North America appeared after the end of World War II. In 1947, Schwartz, Pew, and Schafer of the Forest Products Laboratory, Madison, Wisconsin, reported on experiments to produce insulation board and hardboard from 8 types of western sawmill and logging residues (18). Inclusion of logging residues as potential furnish is of interest, because this aspect of forest utilization has received greatly renewed attention only recently. Most gains to date in residue utilization have been made using material generated at sawmills and plywood and other wood processing plants, not from material left in the woods after logging.

Although Schwartz *et al*. do not specifically mention that bark was present in the residues they used, we can assume that it was included on the slabs and edgings serving as a source of raw material, and also in the logging residues. Their results showed that satisfactorily strong insulation board could be made from western hemlock slabwood, white fir logging residues, or a 50-50 mixture of white fir logging residues and Douglas-fir slabs. Dimensional stability, however, was poor in all combinations. None of the residues produced a hardboard that was acceptable, with the exception of western hemlock slabwood.

In 1948 and 1949, investigators at the Forest Products Laboratory, Ottawa, Canada, reported results of the first preliminary studies where bark alone was the raw material for wet process softboard and hardboard. Three articles were published, each containing essentially the same information. One was by Clermont and Schwartz (19), and the other two by Schwartz (20, 21). Bark from eastern white cedar and western redcedar was used as the entire furnish, or with a mixture of 10% pulp screenings. These species were chosen because of the natural fibrous nature of their barks. Although tests showed the experimental hardboards could not meet standards then in force, eastern white cedar bark showed some promise. For softboards, the eastern white cedar produced acceptable boards.

The next series of reports based on use of bark for wet process hardboard were published over a 7-year span from 1950-1956 by Anderson and Runckel. In the first report (22), the results included laboratory boards made from Douglas-fir slabwood containing bark, which varied between 15 and 45% of the total furnish. Bending tests indicated adequate strengths were obtained, and addition of bark enhanced moisture resistance of the boards. Acceptable boards also were made when white fir and western hemlock barks made up a portion of the furnish. No binder was added to the Asplund-type fiber.

A 1952 report (23) by Anderson and Runckel indicated further experiments had been performed using 100% Douglas-fir, white fir, or western hemlock bark as wet-process hardboard furnish without any chemical additives. Boards made from Douglas-fir bark had by far the best strength and dimensional stability. The report also disclosed that a commercial plant to manufacture a hardboard containing bark was under construction at Dee, Oregon, by the Oregon Lumber Company. A description of the plant after 1 year of operation was given by Runckel (24). In actual production, adding 3/4% phenolic resin to the furnish was necessary, but no sizing was required.

Another interesting report by Anderson and Runckel discussed use of Douglas-fir branchwood in wet-process hardboard (25). This is one of the earliest references to utilizing this type of material as composition board furnish; satisfactory boards were made from 100% branchwood. No binder was added. Actual percentage of bark in the furnish was not disclosed.

Anderson contributed a paper to the 1956 International Consultation on Insulation Board, Hardboard and Particle Board (26). In addition to a good review of the literature, the paper contained results of further experimentation using the barks of several more species as furnish for wet-process hardboard. New species included ponderosa pine, sugar pine, southern pine, pinyon pine, lodgepole pine, noble fir, and red oak. Ponderosa pine bark produced hardboard comparable to Douglas-fir bark in both strength and moisture resistance. The other species varied widely in these properties. A condensed version of this work appears in reference (27).

King and Bender of the Canadian Forest Products Laboratory at Ottawa contributed two studies to the literature on bark composition boards. The first, in 1951, was concerned with producing wet-process insulating fiberboard from western redcedar shingle mill waste (28). Laboratory boards containing up to 100% bark were made from attrition mill prepared furnish. Other raw materials included wood, shingle hay, and sawdust. Some combinations of furnish made boards that were acceptable. The next year (1952), results of using the same type furnish to make laboratory-sized wet-process hardboard were published (29). Boards were made with and without added binder. Those of 100% bark and no binder would not meet specifications. Higher density and additional binder were found to improve properties of the boards.

Up until 1952, all references found on bark in composition board were based upon wet-process softboard or hardboard. The first mention of a dry process came in a one-page article published by the British Columbia Research Council (30). The brief report indicated a dry process had been developed for making an interior wallboard possessing good strength and moisture resistance. No binder was necessary for the cedar mill residues, which included bark. Apparently, the process never was tried commercially.

A report published in 1954 by Cooke (31) was the first reference found giving pertinent details of a dry process for making a particleboard-type composition board containing bark. Ponderosa pine slabs containing 30% bark were chipped and hammermilled. Phenolic resin was added at 4% solids, as was sizing agent. Boards containing bark had lower bending strengths than all-wood boards; addition of a kraft paper overlay helped increase strength of boards that contained bark. Density of the boards ranged from 40.6 to 46.8 pounds per cubic foot. A satisfactory sheathing-type board could be made from ponderosa pine slabs, with or without included bark.

Three years after the report by Cooke, an article appeared describing a commercial particleboard plant operating on unbarked white pine and eastern hemlock slabs and edgings (32). The plant, Granite Board, Inc. at Goffstown, New Hampshire, started in 1955 and operated successfully for several years.

Another report came from Canada in 1959 when Bender published research results of utilizing eastern Canadian barks as furnish for wet-process insulation board and hardboard (33). Bark species included in the study were black spruce and balsam fir; each contained 25-35% wood. A Sprout-Waldron disk refiner was used to prepare the bark fiber, and boards were made with 1¼% wax emulsion but contained no added binder. Physical tests indicated the boards met some commercial specifications; the author believed that addition of more woody fiber would improve the properties. In addition, a few experimental dry-process particleboards were made with addition of some unnamed binder that was a byproduct material.

Burrows in 1959 contributed a study based on making a floor tile from the cork fraction of Douglas-fir bark (34). Added binder in the dry-process tiles was either 5% butadiene styrene or diethylene glycol. Comparison tests were made against tiles from Mediterranean oak cork. Dimensional stability was better in Douglas-fir cork tiles, and most other properties compared favorably. No known commercial application resulted.

One year later, Burrows published results from a comprehensive series of experiments using 100% Douglas-fir bark as furnish for particleboard (35). No binder was used; he relied upon the "self-bonding" properties of Douglas-fir bark. Variables included bark particle size, mat moisture content, pressing pressure, and use of various overlays. Additional boards were made from ponderosa pine, western hemlock, and white fir barks. A pilot-plant-size run was made using results gathered from the study.

Increasing pressing pressure and use of an overlay on the bark core improved properties the most. Burrows concluded that a commercial board product could be made from all-bark furnish without added binder if the bark were overlaid with kraft paper or veneer.

Further work in Canada to produce rigid, wet-process insulation boards from a bark-wood mixture was reported by Branion in 1961 (36). He made boards containing 85% poplar wood and 15% white spruce bark; a few boards were made with added poplar or jack pine bark. White spruce bark worked best; it appeared to cause a significant increase in tensile strength compared to boards made from 100% poplar wood fiber. This effect also was demonstrated in a hardboard. Other boards were made with up to 80% bark. Water absorption decreased as bark content increased. An effort was made to discover the bark ingredient responsible for the strength increase. After a series of extractions, the active component was concluded to be present in the holocellulose.

Lewis in 1961 discussed why composition boards containing bark seldom had found commercial application (37). He believes the reasons are that

bark usually contains dirt and grit, which causes rapid dulling of chipper and flaker knives;

in pulping processes, bark may require different conditions than wood;

increased foam and slime problems occur with bark in wet processes;

maintaining constant bark-to-wood ratios is difficult; and

bark causes darkening of the board surface.

Whatever the reason, a dearth of research in North America on composition boards containing bark is apparent after 1961. The next publication appeared in 1968 when Stewart and Butler reported on making wet-process hardboard from 100% western redcedar bark or mixtures of bark and up to 36% wood by weight (38). Fiber was prepared by Asplund or Sprout-Waldron refiners, and phenolic resin was added as binder. Control boards were made from Douglas-fir woody fiber, either commercially produced or laboratory made. All-bark boards showed a marked reduction in linear expansion and internal bond when compared to boards made partially from wood fiber. Modulus of rupture, water absorption, and thickness swell showed no change. When compared to all-wood boards, the all-bark board lost 25-30% of bending strength and some degree of internal bond.

Renewed interest in bark particleboard was evidenced by a short article written by Murphey and Rishel (39). They reported results of preliminary studies on relative strengths of various bark species compared to aspen flakeboard. Bark species included aspen, black locust, green oak, white pine, oak and locust, poplar, red oak, and mixed oak. Overlaying was suggested as a means of increasing bending strengths.

Results of a comprehensive research project in which three types of particleboard furnish containing bark were included (pole peelings, logging residues, and bark) were reported by Currier and Lehmann in 1970 and 1972 (40, 41). Three-layer boards were made with various percentages (25, 50, and 100) of Douglas-fir, ponderosa pine, western hemlock, or a mixture of all three barks in the core layer. Urea formaldehyde resin at 6% solids was added to core furnish and 8% solids to face furnish. From strength and dimensional stability tests, they concluded that both types of properties were lowered drastically when bark content of the core exceeded 25%. Boards made with logging residue as core furnish were acceptable, because amount of bark occurring in the logging residue naturally did not exceed 25%.

An announcement was made in 1970 that a large producer of particleboard, Boise Cascade Corporation, La Grande, Oregon, planned to initiate full-scale production tests of an underlayment grade board containing 25% pine bark (42). Douglas-fir, white fir, and redwood barks also were tried in preliminary boards made in the laboratory. Scheduling of the plant run was predicated on successful results from the laboratory-made boards. Apparently, the production run did not prove to be successful, even though the laboratory boards met all specifications.

One of today's fastest growing segments of the wood composition board industry is production of medium density fiberboard (MDF) using a dry process similar to that used for particleboard. First mention of the possibility of utilizing bark for MDF came in a presentation by Brooks in 1971 (43). He described a process in which a homogenous board with superior properties could be made from such raw materials as mixed, unbarked hardwood pulp chips; unbarked pine chips, if bark content was less than 30%; forest thinnings, branches, and so on; and hardwood bark. Furnish was prepared by double-disk pressurized refiners. Brooks concluded a plant could be built to operate on 100% hardwood bark.

Dost in 1971 reported on a study where redwood bark fiber was used in three-layer particleboard (44). Amount of bark in the furnish, by weight, was 0, 10, 20, and 30%; hammermilled disk flakes or Pallmann flakes of redwood wood made up the remainder of the furnish. Urea formaldehyde resin was applied at three percentages. Test results showed surface smoothness and strength properties (MOR, MOE, and IB) decreased with increasing bark content in the boards. Water absorption decreased, but thickness swelling and linear expansion increased as the amount of bark increased.

In 1971, Marra and Maloney of Washington State University were interviewed regarding their pilot-plant research on bark board (45). They predicted that a shortage of easy-to-use sawmill and plywood plant wood-type residues would lead composition board manufacturers to seek bark, logging slash, and reclaimed waste paper as a source of furnish. Some potential problems with bark were discussed, especially that bark naturally possesses lower strength properties than wood of the same species.

Two years later, Maloney published results of his comprehensive bark board study (46). Included were barks from western larch, Douglas-fir, ponderosa pine, and western redcedar. Homogeneous boards contained either 7.5 or 10% urea formaldehyde as binder, or 6% phenol formaldehyde. Boards were made at four specific gravities, 0.40, 0.55, 0.70, and 0.85. Strength tests showed homogeneous boards of ponderosa pine or Douglas-fir approached or equalled the internal bond specifications for low-to-medium density particleboard; larch met only the low density criterion. None could meet modulus of rupture specifications. Western redcedar boards met MOR specifications, but had low internal bond. Linear expansion was high, especially for ponderosa pine. Additional work was reported on using impregnating resins to "beef up" bark strength, with some success.

At about the same time as Maloney's research, Chow was attempting to elucidate the so-called self-bonding capabilities of some types of bark when subjected to certain combinations of time, temperature, and pressure (47). Moisture content was not a variable and was kept at 2.5% or less. Boards were made from Douglas-fir bark at temperatures of 180 and 200°C and pressing times from 5 to 120 minutes. Control boards were pressed with bark furnish containing 4.5% phenol formaldehyde resin. Tests indicated that boards made without binder at optimum time-temperature pressing conditions possessed internal bond and bending strengths equal to boards made with 4.5% resin binder. In addition, the best no-binder boards had much better dimensional stability properties than those made with binder. Optimum conditions were a pressing temperature of 200°C and from 40 to 120 minutes press time.

About this time (1973), widespread interest appeared in a further modification of the old basic particleboard products. One example was production of a so-called structural board, usually from furnish consisting of wood flakes or wafers. A successful commercial product had been produced from aspen wood wafers for several years, and Gertjejansen and Haygreen (48) explored the effect of using as furnish aspen that contained bark. Because aspen bark changes drastically in physical characteristics from butt to top logs, its source on the log was a variable. Both wafer and flake-type boards were made, with phenol formaldehyde as binder. Test results showed wafer-type boards with butt-log bark lost 30% in moduli of rupture and elasticity, linear swelling increased 75%, and thickness swelling decreased 28%. Internal bond increased 28%. Upper-log bark reduced MOR and MOE 15% and had little or no effect upon IB or stability properties. Regardless of bark source, flake-type boards lost MOR and MOE; IB did not change. Linear swelling was greater and thickness swelling lower with butt-log bark. Both types of swelling increased with upper-log bark.

Another proposed composite product, this one strictly nonstructural, was a decorative interior panel consisting of a pressed bark overlay on a base material such as plywood, hardboard, or gypsum board (49). Preferred bark species were those chunky in nature, such as fir or pine.

During 1974, Anderson and co-workers Wong and Wu produced a series of three publications dealing with the inclusion of white fir or ponderosa pine bark in particleboard (50, 51, 52). The last two reports also included bark extract as binders. The first study on white fir (50) included addition of 2% paraformaldehyde to react with the polyphenolic extractive components of the 100% bark furnish. Although some improvement was noted, bending strength was too low and linear expansion too high. A better board was made by using bark in the core and wood particles on the faces. Homogeneous boards from a wood-bark mix were satisfactory if the amount of bark was 25% or less.

Results of the later studies showed bark-paraformaldehyde furnish as core of three-layer particleboard resulted in white fir boards meeting medium-density particleboard specifications for bending and internal bond strengths, as well as linear expansion (51).

The use of ponderosa pine bark plus 2% paraformaldehyde as furnish for all-bark boards, or as core of three-layer boards with wood faces was reported in the third publication (52). Tests indicated the all-bark boards had very low bending strength and very high linear expansion. Internal bond, however, was adequate, as was thickness swelling. By going to the three-layer configuration, both bending strength and linear expansion were markedly improved.

Biblos and Coleman investigated another type of potential structural composite product (53). They made and tested panels consisting of a particleboard core from sawdust and bark and faces of veneer. All material was southern pine, and 9% urea formaldehyde served as binder. Strength tests indicated the composite panels were superior to conventional two-layer floor systems of 1/2-inch plywood plus 5/8-inch particleboard underlayment.

In the same year, 1974, Lehmann, Geimer, McNatt and Heebink of the Forest Products Laboratory at Madison, Wisconsin, published three reports, each relating to structural-type particleboard from forest residue furnish (54, 55, 56). Amount of naturally occurring bark in the raw material studied (Douglas-fir, true fir, western hemlock, and lodgepole pine) was low at 7-8%. Various types of mechanically prepared particles were studied; board configurations included structural flakeboard, three-layer boards, homogeneous boards, and overlaid panel siding. Synthetic resin binder and wax size were added to all board furnish. Results of strength and stability tests indicated inclusion of bark or branchwood or both in large amounts results in severe loss of strength and dimensional stability, especially in flakeboards. The recommendation was that bark/branchwood content not exceed 15% of the total furnish. In addition to loss of strength and stability, bark contained about 10 times as much silica as woody residues. Thus, machining problems may occur in boards with high bark contents.

A report from West Virginia University by Koch and Hall (57) discussed the use of hardwood bark as particleboard furnish, with and without added binders. Species included red oak, soft maple, and black birch. Initial studies indicated no-binder boards had to be compressed to 70 pounds per cubic foot to maintain their integrity. Both strength and dimensional stability were enhanced by pressing boards at 400°F instead of 300°F. Longer press times (15, rather than 10 minutes) also helped. Later, boards were made with 5% added starch powder. One potential use of this product was for expandable horticultural planting containers. Both raw and composted barks were tried, with promising results.

The thermal properties of composite boards were the subject of a recent report by Place and Maloney (58). Thermal conductivity tests were made on three-layer boards with surfaces of white pine wood flakes and cores of either Douglas-fir or grand fir bark. Density was varied at 34, 42, and 52 pounds per cubic foot. The composite boards containing bark proved to be better insulators than wood particleboard of comparable density. Douglas-fir bark cores had lower thermal conductivity than did grand fir.

Two articles published last year (59, 60) discussed work at the Vancouver, British Columbia, Western Forest Products Laboratory that amplified the 1972 findings by Chow on the self-bonding of bark and the mechanism responsible for it. High pressing temperatures (200 to 300°C) were found to be the key to activating the chemicals native to bark that will polymerize in the hot press. Pressing times ran from 2 to 60 minutes. Species of bark investigated included Douglas-fir, western hemlock, lodgepole pine, and western redcedar. Board specific gravity was 0.9 to 1.0 grams per cc. Tests showed MOR and IB values were related directly to time-temperature parameters. Higher temperatures, especially above 250°C, and longer pressing times gave the best results. After 2 hours of boiling, the better bark boards retained most of their bending strength, unlike conventional wood particleboard. Dimensional stability also was enhanced by higher temperatures and longer press times. Chow figured the extractive-lignin polymer bond formed was equal to that obtained by addition of 4.5 to 7% synthetic resin binder. Results comparable to Douglas-fir were found for western hemlock and lodgepole pine barks; western redcedar proved to be unpredictable.

So far this year, two reports have been published on the effect of bark in medium-density fiberboard. The first by Woodson (61) covered the bark of three southern hardwoods, sweetgum, southern red oak, and mockernut hickory; percentage of bark in the whole-tree furnish was 13.4, 20.0, and 18.6%, respectively. Urea-melamine formaldehyde binder was added at 8-10%, and wax at 1%. Test results showed inclusion of bark decreased tensile and bending strengths by 16-18%, MOE by 10-14%, and IB by 8%. Linear expansion was not affected significantly by bark. Thickness swelling was improved in two of the three species. Woodson concluded good quality, medium-density fiberboard could be made from barky chips.

The second report on MDF in 1976 came from Chow at the University of Illinois (62). He, too, worked with hardwood barks including cottonwood, red oak, white oak, and walnut. Furnish was prepared in pressurized refiners and by hammermilling. Urea formaldehyde resin percentages were 5.0, 7.5, and 10.0%, plus 1% wax. He concluded that the fiber from the pressurized refiners was superior to hammermilled particles. Cottonwood and white oak furnish gave better boards, exceeding or approaching requirements of present standards for type 1-B-1 commercial particleboard.

The most recent publication reviewed was by Einspahr and Harder (63), who discuss the basic properties of hardwood barks that could be important in the manufacture of any fibrous product. This was a progress report showing results for 16 pulpwood species; work is in progress on 16 additional species. Measured were such bark factors as specific gravity, extractives content, strength, toughness, reaction to hammermilling, and ash content.

Literature Cited

1. Ellis, T. H., Proceedings P-75-13, "Wood Residues as an Energy Source." Forest Prod. Res. Soc. (1975), 17-20.
2. Schuldt, J. P., and Howard, J. O., Special Report 427, Oreg. State Univ. Extension Serv., Corvallis (1974).
3. Marian, J. E., and Wissing, A., Svensk Papperstidning (1956) 59 (21), 751-758; (22), 800-805; (23), 836-837; (1957) 60 (2), 45-49; (3), 85-87; (4), 124-127; (5), 170-174; (7), 255-258; (9), 348-352; (11), 420-424; (14), 522-523.
4. Roth, L., Saeger, G., Lynch, F. J., and Weiner, J., Bibliog. Series 191, Inst. Paper Chem., Appleton, Wis. (1960).
5. Roth, L., and Weiner, J., Bibliog. Series 191, Supplement I, Inst. Paper Chem., Appleton, Wis. (1968).
6. Weiner, J., and Pollock, V., Bibliog. Series 191, Supplement II, Inst. Paper Chem., Appleton, Wis. (1973).
7. Pearl, I. A., and Rowe, J. W., Forest Prod. J. (1960) 10(2), 91-112.
8. Rowe, J. W., Forest Prod. J. (1963) 13(7), 276-290.
9. Gregory, A. S., and Root, D. F., Pulp and Paper Mag. Canada (1961) 62(8), T385-T391.
10. Ross, W. D., Bibliog. Series 6, Forest Res. Lab., Oreg. State Univ., Corvallis (1966).
11. Harkin, J. M., and Rowe, J. W., U.S.D.A. Forest Service, Forest Prod. Lab. Res. Note FPL-091 (1969).
12. Walters, E. O., Proc. Third Texas Industrial Seminar, Texas Forest Prod. Lab., Lufkin (1969), 27-38.
13. Currier, R. A., and Lehmann, W. F., Proc., Conference on Converting Bark into Opportunities, Oreg. State Univ., Corvallis (1971), 85-87.
14. Hall, J. A., "Utilization of Douglas-fir Bark." Pac. N.W. Forest and Range Expt. Station, Portland, Oreg. (1971), 20-23.

15. Scroggins, T. L., and Currier, R. A., Forest Prod. J. (1971) 21(11), 17-24.
16. Moslemi, A. A., "Particleboard Volume 1: Materials," 244 pp. Southern Ill. Univ. Press, Carbondale and Edwardsville (1974).
17. Bhagwat, S. C., Forest Prod. J. (1975) 25(2), 13-15.
18. Schwartz, S. L., Pew, J. C., and Schafer, E. R., Paper Trade J. (1947) 125(4), 37-42.
19. Clermont, L. P., and Schwartz, H., Forest Prod. Res. Soc. Proc. (1948) 2, 130-135.
20. Schwartz, H., Bulletin 25. N.E. Wood Utilization Council, Inc., New Haven, Conn. (1949).
21. Schwartz, H., Paper Trade J. (1949) 128(24), 27-28.
22. Anderson, A. B., and Runckel, W. J., Forest Prod. Res. Soc. Proc. (1950) 4, 301-309.
23. Anderson, A. B., and Runckel, W. J., Paper Trade J. (1952) 134(4), 22-30.
24. Runckel, W. J., J. Forest Prod. Res. Soc. (1953) 3(5), 148, 228.
25. Anderson, A. B., and Runckel, W. J., The Lumberman (1953) 80 (4), 134,136,139.
26. Anderson, A. B., Background Paper 4.2. Internatl. Consultation on Insulation Board, Hardboard and Particle Board. Food and Agric. Org. United Nations, Geneva, Switzerland (1956).
27. Anderson, A. B., Norsk Skogindustri (1956) 10(12), 475-479.
28. King, F. W., and Bender, F., Pulp and Paper Mag. Canada (1951) 52(1), 75-79.
29. King, F. W., and Bender, F., Pulp and Paper Mag. Canada (1952) 53(6), 137-141.
30. Anonymous, Res. Memo 52-3, Brit. Col. Res. Council, Vancouver, B.C. (1952?).
31. Cooke, W. H., Report L-4, Oreg. Forest Prod. Lab., Corvallis (1954).
32. Anonymous, Wood and Wood Products (1957) 62(3), 20-31,78,80.
33. Bender, F., Pulp and Paper Mag. Canada (1959) 60(9), T275-T278.
34. Burrows, C. H., Inf. Circ. 13, Forest Prod. Res. Center, Corvallis, Oreg. (1959).
35. Burrows, C. H., Inf. Circ. 15, Forest Prod. Res. Center, Corvallis, Oreg. (1960).
36. Branion, R., Pulp and Paper Mag. Canada (1961) 62(11), T506-T508.
37. Lewis, W. C., U.S.D.A. Forest Serv., Forest Prod. Lab. Rep. 1666-21 (1961).
38. Stewart, D. L., and Butler, D. L., Forest Prod. J. (1968) 18(12), 19-23.
39. Murphey, W. G., and Rishel, L. E., Forest Prod. J. (1969) 19(1), 52.

40. Currier, R. A., and Lehmann, W. F., Paper, 24th Annual Meeting, Forest Prod. Res. Soc., Miami Beach, Fla. (1970).
41. Currier, R. A., 27th Proc. N.W. Wood Prod. Clinic, Spokane, Wash. (1972), 27-31.
42. Sullivan, M. D., Forest Ind. (1970) 95(8), 42-43.
43. Brooks, S. H. W., Paper, 25th Annual Meeting, Forest Prod. Res. Soc., Pittsburgh, Pa. (1971).
44. Dost, W. A., Forest Prod. J. (1971) 21(10), 38-43.
45. Anonymous, Quest (1971) 9(2), 4-7.
46. Maloney, T. M., Forest Prod. J. (1973) 23(8), 30-38.
47. Chow, S., Wood and Fiber (1972) 4(3), 130-138.
48. Gertjejansen, R., and Haygreen, J., Forest Prod. J. (1973) 23(9), 66-71.
49. Anonymous, Crow's Forest Prod. Digest (1973) 51(10), 18.
50. Anderson, A. B., Wong, A., and Wu, K.-T., Forest Prod. J. (1974) 24(1), 51-54.
51. Anderson, A. B., Wong, A., and Wu, K.-T., Forest Prod. J. (1974) 24(7), 40-45.
52. Anderson, A. B., Wu, K.-T., and Wong, A., Forest Prod. J. (1974) 24(8), 48-53.
53. Biblos, E. J., and Coleman, G. E., Forest Ind. (1974) 101(8), 70-71.
54. Lehmann, W. F., and Geimer, R. L., Forest Prod. J. (1974) 24(10), 17-25.
55. Geimer, R. L., Lehmann, W. F., and McNatt, J. D., Eighth Particleboard Proc., Wash. State Univ., Pullman (1974), 119-142.
56. Heebink, B. G., U.S.D.A. Forest Serv., Forest Prod. Lab. Res. Paper FPL 221 (1974).
57. Koch, C. B., and Hall, C. S., W. Va. Forestry Notes 2, W. Va. Univ., Morgantown (1974), 5-8.
58. Place, T. A., and Maloney, T. M., Forest Prod. J. (1975) 25(1), 33-39.
59. Martin, B., Brit. Col. Lumberman (1975) 60(5), 28-29.
60. Chow, S., Forest Prod. J. (1975) 25(11), 32-37.
61. Woodson, G. E., Forest Prod. J. (1976) 26(2), 39-42.
62. Chow, P., Forest Prod. J. (1976) 26(5), 48-55.
63. Einspahr, D. W., and Harder, M., Forest Prod. J. (1976) 26(6), 28-31.

16

Polyurethane Foams from the Reaction of Bark and Diisocyanate

SEYMOUR HARTMAN

Champion International Corp., U. S. Plywood Technical Center, Brewster, N.Y. 10509

The major areas for bark utilization can be broken down into three groups:

1 - as a source of energy production
2 - as an environment and pollution control product and
3 - as a source for materials or chemicals

Bark, a complex chemical mixture of many organic compounds has been used as a power fuel in the Forest Products Industry - as a means of conserving the burning of natural fuel (oil and natural gas). Now especially, with the increase of fuel prices, the use of bark as a fuel has increased. Its use as an environmental and pollution control product has been for erosion control and slope stabilization. Bark has also been used as a potential oil pollutant scavenger in oil spills as well as a potential odor scavenger in sulfate pulp mills. As a source for materials, bark has found use in horticultural applications as a mulch growth media containing fertilizers, pesticides and herbicides for container growth plants, as well as a soil conditioner. As a source for chemicals, numerous papers have been written [1] [2], [3] and many patents [4], [5] have been issued on the chemistry of bark and on the isolation of its components. One outstanding early work was that of Kurth [6] who classified the principal chemical components present in barks as listed in Table I. Numerous patents [7], [8], [9], [10], [11], have been issued detailing the preparation of alkali bark extract and also the preparation of alkali bark and its use as extenders in phenol formaldehyde resin adhesive systems.

With all the research activity in the areas cited, not one research activity has viewed bark as a base chemical or as a raw material or as a reactive component in a chemical reaction. More specifically, as a monomer in a polymerization reaction. As we see in Table I, bark does contain organic compounds possessing hydroxyl components.

Since diisocyanate is one of the active components of polyurethanes, and unique in that it possesses a high degree of reactivity and will react with any chemical compound containing an

TABLE I

PRINCIPAL CHEMICAL COMPONENTS OF BARK

1. Lignin - the material insoluble in concentrated mineral acids
2. Cork - cutose, suberin, and suberic acid (1, 6, hexane-dicarboxylic acid)
3. Carbohydrates - holocellulose, the total carbohydrate fraction
   A. Cellulose
   B. Hemicelluloses-arabans, xylans, mannans, glucosans, and uronic acid substances
4. Extraneous materials
   A. Volatile acids and oils
   B. Non-volatile fatty oils (fats and fatty acids), higher alcohols, resins, and hydrocarbons
   C. Coloring matters
   D. Tannins and the related water-insoluble phlobaphenes
   E. Polysaccharides, glucosides, pectins, and sugars
   F. Organic nitrogen compounds
   G. Mineral matters
   H. Other organic components-saponins, mannitol, dulcitol, etc.

active hydrogen, I decided to use Bark as a chemical reactant, possessing the active hydrogen to produce a polyurethane foam. Thus my objective in this study was to use bark as the major or sole polyol in the reaction with diisocyanate to produce a polyurethane and more specifically to produce polyurethane foams.

Polyurethane formation, whether it be a foam, a coating, or an adhesive, is the result of a series of complex chemical bonds and linkages other than the urethane group. The basic chemical reactions taking place during foam formation have been cited in the literature [12], [13] and are summarized in Figure 1. All the chemical reactions shown in Figure 1 may occur simultaneously and very likely do. For simplicity the chemical reactions in the equations shown are presented in their monofunctional form, although I am sure you are all aware that all the reactive components must be difunctional or greater to produce a polymeric structure.

The two most important chemical reactions in the preparation of polyurethane foam are the reaction between isocyanate and hydroxyl compound to form a urethane linkage:

$$R\text{-}N{=}C{=}O + R'\text{-}OH \rightarrow \underset{\text{urethane}}{R\text{-}NH\text{-}\underset{\substack{\| \\ O}}{C}\text{-}O\text{-}R'}$$

and the reaction between isocyanate and water. The first step of this reaction is the formation of an unstable carbamic acid, which decomposes to form an amine and carbon dioxide. This re-

$$\text{R-N=C=O} + H_2O \rightarrow \underset{\text{carbamic acid}}{\text{R-NH-}\underset{\overset{\|}{O}}{C}\text{-OH}} \rightarrow R'NH_2 + CO_2$$

action is responsible for foam formation through the liberation of carbon dioxide.

Another reaction taking place is the formation of a disubstituted urea from the reaction of the preceeding amine and isocyanate:

$$\text{R-N=C=O} + R'NH_2 \rightarrow \underset{\text{disubstituted urea}}{\text{R-NH-}\underset{\overset{\|}{O}}{C}\text{-NH-R'}}$$

Other reactions which lead to branching and cross-linking are the formation of allophanate and biuret linkages. The allophanate linkages occurs when the hydrogen on the nitrogen atom of the urethane group reacts with an isocyanate:

$$\text{R-N=C=O} + \text{R-NH-}\underset{\overset{\|}{O}}{C}\text{-O-R'} \rightarrow \underset{\text{allophanate}}{\text{R-NH-}\underset{\overset{\|}{O}}{C}\text{-}\underset{\overset{|}{R}}{N}\text{-}\underset{\overset{\|}{O}}{C}\text{-O-R'}}$$

The biuret linkage occurring when the hydrogens on the nitrogen atoms in the disubstituted urea reacts with isocyanate:

$$\text{R-N=C=O} + \text{R-NH-}\underset{\overset{\|}{O}}{C}\text{-NH-R'} \rightarrow \underset{\text{biuret}}{\text{R-NH-}\underset{\overset{\|}{O}}{C}\text{-}\underset{\overset{|}{R}}{N}\text{-}\underset{\overset{\|}{O}}{C}\text{-NH-R'}}$$

Since the isocyanate reactions in Figure 1 are usually slow, the preparation of polyurethane foams requires besides a diisocyanate and a polyol, a blowing agent, a surfactant and a catalyst. The blowing agents used can be water (as seen in Figure 1) or trichlorofluoro methane and/or dichlorodifluoro methane, commonly known as Freons. The surfactants which act as foam stabilizers and which control the cell size of the foam produced are usually silicones, such as, silicone glycol copolymers, or copolymers of dimethylpolysiloxane and polyalkylene ethers. The catalysts normally used to speed up the reaction are tertiary amines and/or organotin compounds. Tertiary amines included in this group are: triethylenediamine, N,N,N'N', triethylamine, N-methyl or N-ethyl morpholine. The organotin catalysts normally used are dibutyl tin dilaurate, stannous octoate, dibutyl tin di-(2 ethyl) hexoate. Combinations of tertiary amines and organotin compounds can also be used.

Since the properties of a polyurethane foam in the reaction between a polyol and an diisocyanate depends somewhat on the

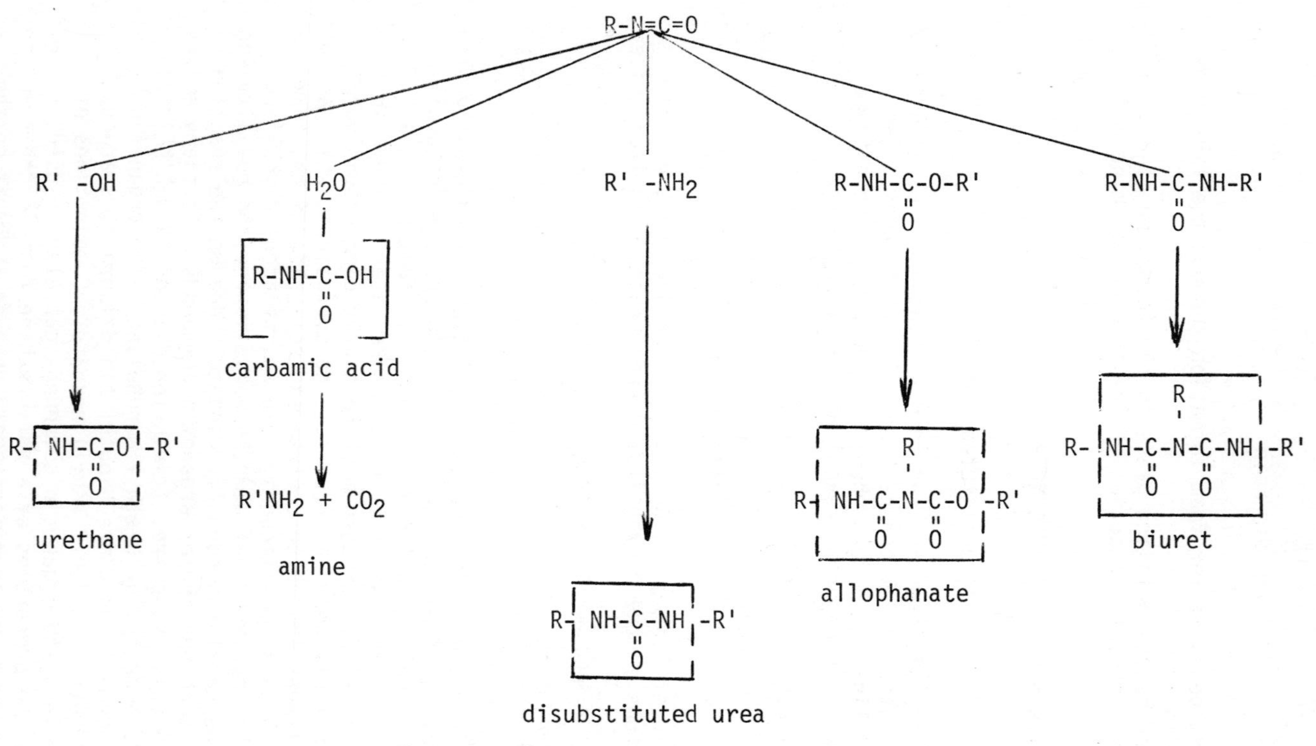

Figure 1. Chemistry of urethane foam formation

stoichiometric relationship of these components, the hydroxyl number of the bark used had to be determined. Results of the analysis of the two bark materials studied, namely, Ponderosa Pine and Douglas Fir are given Table II. The amount of diiso-

TABLE II

BARK ANALYSIS

Bark	% Hydroxy	Hydroxyl Number
Ponderosa Pine	12.56	414.48
Douglas Fir	13.35	440.55

cyanate required to react with the bark was then calculated on the basis of the total concentration of the hydroxyl number of the bark and water when used. An excess of isocyanate over the amount required by stoichiometric calculations was used. After much experimentation with various catalysts and surfactant systems, a typical formula for our foam preparation is set forth in Table III.

TABLE III

GENERAL FORMULA OF A BARK - DIISOCYANATE POLYURETHANE FOAM

Reactants	Parts in Grams
Bark	50
Catalyst	1-2
Surfactant	.5-3
Diisocyanate (crude) MDI	100
H_2O	2-10

The foams were prepared by first mixing the bark and the isocyanate, the other components (catalyst, surfactant, and blowing agent) were pre-mixed and added to this bark - isocyanate mix, and the mixing was continued to yield a homogeneous blend which began to rise and yield a foam product. The rate of foaming was a function of the catalyst used.

A number of bark-diisocyanate polyurethane foams were prepared. An area which I felt had to be explored was whether I was producing a polyurethane foam from the reaction of bark and diisocyanate. Initial attempt to elucidate or confirm a urethane linkage in the prepared foams, was made by the Infra-red analysis on a series of three prepared polyurethane foam formulas, wherein the bark was the varying factor. The three prepared foam formulations studied are shown in Table IV. Infra-red analysis was conducted on the three foam formulations listed in Table IV. The resulting curves are shown in Figure 2.

The three IR curves shown in Figure 2, when interpreted revealed the presence of urea linkages in Formula 1. The pres-

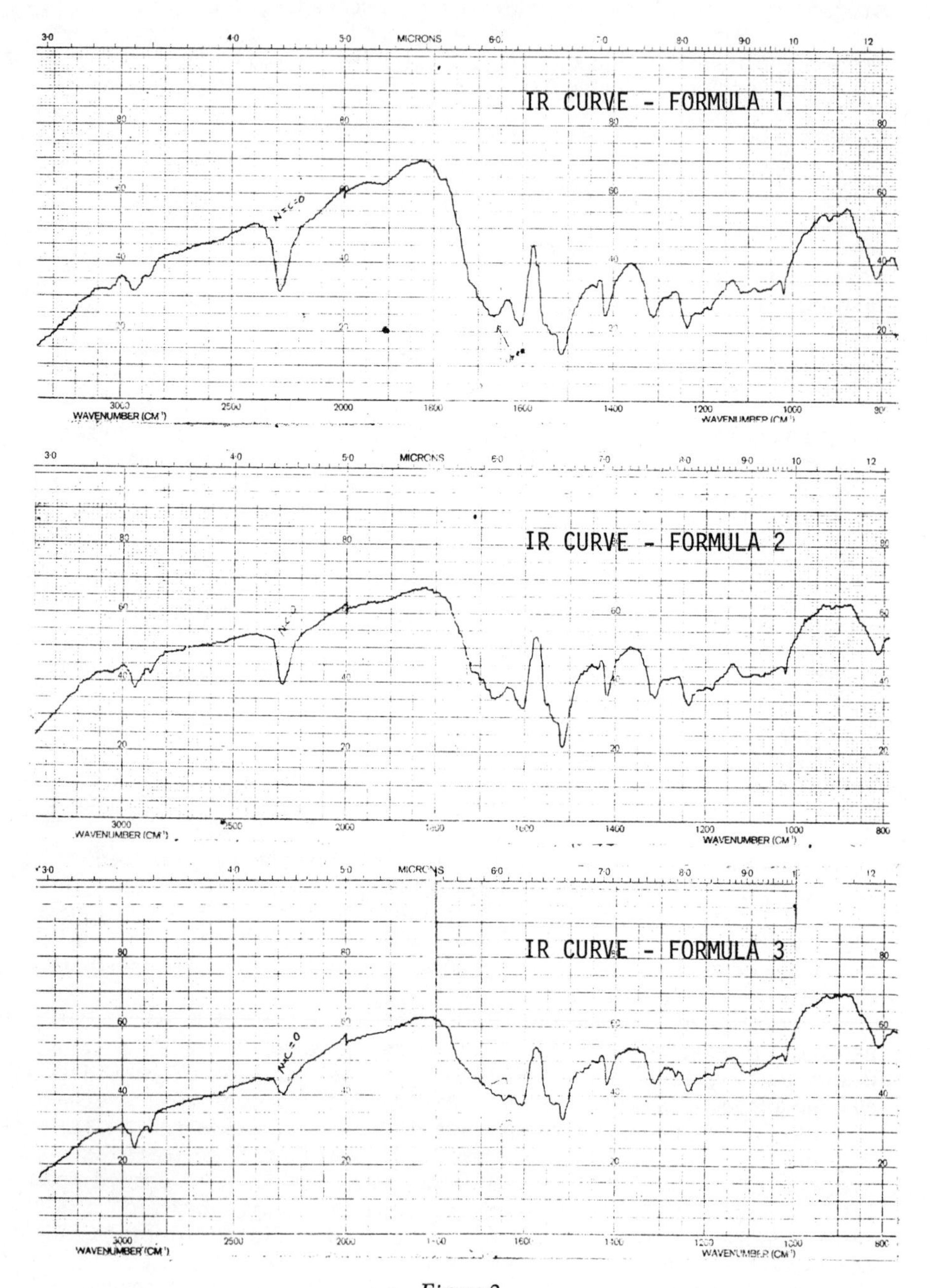
IR CURVE - FORMULA 1
IR CURVE - FORMULA 2
IR CURVE - FORMULA 3
MICRONS
WAVENUMBER (CM⁻¹)

Figure 2

TABLE IV

BARK-DIISOCYANATE - POLYURETHANE FOAM FORMULATIONS USED FOR INFRA-RED (IR) AND FOR THERMOGRAVIMETRIC ANALYSIS (TGA)

Formulations	#1	#2	#3
Ponderosa Pine Bark	50	25	-
Douglas Fir Bark	-	25	50
Catalyst - T-6	2	2	2
Surfactant L-5420	1	1	1
Freon	8	8	-
Water	-	-	7
MDI (crude)	80	80	100

ence of urethane and urea linkages were found in Formula 2, which contained a mixture of Douglas Fir and Ponderosa Pine bark. While Formula 3, which contained Douglas Fir bark, revealed the presence of urethane linkages. All three foams showed the presence of isocyanate linkages.

The lack of distinct urethane linkage resolution in the formula containing Ponderosa Pine bark could well be due to two factors:

1 - the inherent highly rigid polymeric system formed and
2 - the position of the urethane linkage absorption band.

Since the absorption bands in IR spectrum are at frequencies related to the vibration of the functional groupings present, highly rigid polymeric materials do not vibrate as much as non highly rigid structures, therefore, distinct resolution of specific groupings (urethane) may not be that visible in the IR spectrum. Furthermore, the urethane linkage absorption band is at about 5.8 microns, and the other groupings or linkages associated with polyurethane formation, namely, the disubstituted urea, biuret, allophanate, uretidione, and isocyanurates, are all in close proximity to this absorption band. This can be seen in Table V, [14]. Any one of these groupings or linkages and more particularly the disubstituted urea, which yields a strong shoulder, can wash out the urethane linkage. Niederdellmann, et al. [15], in their study of polyurethane and polyurea foam mixtures by IR analysis, found that small concentrations of polyurethane linkages cannot be easily determined or detected in the presence of large amounts of polyurea linkages. Thus we see that due to the possible wash out of the polyurethane linkage and also due to the possible highly rigid polymeric system formed, when Ponderosa Pine bark was used as the polyol, more definite confirmation of the polyurethane grouping was sought. In this behalf, Thermogravimetric Analysis (TGA) was used, since it would provide (a) information on the composition of the initial sample, (b) information on the composition of any intermediate compounds that may be formed and (c) information on the composition of the residue if any was formed. This analytical method,

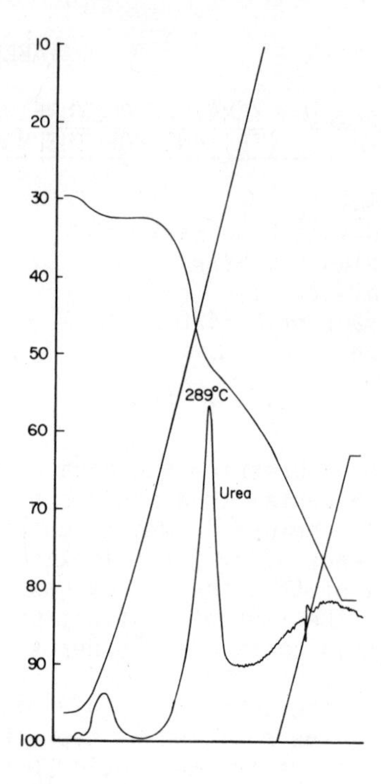

Figure 3. TGA—Thermogram—Formula 1

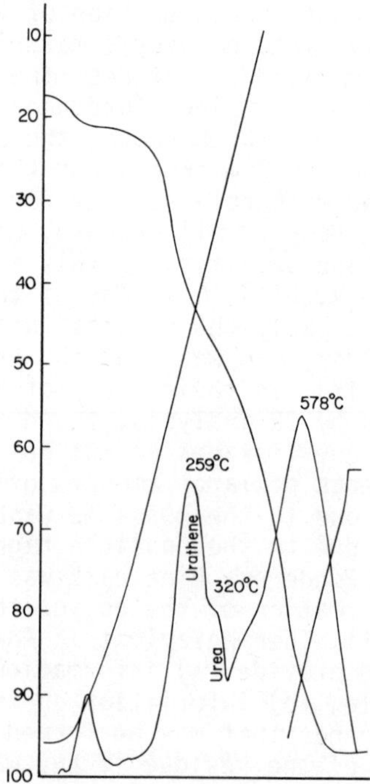

Figure 4. TGA—Thermogram—Formula 2

TABLE V

WAVELENGTHS FOR CHARACTERIZATION OF ISOCYANATE DERIVITIVES

Group	Wavelength (micron)	Wave Number (cm^{-1})
-NCO - in isocyanates	4.40-4.46	2270-2240
-N=C=N - in carbodiimides	4.72	2120
-C=O - in uretidione ring (dimer)	5.61-5.65	1783-1770
-C=O - in isocyanurate ring (trimer)	5.85-5.92	1709-1689
-C=O - in carbamates (urethane)	5.75-5.88	1739-1700
-C=O - in allophanates	5.71-5.81°	1751-1721
	5.81-5.90	1721-1695
-C=O - in ureas	5.99-6.13	1670-1630
-C=O - in biurets	5.81-5.92°	1720-1690
-NH - deformation (Amide II bond)	6.41-6.58	1560-1520
-NH - stretch (bonded)	2.94-3.13	3400-3200
-NH - stretch (free)	2.86-2.94	3500-3400
-C=N - in carbodiimide dimer	5.95	1681
-C=N - in carbodiimide trimer	5.99	1669

°= Higher frequency band is more intense

when compared to Infra-red analysis, is also useful in the identification and characterization of materials.

In TGA, the mass of a sample is continuously recorded as a function of temperature. The thermogram produced represents the temperature at which the mass changes, with a peak temperature representing the maximum mass change. TGA was run on the same prepared foam samples shown in Table IV, using a Fisher Thermal Analyzer #442, in a Helium atmosphere at 50 ml./min. and at a rate of 10°/min. The thermograms produced are set forth in Figures 3, 4 and 5. The interpretation of these thermograms were based on comparing them with confirmed polyurea and polyurethane foam thermograms. They showed, as did the IR analysis, the presence of urea linkages in Formula I; the presence of urea and urethane linkages in Formula 2; and the presence of urethane linkages in Formula 3. A summary of the instrumental analysis (IR and TGA) is presented in Table VI.

From the three formulations studied, I would have expected to find the presence of urea linkages, if anywhere, in Formula 1, since this foam was prepared with water as the blowing agent, and as seen in Figure 1, the production of disubstituted urea structures can be produced from the reaction of water and diisocyanate. Why the appearance of urea-linkages when using Ponderosa Pine bark as the polyol with Freon, as the blowing agent, is difficult to explain at this time.

Compressive strengths were performed on a number of foam samples in accordance with ASTM #1621. The results obtained were comparable to other polyurethane foams prepared from conventional polyols and diisocyanates.

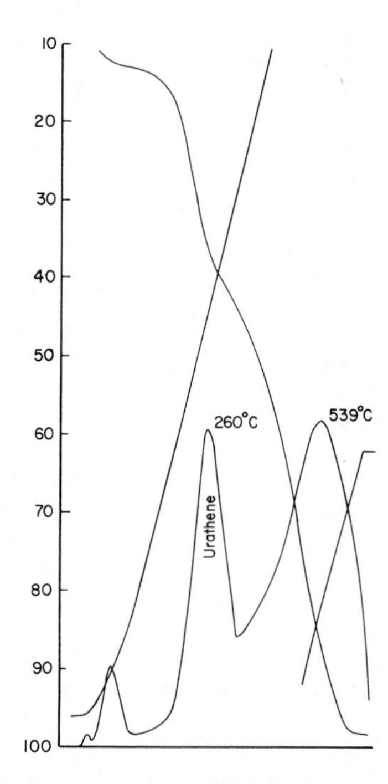

Figure 5. TGA—Thermogram—Formula 3

TABLE VI

INFRA-RED (IR) THERMOGRAVIMETRIC ANALYSIS (TGA) OF DIFFERENT BARK-POLYURETHANE FOAM FORMULATIONS

Formulations	#1	#2	#3
Ponderosa Pine-bark	50	25	-
Douglas Fir-bark	-	25	50
Catalyst - T-6*	2	2	2
Surfactant - L-5420**	1	1	1
Freon	8	8	-
Water	-	-	7
MDI (crude)	80	80	100
Infra-red analysis	unreacted NCO plus high concentration of urea	unreacted NCO plus urethane and urea	urethane
TGA analysis	urea - no urethane	urea - plus urethane	urethane

* M&T - stannous salt of long fatty acid
** Union Carbide Silicone

When the Flammability Test of Plastic Sheeting and Cellular Plastics in accordance with ASTM #D-1692-68 was performed on a series of foams, the foams were found to have (a) a high degree of thermal stability and to possess (b) a high degree of fire retardancy (self-extinguishing). The average result obtained utilizing this procedure on a series of prepared foams is given in Table VII.

TABLE VII

ASTM - D-1692-68
FLAMMABILITY OF PLASTIC SHEETING AND CELLULAR PLASTICS

burning extent - - - - - - - - - - - - - - - - - 2 inches
burning or extinguishing time - - - - - - - 70 seconds
burning rate (BR) - - - - - - - - - - - - - - 1.71 inches/min.

$$BR = \frac{\text{distance burned (in.)} \times 60}{\text{burning time (sec.)}}$$

Burning Characteristics of Foam
intumescence
self-extinguishing

The low burning result and the self-extinguishing characteristic obtained were not surprising. I say this in hindsight, for the inherent flammability of polyurethane foam, in general, is almost entirely a function of the chemical composition of the solid polymer; and one route used to produce flame retardant polyurethane foams has been to alter the structure of the urethane molecule. Stepniczka [16] has reported that the flammability of polyurethane foams can be reduced by using:

1 - components with a high degree of aromaticity.
2 - high molecular weight polyols.
3 - polyols with high functionality - at least 4.
4 - aromatic isocyanate with functionality of 2.3 to 3.2.
5 - cyclic rather than open chain polyols.

From Table 1, we can see that bark components do possess many of the above inherent fire retardant features. Thus the fire retardant properties obtained should not have been a surprise but rather an expected result which I can attribute to the design of the components used.

In conclusion, I have attempted to demonstrate that bark can be used as the sole polyol component in the production of rigid polyurethane foam. These foams possess good physical as well as inherent fire retardant properties. With the availability of bark as a reactive polyol, I can see a means of utilizing bark as a cheap raw material in producing economical polyurethane foam products.

Acknowledgement:

I wish to thank Mr. David L. Williams of Upjohn Chemical Co., for his analytical help in performing the IR and TGA analysis, and for his interpretations of the resulting curves. Furthermore, I wish to thank Mr. R. R. Kirk, Mobay Chemical Co., for his IR interpretations.

"Literature Cited"

1 - Kurth, E. F., Tappi, (1953), 36 (7).
2 - Hergert, H. L. and Kurth, E. F., Tappi,(1953), 36 (3).
3 - Graham, H. M. and Kurth, E.F., Industrial and Engineering Chemistry, (1949), 41 (2).
4 - Brink, D. L., Dowd, L. E. and Root, E. F., U. S. Patent No. 3,234,202, (1966).
5 - Dowd, L. E., U. S. Patent No. 3,255,221, (1966).
6 - Kurth, E. F., Oregon State College Research Paper #106, School of Science, Dept. of Chemistry, (1948).
7 - Herrick, F. W. and Bock, L. H., U. S. Patent No. 3,025,250, (1962).
8 - Herrick, F. W. and Bock, L. H., U. S. Patent No. 3,053,784, (1962)
9 - Herrick, F. W. and Bock, L. H., U. S. Patent No. 3,223,667, (1965).
10 - Klein, J. A. and Poletika, N.V., U. S. Patent No. 3,213,045, (1965).
11 - Heritage, C. C., U. S. Patent No. 2,574,785, (1947).
12 - Saunders, J. H., and Frisch, K. C., "Polyurethanes": Chemistry and Technology (Part 1: Chemistry), Interscience Pub-

lishers, New York, (1962).
13 - Ferrigno, T. H., "Rigid Plastics Foams", 2nd Edition, Reinhold Publishing Company, New York, (1967).
14 - David, D. J. and Staley, H. B., "Analytical Chemistry of the Polyurethanes", Part III, Interscience, John Wiley, New York (1969).
15 - Niederdellman, G., Lauerer, D. and Merten, K., Central Research Lab., Farbenfabriken Bayer, A. G., private communications, November, 1965.
16 - Stepniczka, H. E., JFF/Fire Retardant Chemistry, (1974), 1 (5).

17

A Plywood Review and its Chemical Implications

TERRY SELLERS, JR.

Reichhold Chemicals, Inc., P.O. Box 1610, Tuscaloosa, Ala. 35401

The various operations for converting round wood into veneer and finally into plywood are chiefly mechanical. However, there are many areas in which chemicals, both common and uncommon to us, are involved in substantial quantities. As the various manufacturing variables are viewed, these chemical implications will be discussed.

Plywood is normally defined as an assembly of an odd number of layers of wood joined together by means of adhesive. In most cases, the grain direction of each layer or ply is at right angles to the grain of the adjacent ply or layer. In recent years this definition has been modified slightly.

The outer two surface plies are called faces (often one surface ply will be called the face and the other surface the back). The innermost ply is called core in a three-ply configuration. Other layers are called centers and in some cases crossbands.

Plywood construction refers to the composition of the inner plies or layers. The faces are usually always veneer but the cores may consist of four different types of wood materials: all inner plies of (1) wood veneer; (2) lumber (strips of lumber bonded together); (3) particleboard (chips or flakes combined with an adhesive binder); and (4) hardboard (wood fibers combined with an adhesive).

Hardwood plywood is distinguished from softwood plywood in that the former is generally used for decorative purposes and has a face ply from a deciduous or broad leaf tree. Softwood plywood is generally used for construction and structural purposes, and the veneers typically are of wood from evergreen or needle-bearing trees.

Douglas Fir and Southern Pines make up the primary domestic softwood species in this country used for softwood plywood. Hardwood species consisting of Oak, Walnut, Hickory, Gum, Birch, Elm, Cherry, Pecan, Maple, etc., from the United States and such tropical woods as Philippine Mahogany (Lauan), and at least 200 other species from all parts of the world are used in making hardwood plywood.

Plywood is a multi-use material characterized by its ability to be designed and engineered for both construction and decorative purposes in flat and curved shapes. The first plywood manufactured by man was used in overlaid and inlaid furniture. The objective was to achieve beautiful and decorative effects by the use of highly figured veneer. The earliest record of veneer from wood is in pictorial murals which archaeologists unearthed in Egypt and have dated as early as 1500 B. C. A cedar wood casket was found in the tomb of King "Tut" of Egypt bearing ebony veneer and ivory inlay. No doubt the adhesive used in this period was of animal nature. Thin sheets of wood veneer were spread with this animal glue and pressure applied by sand bag weights.

Historians have described how Greeks and Romans used plywood in their culture to achieve beautiful decorative effects. Articles of furniture overlaid with veneers in these times were highly treasured articles. There is little reference to the use of veneer in the years that followed these cultures until the 17th and 18th centuries, when a revival of the art occurred in Western Europe.

In 1830, the piano industry became the first industry to use plywood. The second half of the 19th century saw plywood being used to make sewing machines, chairs, organs, desk tops, and other articles of furniture as well as panelled doors. The early years of the 20th century brought the inception of the stock panel idea: A panel made in standard size that can be cut into smaller sizes if desired. The sizes originally were 3 feet by 6 feet, but today are 4 feet by 8 feet.

Advantages of plywood are many fold. The chief purpose in overlaying the wood first practiced by the ancients and later by artisans in the 18th and 19th centuries was to produce a surface of beauty over a less aesthetic surface. Today plywood is a mixture of this same objective of aesthetics and the functional period. The inherent property of veneer-cutting utilizes the valued woods in full-size sheets to its greatest potential, whereas other methods of utilization would be too costly or impossible.

Plywood has many structural advantages. It is stronger along the grain than it is across the grain and by altering the direction of the grain 90 degrees with each successive wood layer or ply, the strength properties are equalized. Pound for pound, wood (plywood) has been proven to be stronger than steel. A property of plywood familiar to us all is its splitting characteristics. Wood easily splits along the grain, but plywood by design cannot. The cross-layering concept of plywood creates a more stable product.

In many ways, plywood manufacturing achieves a more complete utilization of the log than does lumber manufacturing. No sawdust results from rotary cutting or slicing of veneer. Valuable woods can be used as face veneers and less valuable woods for the inner plies. Plywood manufacturing makes for a more efficient utilization of specific species. Low grade veneers within a species may be utilized within the panel with the more perfect sheets of veneer utilized on the face. This has both aesthetic and structural design implications.

As the timber in the forest is harvested, it is delivered to storage yards in tree length form or in log lengths. These logs are stored for periods of days to months and are commonly sprayed with water to prevent bacteria and fungi deterioration. Timber cut from the forest contains moisture contents in the range of 30 percent to 200 percent based on its oven dry weight. Usually, the heartwood of the tree is much lower in moisture than the growth areas around the perimeter (or sapwood portion) or the tree.

Veneer logs are usually fed through debarkers of various types. One is called a ring debarker. Most often the trimmed debarked logs are heated in hot water vats or in steam chambers. Heated logs are easier to peel into veneer than are cold logs. Remember in many areas where the forest is harvested and plywood is made, there are very cold periods of time including freezing weather. Heated debarked logs are transferred to the barker deck and led to the infeed chain of the lathe charger.

Most veneer in this country is rotary lathe peeled. The log is centered and held on each end and while it rotates against a knife producing veneer similar to unwinding a roll of paper. There are some higher priced hardwood veneers which are sliced. Most slicers consist of a stationary knife. The segment of the log prepared for slicing is called a flitch and is mounted on what is called a log bed which passes up and down against the knife. One slice of veneer is produced on each downward stroke of the log bed. These techniques are particularly useful in cutting veneers for various bond figures which are found in different sections of trees depending on their rough pattern. These sliced veneers are matched to create designs for the fine furniture manufactured in this country.

After the veneer is peeled and is being transferred along belts, clippers cut out defects in the veneer. The veneer then proceeds along what is called the green (undried veneer) chain to forklift loading docks for transfer to the dryers. Most all green wood waste is chipped and shipped to paper mills via rail cars.

Looking at this initial phase in veneer preparation in the plywood manufacturing process, we see that water spraying of the logs creates a situation subject to water pollution. Also, heating logs in hot water vats and in steam chambers creates additional contaminated water subject to pollution control. Various means of compounding this water, treating and reusing the water or using the solution for fertilizer have been attempted. In essence, due to water leaching the tannins from the bark and other chemicals in the trees themselves, these water solutions must be subjected to zero discharge by 1977, according to the new guidelines by the Environmental Protection Agency.

Debarking these logs creates millions of tons of bark, a material which today is not fully utilized. Some bark is used as mulch around flowering shrubs. One Oregon plant finely grinds Douglas fir bark after removing its waxes by petrochemical extraction, yielding 24,000 tons of extender for adhesive mixes and 1,000 tons of vegetable-type wax per year. Bark is increasingly becoming a source of fuel, but there remains a tremendous opportunity for utilization.

The veneer dryers of a plywood mill are large chambers equipped with heating elements and fans utilizing automatic conveying systems on which the veneers are transported. Some dryers are located behind the rotary lathe, which allows the cut veneer to be dried in a continuous sheet or ribbon and then clipped dry afterwards.

Drying time of veneer depends on the wood species, veneer thickness, moisture content of the veneer, dryer temperature and the desired final moisture content. Usually, veneer moisture must be no higher than 7 percent for softwoods because the typical phenol-formaldehyde resin adhesive will not properly cure at substantially greater moisture contents under highly automative productive cycles. There is the danger of rupturing the panel in the hot press due to trapped steam pressure if veneer moisture is too high. Ordinarily, when this occurs, lower hot press temperatures are used with longer press times involved. Dryer temperatures for coniferous veneer should be 350° $\pm$ 50° Fahrenheit, the point where wood resin vaporizes. If the veneer is overdried, irreversible chemical changes may take place where moisture content is reduced below the fiber saturation point.

Remember the undried (green) veneer may average 60 - 120 percent moisture content and softwood veneer is usually dried to 4 $\pm$ 2 percent and hardwood 6 $\pm$ 2 percent moisture content, oven-dry basis.

Impurities in the drying atmosphere are exhausted in the air and have been referred to as "blue haze." One company has made substantial progress in collecting the blue haze emissions in a duct where they are directed through a series of water showers, filters and other equipment. Then the water is evaporated leaving a heavy liquefied residue which is collected on a stainless steel belt and deposited in storage vats. Currently this residue is being used as a fuel supplement, but some feel that it may be a future source for development of new chemicals. Apparently, this process meets Environmental Quality Standards with reference to air pollution.

Dryers are tremendous consumers of energy such as natural gas, fuel oils and steam heat. Heat conservation is quite important and considerable attention has been given to improving the efficiency of plywood mills to generate most of their own steam heat energy with residues such as bark generated from the process variables. With the shortage of natural gas in recent years, it has become imperative that older mills be converted as quickly as possible in this direction.

Boiler steam is used to dry veneer, heat the hot presses, and heat log conditioning vats or chambers. Over a three-shift day, a softwood plywood plant requires 60,000 - 70,000 pounds of steam per hour for production of 15,000 square feet per hour, 3/8" basis. By the use of residues, it is possible to reduce oil and natural gas requirements to about zero, using them only to even out changes in fuel demands.

Some contaminated water results in the drying process and is a water pollution item. Dried veneer residues collect on the roller bars in the dryer and air tubes of the cooling sections of the dryers which involves certain cleaning chemicals and all of these items must be controlled in terms of pollution. Imagine the quantity of water involved in drying veneer. It approximates the weight of the 18 million cubic meters of plywood produced in the United States. Some calculate this to be 28.6 billion pounds of water.

Some green or wet veneer is stitched together as if with a sewing machine prior to transferring through veneer dryers. Most veneer is sent through the veneer dryer in a clipped form of various widths, but stacked adjacent to each other to form a layer and subsequently unitized into larger uniform pieces. Strips are dry clipped to remove defective edges and subsequently taped or glued with contact adhesives which may be water, petrosolvent or rubber base types. They may be connected along the width by 1,000 - 2,000 denier strings composed of dacron, nylon, polyester or fiberglass filaments coated with hot melt adhesives.

All dry veneer waste and plywood trim waste not used for fuel is commonly chipped and sold to particleboard plants as furnish.

After the veneer has been properly unitized and dried, it is transported to the gluing operation. It is here that the greatest proportion of chemicals other than water are involved in the plywood process. To better understand the adhesive resins involved, perhaps it is best to review the quantities of softwood and hardwood plywood manufactured in relation to the adhesive needs required. The 1972 - 1973 era were years of peak production in the United States for plywood. Data will be extracted from various reports for presentation purposes. Hardwood and softwood plywood production is normally reported in different manners and it is difficult to compare. While the values reported are not intended to be exactly accurate, they will give some comparison, relatively speaking, and should give some concept of the volume of plywood produced and in turn the volume of adhesives used.

In the April, 1975, issue of Plywood & Panel magazine, we find reported square feet, quarter inch basis, domestic hardwood plywood manufactured in this country was approximately 3.5 billion and softwood plywood was 27.5 billion. This means 89 percent of the total plywood produced was softwood plywood, and 97 percent of this production was glued with phenol-formaldehyde resin adhesives, according to the American Plywood Association. This leaves 11 percent domestic hardwood plywood which was 95 percent glued with urea-formaldehyde resin adhesives, according to the Hardwood Plywood Association. Some protein-type glues were used in gluing West Coast softwood plywood production and some melamine and phenolic was used in gluing domestic hardwood plywood. It may be of interest that the imported hardwood plywood into this country exceeded 5 billion square feet, quarter inch basis, mostly thin wall panelling with Lauan faces. Little softwood plywood was imported and the United States is becoming an exporter of softwood plywood.

In 1973, over 500 million solid pounds of phenol-formaldehyde resin were used in the manufacture of softwood plywood in the U. S. A. About 60 million solid pounds of urea-melamine-formaldehyde resin were consumed in the hardwood plywood industry.

J. T. White reported at the Washington State University Symposium on Particleboard in 1973 that 18.7 percent of the formaldehyde manufactured in the U. S. A. went into wood binders; likewise, 8.9 percent of the urea chemical produced in the U. S. A. went into wood binders.

Some report that over 50 percent of the urea-formaldehyde resins consumed went into particleboard. This is brought out because there may be a shift away from urea resin for certain types of oriented particleboard used in structural plywood constructions. Historically, particleboard has been used for inner plies as previously mentioned in some hardwood plywood. There is now one plant in production in Idaho which produces mechanically oriented strand particleboard for use specifically as core for softwood plywood production. It is anticipated that this trend to some degree will increase in the future, and phenolic resins appear to be the mechanism with which this particleboard will be bonded.

Most phenol-formaldehyde resin in this country is a water-soluble, 40 percent non-volatile type, when used for softwood plywood production. In this solution, the volume of phenol-formaldehyde resin exceeds 1,250 million pounds per year. Most plywood adhesive resins are shipped in bulk quantities of 4,000 to 10,000 gallons via tankwagon or tankcar to the plywood mills and stored in storage tanks. Other ingredients are collected in a glue mix area and combined with adhesive resin.

In the past, phenolic mixes were simply a combination of the phenolic resin with a filler such as walnut shell flour or pecan shell flour. These type mixes are still used in some hardwood exterior plywood. Today, phenolic glue mixes for softwood plywood involve mixing phenolic resin with water, filler, extender and sodium hydroxide (usually 50 percent). The filler is usually "furafil," (a ligno-cellulose by-product of furfural production of Quaker Oats Company), Douglas fir or alder bark, wood particleboard sander dust and/or attapulgite clay. A sulfite paper mill lignin is also used as a filler in the Northwest. Between 50,000 and 60,000 tons of filler were used in phenolic glue mixes per year.

Furfural is made from corn cobs, rice, cottonseed and oat hulls or other agricultural residue. The by-product "furafil" is principally a slightly acidic, dark brown solid product pulverized similar to coffee grounds. It is further processed to various particle size, but the most common has at least 98 percent passing through a 100-mesh screen. It is typically composed of 38 percent cellulose and 42 percent residue from saccharification (lignins and resins).

Douglas fir bark is a ligno-cellulosic material, too, containing up to 20 percent cork, 20 percent amorphous tannin powders and 5 percent vegetable wax (carnauba type). As previously mentioned, the wax is extracted before use as an extender. The balance is dried and ground into powder form similar to particle size of furafil before use as an extender. Alder bark is dried and ground directly without further treatment and used as a filler.

Clay mineral attapulgite is an acicular-shaped hydrous magnesium aluminum silicate. Large deposits occur in South Georgia and North Florida. It is receiving increasing use in partial replacement of furafil.

All of the products, with the possible exception of particleboard wood sander dust, yield varying degrees of thixotropic properties which permit easy application, yet fixed in a consistent pattern after application.

Other extenders of proteinaceous-amylaceous products such as flour from wheat, soy, rye and milo-sorghum grains are used. One of the most highly regarded protein-starch extenders is the endosperm of selected soft winter wheats. About 25,000 tons of proteinaceous-amylaceous products are used each year in phenolic mixes for softwood plywood. Perhaps an equal quantity is consumed by the hardwood plywood industry.

The objectives of the extender-filler in glue mixes are to improve the adhesive performance, to help control the mix viscosity and to conserve the phenolic and urea adhesives. These fillers and extenders serve specific needs in these glue mixes. They extend the resin solution from its original non-volatile state to lower levels, such as 26 percent resin solids in the mix. They also function as a means of maintaining the very low resin molecular weight and size molecules on the surface when they are applied to the veneer to prevent over-penetration prior to hot press pressure and temperature which gels the resin to a permanently set condition.

In hardwood plywood, the glue mix is typically formed by adding water and soft wheat flour at rather high proportions with a minor addition of filler, such as nut shell flour to the urea-formaldehyde resin. Urea-formaldehyde resins are usually shipped in 60 - 65 percent non-volatile solids form. Acid salts such as ammonium sulfate are added to increase the rate of cure of the urea-formaldehyde resin when under pressure and subsequently heated in hot presses. Some acid salts cure the urea-formaldehyde sufficiently well under ambient "cold" press conditions. It is evident this type of plywood would be oriented toward interior use such as furniture. If water-proof type bonds are required of hardwood plywood, then melamine-formaldehyde resins are used in similar mix form but with higher resin solids levels. Sometimes phenolic resins with shell flour are used to obtain the desired grade of exterior glue line.

After veneers have been unitized from small strips to full sheets of the desired width, usually 4 feet by 8 feet in softwood plywood and various widths and lengths for hardwood custom manufactured plywood, they are ready to be fed through a glue application which coats one or both sides of veneer sheets with liquid adhesive. These applicators control the amount of adhesive transferred to the veneer. The amount of glue mix applied per 100 square feet of surface area (single glue line basis) will approximate 4 - 5 pounds and the amount of resin solids 1 - 1.5 pounds.

In hardwood plywood, almost all applicators are roll coater types. In softwood plywood, the large production mills used roll coaters for 50 years or more, but are now moving more and more toward spray applicators and curtain coater applicators. In the southern section of our country where 57 softwood plywood mills are now located and have been built since 1963, approximately one-third of the mills use curtain coaters, one-third use spray applicators and the other third use roll coater applicators.

The curtain coater and spray applicator are innovations which can conserve adhesives in the range of 20 percent and permit recycling of glue which is not applied to the veneer.

The combined panels of the desired number of plies or layers are now transferred either directly to the hot press or to a cold press and then to the hot press. The cold prepress tacks the veneer adhesive glue line together so the panels require smaller openings in the hot press, thus permitting larger numbers of hot press openings per cycle of production.

Now comes one of the most important steps in the process — the pressing together of the veneers under heat and pressure, thus setting the phenolic or urea adhesive, whichever may be applicable.

Hot press plates heated to about 250° Fahrenheit for urea hardwood plywood and 300° Farenheit for phenolic bonded softwood plywood are closed under pressure at 150 - 200 pounds per square inch. The hot presses may vary from ten openings to as many as fifty openings capable of pressing one or two thin panels per opening.

In hardwood plywood, cold presses without heat application are occasionally used where curved plywood in particular is desired. Dielectric heat curing of the plywood glue lines as a means to increase productivity is sometimes used in conjunction with this process. The acid salt becomes critically important in these cases.

After hot pressing, the panels are stacked for conditioning, sawed to dimension and sanded. They are ready for inspection, grading, strapping and shipping.

Returning to the gluing operation and expanding the discussion on the chemical aspects of plywood gluing is desirable. James A. Klein gave an excellent presentation (at the 1975 Forest Products Research Society Annual

Meeting) on the nature of the bond between phenolic resin and veneer. This discussion is quite comprehensive and is available through a "separate" publication from the FPRS office in Madison. His review indicates microscopic studies of plywood glue bond show penetration of 4 - 8 cells on each side of the glue line in the wood. If an adhesive penetrates and completely wets the surface of the wood four cells deep on each side of the glue line, the actual bonding area can be twenty-four times as great as the surface area of the glue line. Controlled wetting and penetration of the wood by the adhesive is necessary for good bonding. An adhesive that does not wet and penetrate results in discontinuous bonds which are unsatisfactory. The resin synthesis or adhesive formulation can be adjusted to change the wetting and penetration of the adhesive as applied in a particular plywood plant and for particular veneer species. D. L. Gumprecht discussed the tailoring of phenolic adhesives in 1969 in his article published by the Forest Products Journal. Because each sheet of veneer represents a different combination of available capillaries and surfaces of available wettability, formulating a phenolic adhesive to obtain acceptable or desirable bond qualities within the operating conditions of the plywood plant becomes as much an art as a science. Changes in seasons, weather conditions, raw materials, equipment or operating procedures very often result in changes in adhesive recommendations.

The nature of adhesive to wood bonds has been variously described as physical bonding, chemical bonding and secondary chemical bonding. Physical bonding, or the interlocking of the cured adhesive in the internal capillary structure of wood, occurs in all good glue bonds but contributes only slightly to the overall strength of the bond. Some studies have shown that true chemical bonding may occur between phenolic resins and wood. True covalent bonds between the resin and wood molecules would be extremely strong. Although this type of bond may and probably does occur to a limited extent, its contribution to the overall strength is believed very small in most plywood bonds.

Most of the strength of a phenolic resin to wood bond is generally concluded to be the result of hydrogen bonding between the phenolic resin and the wood chemical molecules. Hydrogen bonding is one type of secondary chemical bonding in which the hydrogen ions are mutually shared by reactive polar groups, usually hydroxyl or carboxyl. This type of bond is very similar to the manner in which wood holds water for the first few percent of moisture above the oven dry level.

The potential for hydrogen bonding between a phenolic resin and wood depends upon: (1) The number of reactive groups on the resin and wood molecules at which hydrogen bonding may occur; (2) The ability of the adhesive to wet and penetrate the capillary structure of the wood; (3) The mobility of the phenolic molecule to align reactive sites on the resin molecule to reactive sites on the wood molecule, and (4) The ability of adhesive to cure in place with a minimum of shrinkage and internal stresses while reactive sites on the resin and wood molecules are sharing hydrogen ions. It is agreed that this concept is not universally or uniformly in agreement.

Most plywood adhesives contain from 5 - 7 percent sodium hydroxide or caustic. The caustic can cause partial pulping and excessive swelling of the wood adjacent to the glue line, allowing some limited penetration by the larger phenol molecules. Most of the bonding between the resin and veneer, however, occurs at the surface of the larger capillaries near the glue line.

Veneer is easiest to bond when it is dried to a moisture content above zero percent (oven dry method). Overdrying can result in thermal decomposition or the removal of some of the chemically bound water normally remaining in the wood fiber at zero percent moisture. The removal of the chemically bound water converts reactive hydroxyl groups to relatively unreactive ether linkages. This and the possible migration of wood resin to the surface of the wood in overdrying results in a surface which is extremely difficult to wet and is referred to as surface inactivated. This in itself does not make the veneer impossible to glue, but it does increase the difficulty when this type of veneer is mixed with properly dried veneer. Adhesives can sometimes be adjusted to obtain satisfactory wetting and penetration of inactivated surface veneer, but when the degree of surface inactivation varies considerably within units of veneer for bonding, it may be difficult or impossible to obtain a satisfactory adhesive formulation for any length of time which would result in excellent glue bonds.

Most components of adhesive will not penetrate a cell wall in wood. The movement of adhesive through four to eight cells away from the glue line required for good bonding depends on the capillaries available near the glue line. Unless drying or mechanical checks or cracks are present, the effective capillary area in uncut summer wood cells is less than 0.1 percent when the break occurs between cells or in outer cell walls. Wood cells are not all parallel to the veneer surface, so many are cut across the grain when peeled. Cut cells will vary from 60 - 85 percent depending on the summer or spring wood growth. These cut cells can form a large passage for the rapid movement of adhesive away from the glue line. There are also crushed cells due to the mechanical surface pressure of the lathe nose or roller bar, etc. Other openings such as resin ducts are quite prevalent in southern pines and permit adhesive penetration 50 - 70 cells from the glue line. The capillary area in spruce, larch and Douglas fir is much less and the heart wood resin ducts are usually closed or occluded with natural wood resins.

Obtaining uniform penetration of adhesive in all of the capillaries present on the veneer surface presents an interesting problem to the resin chemist and adhesive technologist. Wood resins and extractives may also affect the wettability of veneer and accessibility of reactive sites on the cell walls. This is especially true of certain imported woods from Malaysia, Indonesia and other areas of Southeast Asia.

Phenolic resins for plywood are typically caustic-catalyzed phenol-formaldehyde resoles. Typical resins are water solutions containing 40 - 44 percent resin solids composed of 23 - 25 percent phenol, 5 - 7 percent sodium hydroxide and 10 - 12 percent formaldehyde. The ratio of formaldehyde to phenol in a resin has a decided effect on performance characteristics as does the synthesis procedure used in manufacturing the resin. Various "chemical modifiers" may be added to the resin to improve certain features such as prepressing, dryout resistance or for specific application methods such as curtain coating where surface tension is important.

As previously mentioned, phenolic resin is usually mixed with extenders, additional caustic, water and certain other chemicals. The extenders must be fine enough to stay in suspension and must be compatible with the phenolic resin molecules to avoid filtering or extender separation. Most successful extenders have an abundance of polar reactive groups allowing them to also bond to the resin and be included in the polymerization matrix. Some extenders may also have reactive groups which contribute to the total bond strength, but most act as a bulking agent to give the adhesive mix the desired viscosity and surface retention characteristics. They may also hold water and the amount of water held in the adhesive prior to hot press heat application affects the assembly time tolerance, penetration, wettability and the flow and cure of the resin. The amount of additional caustic added to the adhesive mix affects the digestion of the filler-extenders, the penetration, assembly time tolerance and cure rate of the adhesive. Sometimes defoamers are added if there is a foaming problem in mixing or application. Chemical modifiers may be added to the mix separately for curtain coating, etc. Small percentages of chemical modifiers affect the surface tension of the adhesive but the effect on glue bond is usually masked by the high caustic content of most softwood plywood phenolic adhesives.

While some moisture starts to leave the adhesive as soon as it is applied to the veneer, enough water must be absorbed by the wood or retained on the glue film sufficient to keep the resin molecules mobile. Otherwise, the resin will not flow properly in the bonding set time. The adhesive must gel in a continuous film. It must also cure in place. The adhesive should form a permanent durable bond to the wood.

The exterior durability of softwood veneer species in this country has been demonstrated. There have been some difficulties encountered in the long term exterior durability of some Asian veneer species when bonded with phenolic resins. Extractives interfering with the cure of the resin directly or the resin bonded to the extractives rather than the wood cause failures along the glue line. These species shrink and swell more than native softwoods. Stresses are greater and breaks in the wood hydrogen bonds may be occuring. Extensive research on this subject is being performed at the present time and substantial knowledge has already been gained in this regard.

When involved with gluing, there is the opportunity for water control. Since the plywood industry is one of the industries subject to zero water discharge by 1977 under the Environmental Protection Agency Effluent Limitation Guidelines, this area receives substantial attention. For phenolic gluing in the softwood industry, water discharges from the gluing operations may be recycled and reused in the adhesive mix formulation. Concentrations of from 1 percent to greater than 20 percent non-volatile solids in the wash water may be concentrated, gathered and used in the adhesive mix formulations. Usually these solids are not incorporated in the adhesive resin solids figures but are considered as part of the total extender solids in the mixture. This wash water does affect the digestion of the glue mix.

Many mills have successfully controlled their wash water in the softwood industry and there are some in the hardwood plywood industry that are recycling wash water. Wash water from hardwood plywood is an acid-catalyzed solution or mixture and presents another problem. Usually, the resin solids are allowed to settle and the water above reused and the sump solids removed and disposed of in another manner, such as burying the material or burning the solids in a boiler. However, the latter has presented problems within boiler fire chambers. Zero water discharge philosophy is a debatable subject but its application is practical and reasonable in most cases. By recycling or transfer from one source to another, or with pools for evaporation, the water waste materials are contained to prevent discharge into navigable streams.

Another area of real concern deals with HEW Occupational and Health Standards on hazardous substances listing formaldehyde and phenol. With phenolic adhesives, this is not a problem since all of the resin solids and non-volatile solids are usually tied up in the set glue line.

Well over 95 percent of the hardwood plywood production in the United States uses urea-formaldehyde as the adhesive bonding agent between veneers for reasons of quality, intended use and economics. In this industry, formaldehyde evolution is a subject of concern because normally all of the adhesive formaldehyde released under elevated temperatures is not tied up in the set glue line. This is an area that is of real concern to the particleboard industry which uses urea-formaldehyde as its binder. Some of the formaldehyde is given off as a gas when the resin is cured in the hot press in the manufacturing cycle. Most of this gaseous formaldehyde escapes during the pressing or during subsequent storage, sanding and trimming. However, in a few cases, minute amounts are given off by the hardwood plywood for a period of time after shipment. The free formaldehyde can diffuse out of the panels into the surrounding atmosphere and occasionally there are complaints about formaldehyde odor in mobile homes or rooms where hardwood panelling has been installed.

Overlays for plywood deserves mentioning. Overlays for plywood may be anything that conceivably can stick to the panel and have end use utility. Overlays include polyester or phenolic impregnated paper (either medium or high density types), fiberglass-reinforced plastic, fabric, high pressure laminates, aluminum, lead, polyurethane insulation and pebbles. The uses for plywood are extensive and marriage with other overlay materials expand these uses tremendously.

Phenolic bonded plywood may be treated with fire-retardant chemicals where desired or required.

Another item facing plywood and its chemical usage is the conversion to the metric system of measurements, weights and volumes.

We trust this review of the chemical implication of plywood production stimulates interest and understanding of this important industry.

Abstract

Plywood is normally defined as an assembly of an odd number of layers of wood joined together by means of an adhesive. The various operations for converting round wood into veneer and finally into plywood are chiefly mechanical. However, there are many areas in which chemicals are involved in substantial quantities. A weight of water equal to the oven dry weight of wood used to manufacture plywood must be removed from the veneer of the forest grown tree before adhesive application to assure cure of the adhesive and to meet the needs of plywood use conditions. A billion pounds of liquid synthetic phenol- and urea-formaldehyde resins are consumed yearly by the plywood industry in the United States. One hundred million pounds of proteinaceous grain flour extender and another one hundred million pounds of ligno-cellulose and clay filler are consumed in plywood manufacture each year. Zero water discharge philosophy is a debatable pollution subject in the plywood industry but its application is practical and reasonable in most cases. Air omission control is receiving considerable attention and presents particular problems with free formaldehyde where urea-formaldehyde resins are utilized. Debarked veneer logs create millions of tons of bark which today is not fully utilized. Using this bark and other wood residues for energy generation is a primary objective of plywood mills to essentially free them of petrochemical requirements for energy. These and other areas are reviewed with reference to eighteen million cubic meters of plywood made in the United States yearly.

18

The Bonding of Glued-Laminated Timbers

ROY H. MOULT

Koppers Co., Inc., Monroeville, Pa. 15146

Many examples of bonded wood constructions have been known from ancient times for their usefulness in the building of boats and ships. However, the important technology of combining small pieces of wood to form large timber sections, with the strength and durability of intact lumber, is certainly a more recent product of the last forty years. The early stages of this modern development have been reviewed by Selbo and Knauss (1).

The term "glued-laminated timber" is used to classify those beams, columns, or arches, which are produced by adhesively joining boards together, in face-to-face, parallel-grain alignment. This type of construction is thereby distinguished from plywood, where the wood layers, or plies, are placed together with the grain at right angles, and where an odd number of plies must be used to give a balanced, non-warping system. Although articles concerned with wood adhesion frequently assume that the same factors apply to both types, it is preferred here to contrast them on the basis of adhesives and bonding methods most commonly used for their production.

Adhesives Used

The types of adhesives suitable for laminating beams are restricted by the conditions of application and by their end-use requirements. A wider choice of adhesives for plywood depends on whether softwoods or hardwoods are used, whether they are required for internal or external exposures, or whether they are to be used for ornamental or structural purposes. Thus phenol-formaldehyde types would be used for marine or exterior construction uses; urea-formaldehyde types would be advantageous for cold pressing, or melamine-urea adhesives might be preferred for hardwood plywood, or lumber-core panels used in furniture production.

None of these adhesives would be suitable for glue-laminating using present procedures. Room temperature curing glues, with good exterior durability, are now almost exclusively used. Resorcinolic adhesives are the only type which meet the industrial and military requirements commonly accepted by this industry. Modern laminated beams are too large or too cumbersome in shape to be conveniently transferred to heated areas or ovens. Therefore, the glues must be capable of setting without the application of external heat. In order to meet economical production schedules the adhesives are now formulated to cure more rapidly at the ambient temperature of non-heated plant buildings so that clamping pressures may be removed within four hours. For this purpose, the glues are expected to attain at least half of the mature shear strength and wood failure within that time.

Resorcinolic adhesives are either straight resorcinolic resins, or phenol-resorcinol copolymers. When military requirements dominated the field, during and after World War II, most of the beams were produced from oak, and the straight resorcinol resins were preferred. By the time of the Korean War, excellent phenol-resorcinol types had been developed. These glues were used in the production of the all-wood minesweepers used in naval operations. They demonstrated a high degree of durability in their many years of exposure to all the rigors of the Seven Seas.

Resorcinol differs from other phenols in that it reacts readily with formaldehyde under neutral conditions at ambient temperature. To make stable adhesives, which can be cured at the point of use, they are prepared with less than a stoichiometric amount of formaldehyde. About two thirds of a mole of formaldehyde for each mole of resorcinol will give a stable resinous condensation product. The resin is formed into a liquid of convenient solids content and viscosity. Such solutions have infinite stability when stored in closed containers. Glue mixes formed at the point of use from these solutions, on addition of paraformaldehyde-containing hardeners, will have a useful life of several hours due to two principal factors: (1) the paraformaldehyde depolymerizes to supply monomeric formaldehyde at a slow rate, as determined by the pH; (2) the availability of the formaldehyde is also controlled by the kind and amount of alcohol in the solvent. Formaldehyde reacts with the alcohol to form a hemiacetal. This reaction is reversible and forms an equilibrium which exerts further control on the availability of the formaldehyde.

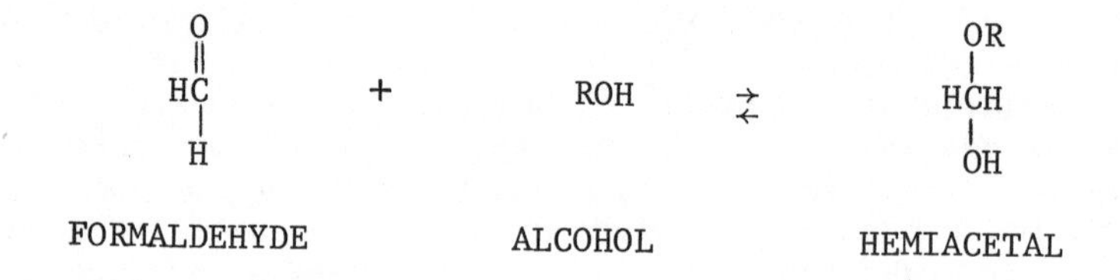

Resorcinol-phenol-formaldehyde resins are produced by combining phenol with formaldehyde, under mild alkaline conditions, before the resorcinol is introduced. Thus these resins may be considered to be a resorcinol-modified resole type and would be a mixture of oligomers something like that shown in Figure 1.

The fact that a major proportion of the resin consists of phenol serves a two-fold purpose: (1) it reduces cost considerably because of the lower cost of phenol; (2) it controls the extent of final cross-linking in the glue film thus improving flexibility and strength.

Wood Surface Factors

The boards used in laminating beams are planed smooth just prior to use to insure fresh, clean, flat surfaces. If roughsawn lumber were used, these boards would be separated by large gaps when placed face-to-face due to surface asperities, cupping, twisting, and end-joint bulges. It would not be economical to fill these gaps with an expensive adhesive. Besides, this planing removes substances from the surface that interfere with proper wetting. Wood surfaces become oxidized during storage, extractives migrate to the surface, molds frequently form stained areas, oily films are deposited from the atmosphere, and dirt may be picked up from handling.

On the other hand, great care must be taken to assure proper planing. Surface problems can result from the use of dull planer knives, or improperly adjusted pressure rolls; burnishing, crushing, or loosening of the surface fibers may occur. Densification of the summerwood areas may seal off the cut lumens, thus reducing the effective amount of active area available. Loose fibers on the surface, or weak subsurface layers, may seriously reduce the strength of the bond.

The moisture content of the wood exerts a very important influence upon the bonding processes and upon the adhesives used for both laminated beams and plywood. Once the beams are clamped up, moisture changes become impractical due to the large cross sections, so it is more convenient to prepare the beams from lumber which is dried to normal equilibrium ranges of eight to 15 percent. Resorcinolic adhesives, when cured at room temperature, give their highest adhesion to wood in that range. At higher moisture levels, the absorption of moisture from the glue film is retarded and this delays or prevents proper cure. At lower levels excessive shrinkage, drying checks and glue film crazing reduce bond strength. Thus the characteristic curing properties of resorcinolic adhesives are admirably suited to the conditions imposed when beams are laminated at room temperature.

In the hot pressing of exterior grade plywood, phenolic glues behave quite differently. Veneers must be dried to less than 5% moisture content and glueline temperatures must be 100 to 150°C.

to insure optimum cure at short press times. Because water is converted to steam at such temperatures the water must be removed rapidly; this prevents over-penetration of the glue, steam blows, or disbonding. Dry veneers act as desiccants to absorb the moisture during the pressing cycle. However, it is also important to note that complete removal of water (i.e., "bone dry veneer") should be avoided or other problems arise, such as poor wetting, underpenetration, and precuring.

Glue Application

Two methods of glue application are currently being used in production of glue-lams. The older method is to pass the boards through rubber-covered roll spreaders so that both sides are coated simultaneously. The boards are then stacked horizontally until all are spread and clamps are in proper position for pressure application. During this assembly time the wet glue film is exposed to the atmosphere for only a short time. This minimizes loss of solvents and water to the atmosphere.

A more recent method for spreading glue is to extrude it from a pipe in a curtain of thin, falling streams onto a horizontally disposed board as it passes under the extruder. In this case, one side of each board is coated with ribbons of extruded glue extending full length. The glues used for this purpose are modified by certain agents to give thixotropic behavior. That is, when they are stirred rapidly, the viscosity is low enough to permit pumping. When at rest, they have a gelatinous consistency. The boards are set on edge but the adhesive does not sag or run off. Because the adhesive is in ribbon form instead of a flat-film, the wet surface area is small and solvent loss is reduced. This permits open assembly times of up to one hour for convenience of production. When all of the boards are spread they are then placed face-to-face and clamped.

The Importance of Solvent Composition

The room temperature performance of resorcinolic adhesives is highly dependent upon the presence of the alcoholic solvents. In order to perform their proper functions, they must be able to flow easily into all of the interstices of the faying surfaces and to achieve adsorption to the maximum amount of area available. The alcohols are good solvents for resorcinolic resins. They reduce the viscosity and its rate of increase during the time the glue is being mixed and applied. This promotes good flow and transfer performance. Alcohols maintain the mobility of the resin molecules during gel formation so that cross-linking can take place more uniformly and to a greater extent. These adhesives, when diluted only with water, produce sandy, friable and weak glue films.

Alcohols also promote wettability and penetration of the wood surface. This may easily be shown by the following simple experiment. When equal sized drops of distilled water were placed on the surface of a freshly planed piece of southern yellow pine, the times for the drops to completely soak into the wood were observed. On the early wood it took 65 seconds and on the latewood 179 seconds. When similar drops of 50% ethanol solution were used instead of pure water, it required only six seconds to disappear into the earlywood and 26 seconds into the latewood. However, if a small drop of adhesive syrup, with no hardener added, was placed on the wood surface, no adsorption took place at all. It was surmised that the viscosity prevented its permeation. When the adhesive was diluted with 50% alcohol it was readily absorbed and produced a red stained spot on either earlywood or latewood areas. This showed that the low molecular weight adhesive molecules could readily permeate the wood structure before condensation with the curing agent.

Assembly Factors

When the glue mix is freshly applied to a board surface, it is in the best condition for exhibiting proper mobility and adsorption. Thus, a properly adjusted adhesive will give higher strength bonds after short assembly time periods, particularly when both faying surfaces are coated. However, during longer periods of open assembly when the wet glue film is exposed to the atmosphere, alcohols will volatize and simultaneously promote dehydration of the glue film. At higher ambient temperatures prevailing during the summer, this action is greatly accelerated, so that permissible open assembly times are reduced. As noted previously, when the glue is applied in the form of ribbons, the escape of alcoholic solvents is reduced so that longer open assembly times are attainable.

When the boards are placed together in closed assemblies ready for clamping, the differences between single and double spread methods play a part in the nature of the adhesive bond. If both surfaces have been spread with a thick layer of glue, the adhesive will readily perform its functions of wetting and permeating both of the faying surfaces. Even though considerable alcohol has evaporated, the films will flow together to form a continuous interlayer. During the closed assembly period there is no further evaporation, but solvent is absorbed into the substrate, promoting better bond formation. When clamping pressure is applied, the board surfaces are forced into close contact so that gaps are closed, air spaces and excess glue are squeezed out, and thinner, stronger adhesive films are obtained.

When the adhesive is applied in ribbon form to only one of the surfaces, a minor proportion of the total surface area will be wetted during the open assembly period, that part which is

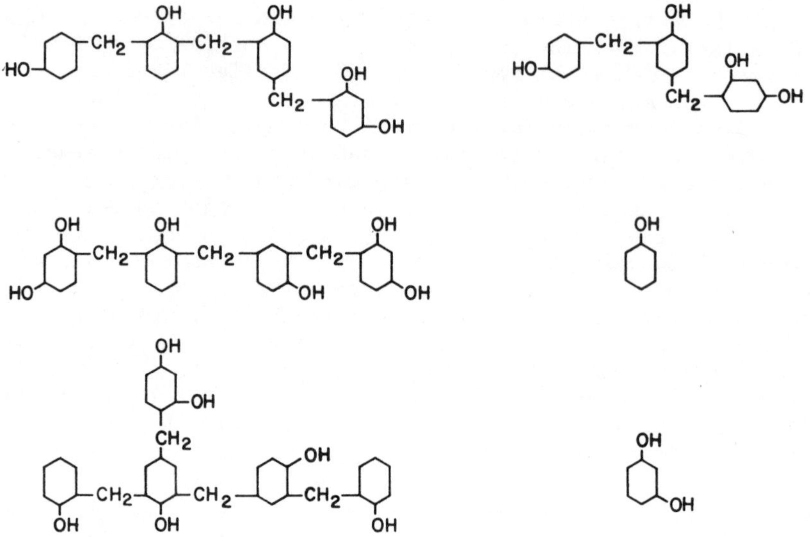

Figure 1. Resorcinol–phenol–formaldehyde resin oligomers

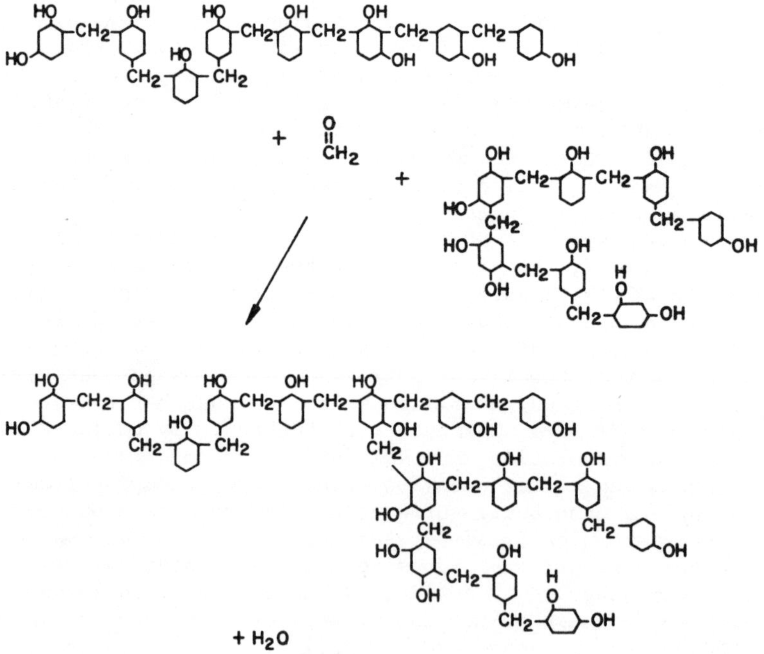

Figure 2. Cross-linking of resorcinolic resins

immediately under the ribbons. The unspread portion between the ribbons, and the entire face of the unspread board, will be wetted only when the boards are pressed together. Penetration of these latter areas is therefore controlled by the solvent composition which remains in the ribbons of glue after this exposure period. As long as this is done at normal room temperature, with properly surfaced boards, there is very little tendency for the solvent to escape and good results are usually obtained.

However, when higher ambient temperatures prevail, such as those often experienced during the summer months (80 to 110°F), the loss of active solvent becomes critical. The glue film after extended assembly time may become deficient. Its ability to wet and penetrate the surface is then severely reduced. Thereafter the probability that this film will produce an adequate bond soon falls to zero. Consequently, the allowable assembly time may be quite limited.

The Relationship of Adhesion to Resin Cure

Resorcinolic adhesives are outstanding in durability because they can form relatively high proportions of cross-links without application of heat or strong catalytic agents. The rate at which these cross-links form, and the amount of conversion, is highly dependent upon the removal of water from the film. This includes not only the water present in the solvent system but also that which is formed by the condensation reaction with the hardener. This reaction is shown in Figure 2.

The importance of water removal may be shown by allowing a sample of glue mix to gel in a test tube. It is quite apparent that the hardening speed is considerably reduced, and that its full strength does not develop until it is allowed to shrink on drying in the open air.

When the glue spread is enclosed between two wood surfaces, the film can no longer lose moisture to the atmosphere; water must then be lost by absorption into the wood. Thus the curing of the glue film is dependent upon the relative ability of the faying surfaces to remove moisture. Hence, any condition of the wood surface that reduces the absorption of moisture lessens the rate of hardening and decreases the strength of the glue film.

From the many articles I have studied which discussed relationships between various wood factors and adhesion, I have selected those most frequently considered to be responsible for low bond strength values. These I have listed in Table I. In the upper column I have indicated those which repel moisture absorption, and in the lower those which promote absorption. Although all of these factors do not carry equal weight, it is interesting to note that 12 of the 20 factors will inhibit penetration while eight will promote it.

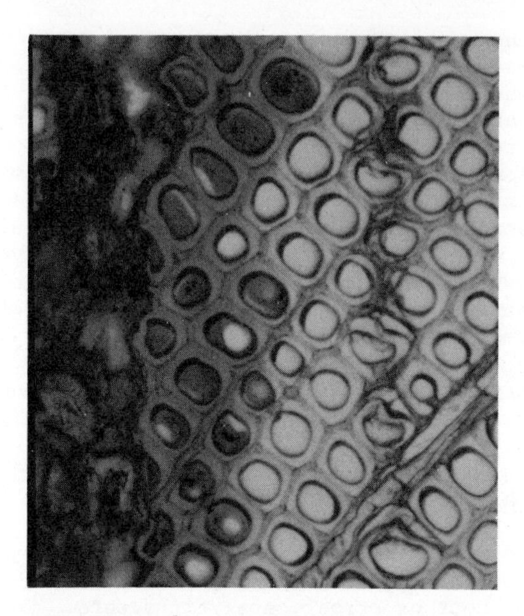

Figure 3. The absorption of alcohol–water solvent by wood near glue line (0 min open assembly time and 15 min closed time)

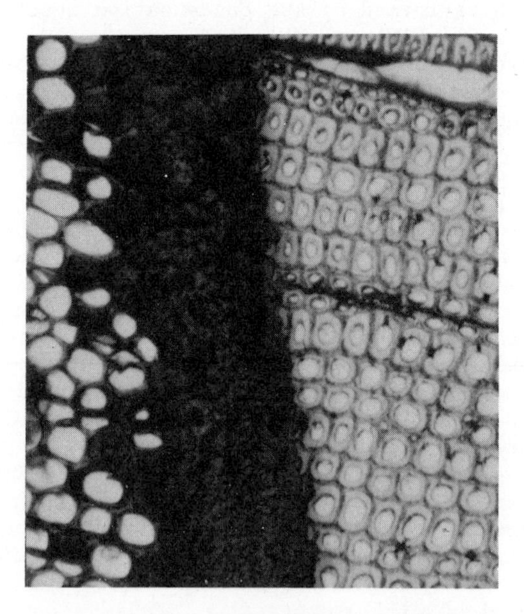

Figure 4. Lack of absorption of solvent by wood near glue line (60 min open assembly time and 0 min closed time) (240×)

The absorption of the alcohol-water solvent by the wood is shown in Figure 3. Latewood and earlywood tissues are both seen near the glue-line. The penetration of adhesive into the ducts and into the cell lumens presumably through the cut fiber ends may be noted. However, the coloration of the cell walls throughout all of the tissue adjacent to the glue-line and extending into the substrate three to five cells deep is particularly significant. This is apparently due to the penetration of the solvent; the color being due to the presence of soluble, low-molecular-weight components. Beyond this permeation region the tissue is practically colorless. This picture was made of a southern yellow pine glue joint after 0 minutes of open assembly time and 15 minutes closed time.

In contrast, compare with the picture of a joint made with the same glue and wood but after 60 minutes open time and 0 minutes closed, Figure 4. The very thick glue line and almost total lack of solvent absorption are quite apparent. The wood failure in testing this joint was very low, while that of the previous figure was very high.

It therefore appears that reduction of moisture absorption by the wood accounts for much of the variation noted in the bond strengths of glued laminated timbers. For example, it is often noted that in the evaluation of bonds of southern yellow pine the wood failure is high in those areas of the sheared bond where earlywood faces earlywood, intermediate where earlywood is opposite latewood, and very low where the latewood areas adjoin. Many reasons for the poor adhesion to latewood surfaces have been proposed in the literature including: poor wettability, high density, highly crystalline cellulose content, high tensile strength, presence of microchecks, and low surface area. However, these are relationships which have been observed which do not completely explain why the adhesion is low.

It is quite probable that the resin is insufficiently cured at the junction between latewood surfaces because the water has not been efficiently removed from the film. One observation which substantiates this is that when the specimens are heated, the sheer strength and wood failure values both increase, and the distinction between earlywood and latewood disappears. This definitely indicates that the problem is due to under cure.

In those cases where water is absorbed very rapidly it would be expected that adhesion values would be very high. They are. That is, if the open assembly time is low. But, if the film is applied to a very absorptive surface and then held for 30-60 minutes before closing with an unspread surface, the solvent will have been removed prematurely, and the bond subsequently obtained will be poor due to low penetration. Thus, all of the factors listed in Table I can potentially cause poor curing.

Resorcinolic adhesives are almost uniquely suitable for the glue-laminating of beams because of their ambient curing characteristics, structural strength and great durability. Their

TABLE I

WOOD SURFACE PROPERTIES THAT AFFECT ADHESION

Moisture Repellent

Heartwood (2), (8), (13)

Latewood (3), (4), (5), (7), (8).

Burnishing (8)

Density (5), (8), (12)

High Moisture Content (8), (9), (10), (13)

Dirt (8), (14), (15)

Extractives (10), (11), (15)

Incipient Rot (8), (14)

Aging (10), (13), (14)

Preservative Treatment (16)

Fire Retardant Treatment (17)

Refractory Wood (5)

Moisture Absorbent

Sapwood (2), (8), (13)

Earlywood (3), (4), (5), (7), (8)

Roughness (7), (10), (15)

Sloped Grain (7), (8)

Porosity (5), (7), (8), (12)

Low Moisture (8), (9), (10), (13)

Sanding (7), (10), (13)

Planing (7), (10)

mobility and curing properties are highly dependent upon the characteristics of their solvent composition. Changes in this composition during the application and assembly periods can have important effects on the laminating process and on the eventual strength. It is now realized that the solvent exerts a powerful influence on the curing rate of the glue bond, due to its ability to promote penetration of the substrate and assistance in promoting absorption of water from the glue line. This seems to fit almost all of the observed effects on adhesion to wood previously reported in the literature.

Literature Cited

1. Selbo, M.L. and Knauss, A.C., For. Prod. J. (1954), 4(4), 69.
2. Huffman, J.B., For. Prod. J. (1955), 6(4), 135.
3. Browne, F.L., I & E Chem. (1931), 23(3), 290.
4. Lutz, J., For. Prod. J. (1964), 14(3), 97.
5. Krahmer, R.L., For. Prod. J. (1961), 11(9), 439.
6. Troop, B.S. and Wangaard, F.F., Yale University Tech. Report No. 4 (1950).
7. Suchsland, O., Holz als Roh-und Werkstoff (1958), 16(3), 101.
8. Hemming, C.B., "Handbook of Adhesives," p. 505, Reinhold Pub. Corp, New York, 1962.
9. Marra, A.A., For. Prod. J. (1962), 12(2), 81.
10. Gray, V.R., For. Prod. J. (1962), 12(9), 452.
11. Blomquist, R.F. and Olson, W.Z., For. Prod. Lab. Report No. 032, Madison, Wisconsin.
12. Freeman, H.A., For. Prod. J. (1959), 9(12), 451.
13. Northcott, P.L., Colbeck, B.G.M., Hancock, W.V. and Shen, K.C., For. Prod. J. (1959), 9(12), 442.
14. Stumbo, D.A., For. Prod. J. (1962), 14(12), 582.
15. Bikerman, I.I., "The Science of Adhesive Joints," Academic Press, New York, 1961.
16. Raknes, E., J. Instit. Wood Sci. (1963), 11, 24.
17. Schaeffer, R.E., U.S. For. Serv. Res. Note FPL-0151.

19

Finishing Wood for Exterior Applications—Paints, Stains, and Pretreatments

WILLIAM C. FEIST

Forest Products Laboratory,[1] P.O. Box 5130, Madison, Wis. 53705

Weather affects the performance of wood finishes adversely. Both types of finish the film-forming (paints, varnishes) and the penetrating (stains, preservatives) are adversely affected by a combination of the following factors:

1) Photodegradation by ultraviolet light,
2) Leaching, hydrolysis, and shrinking and swelling by water,
3) Discoloration and degradation by micro-organisms.

Wood surfaces exposed to weather without any finish change color, are roughened by photodegradation and surface checking, and erode.

Film-forming finishes on wood provide the most protection against sunlight and moisture. They offer the widest selection of colors, but are susceptible to failure by cracking, blistering, and peeling. These types of failures are typical of coating systems for wood exposed to a varying moisture as well as to sunlight.

Finishes or stains that penetrate wood surface and form no coating also provide protection from weather. Because these types of finishes do not fail by peeling and blistering, they are more easily maintained and refinished.

The adverse effects of weather on wood and wood finishes and resultant chemical and physical changes on wood surface will be reviewed and discussed in this work. In addition, recent research findings on surface pretreatment of wood for enhancing durability of applied finishes will be presented.

Paints

Film-forming finishes such as paints, lacquers, and varnishes have long been used to protect wood surfaces. The film-forming finishes are not preservatives; they will not prevent decay if conditions are favorable for growth of micro-organisms.

[1]Forest Service, U.S. Department of Agriculture. Maintained at Madison, Wis., in cooperation with the University of Wisconsin.

Paints are commonly divided into oil-base, or solvent systems, and latex, or water-base, paints. These paints are essentially a suspension of inorganic pigments and a resin, or vehicle, in a suitable "solvent." Paints protect wood surfaces by forming an ultraviolet absorbing film that adheres strongly to the surface.

Much has been written about paints and their mode of failure on wood surfaces (1-6); thus, this subject will not be discussed in detail here.

Stains

Penetrating finishes such as stains and water-repellent preservatives (7,8) leave little or no continuous coating on wood surface. Because there is no coating, there is no failure by blistering or peeling. These finishes are ideally suited for rough-textured surfaces difficult to paint effectively.

The oil-base, semi-transparent, penetrating finishes are generally comprised of a resin, a solvent, a UV-absorbing pigment, a wax, and a preservative (7).

The water-repellent preservatives are generally comprised of a resin, a solvent, a wax (as the water-repellent), and a preservative (generally pentachlorophenol) (8). Wood surfaces treated with these finishes are water repellent and mildew and decay resistant. The water-repellent preservatives are excellent natural finishes for wood.

Weathering of Wood

The appearance of wood exposed outdoors without protective paint, stain, or treatment changes markedly in a few months; then the wood remains almost unaltered for years (9-11). Color of the wood is affected very early (11,12). Generally, all unprotected woods change in color to a brown. After prolonged weathering, all woods become gray (9). Recent research has been concerned with many aspects of weathering of wood (13-18).

Chemical changes in wood during weathering cause the changes in color (9,10,19). Only exposed surfaces of wood are affected. The brown of the wood surface (0.02-0.10 in.) is caused chiefly by decomposition products of wood lignin. As rain slowly leaches the brown decomposition products, a gray layer (0.003-0.01 in. thick) of a disorderly arrangement of loosely matted fibers, develops over the brown layer. The gray layer is composed primarily of the most leach-resistant parts of wood carbohydrates. Of these, xylan and arabinan are leached most extensively from weathered wood; galactan and mannan, to a lesser degree; and the more resistant glucosan, least (9,19). As wood substance is leached from the weathered surface, the brown and the gray layers move into the wood, but the process is so slow that only approximately 1/4 inch of thickness is lost in a century (9,10,20).

The color and the appearance of weathered wood can be affected, to a marked degree, by dark-colored spores and mycelia of fungi (4,10). When these grow on a wood surface, they contribute significantly to the dark gray appearance of weathered wood.

Weathering produces physical as well as chemical changes. Miniutti, in a series of papers, examined microscopic changes in cell structure of softwood surfaces during weathering and ultraviolet irradiation (21-25).

Wood near an exposed surface of a board fluctuates between wet and dry conditions more rapidly than does wood in the interior of the board. Swelling and shrinking stresses roughen the surface, raise the grain, and minute checks or cracks become visible. The cracks steadily increase in number and size. The chemical changes add to the roughness and soften the surface of the board (9,25). In addition to the minute checks on the surface, most woods soon develop larger and deeper checks or cracks that are easily visible. As the weathering process continues, boards tend to cup, warp, and pull at their fastenings (5). Woods of moderate-to-low density acquire fewer checks than do woods of high density.

Pretreatments

Serious problems are encountered in the exterior use of wood due to the influence of moisture, ultraviolet light, fungi, and erosion. New, improved, exterior finishes such as silicone-modified alkyds, polymers of vinyl fluoride, and acrylic and related polymer latex paints are continually being introduced. Yet, problems of erratic bonding, peeling, discoloration by wood extractives, and the rapid deterioration of transparent finishes remain.

Recent work at the Forest Products Laboratory (26-30) has shown that certain inorganic chemicals when applied as dilute aqueous solutions to wood surfaces provide the following benefits:

1) Retard degradation of wood surface by ultraviolet irradiation.

2) Improve durability of ultraviolet transparent polymer coatings.

3) Improve durability of paints and stains.

4) Provide a degree of dimensional stabilization to wood surface.

5) Provide fungal resistance to wood surface and to coatings on the surface.

6) Serve without further treatment as natural finishes for wood.

7) Fix water-soluble extractives in woods such as redwood and cedar, thereby minimizing subsequent staining of applied latex paints.

The most successful treatments investigated were those containing chromium trioxide, copper chromate, or ammoniacal solutions of copper chromate (27,29). Successful results using zinc-containing compounds have also been reported (31-35).

The effectiveness of several surface treatments is illustrated in Table I. The springwood erosion (in microns) is determined using a compound reflected-light microscope at 140X. End-matched samples are used for comparison. Accelerated exposure was in a weatherometer with exposure cycles of 20 hours light, 4 hours water spray, 5 days per week.

Table I. Springwood Loss (In Microns) of Western Redcedar Treated with Eight Aqueous Chemical Solutions

Treatment	Hours of Exposure			
	440	840	1240	1700
	Microns			
None, control	20	80	155	310
Copper chromate	10	15	25	115
Ammoniacal copper chromate	5	10	15	90
Ammonium chromate	5	10	40	120
Sodium dichromate	5	10	35	130
Chromium trioxide	0	0	5	20
Stannous chloride	25	80	145	250
Ammoniacal zinc oxide	20	40	130	260
Copper sulfate	--	--	--	250

The most effective treatments were those containing chromium. Of these, the chromium trioxide and the ammoniacal copper chromate treatments were the most effective.

The chromate treatments were found effective on both new wood surfaces and on surfaces exposed to natural and artificial weathering. The treatments, particularly chromium trioxide, will retard the erosion of weathered wood if applied to a surface as a 5% aqueous solution.

Little is understood of the mechanism involved in the improvement of a wood surface after treating with chromium-containing chemicals and the subsequent durability of applied finishes (5). The data in Table II show that the reduction of the rate of weathering (as measured by loss of springwood) is directly related to chromium concentration.

Table II. Springwood Loss of Western Redcedar After Accelerated Weathering (480 Hr In The Weatherometer) As Related to Chromium Metal Concentration

Treatment	Springwood Loss, Microns
Sodium dichromate	
1% Chromium	336
3% Chromium	82
5% Chromium	59
Chromic acid	
1% Chromium	130
3% Chromium	47
5% Chromium	18
None	313

Surface dimensional stabilization also apparently is involved in the mechanism of improvement (27).

Summary. Treatment of wood surfaces with chromium-containing chemicals such as chromium trioxide is an effective method to retard weathering of wood and improve performance of subsequently applied finishes. Continuing work is concerned with determining the mechanisms responsible for improving durability of wood surfaces and finishes after treatment. Performance of wood-derived products (plywood, hardboard, fiberboard, particle board) after surface treatment with inorganic chemicals is also being investigated. The overall objective of the continuing research is to investigate new environmentally safe procedures to stabilize wood surfaces and to improve performance of applied finishes.

Literature Cited

1. Browne, F. L., Forest Prod. J. (1959), 9 (11), 417-427.
2. Banov, A., "Paints and Coatings Handbook," Structures Publi. Co., Farmington, Mich., 1973.
3. Hess, M., "Paint Film Defects," 2nd ed., Chapman and Hall, Ltd., London, Eng., 1965.
4. Duncan, C. G., Off. Digest (1963), 35 (465), 1003-1012.
5. Forest Products Laboratory, "Wood Handbook: Wood as an Engineering Material," USDA Agriculture Handb. No. 72, revi., Gov. Print. Off., Washington, D.C., 1974.

6. Forest Products Laboratory, "Wood Finishing: Blistering, Peeling and Cracking of House Paints from Moisture," 7 pp., USDA For. Serv. Res. Note FPL-0125, For. Prod. Lab., Madison, Wis., 1970.
7. Black, J. M., Laughnan, D. F., and Mraz, E. A., USDA For. Serv. Res. Note FPL-046, For. Prod. Lab., Madison, Wis., 1975.
8. Forest Products Laboratory, "Wood Finishing: Water-Repellent Preservatives," 7 pp., USDA For. Serv. Res. Note FPL-0124, For. Prod. Lab., Madison, Wis., 1973.
9. Browne, F. L., South. Lumberman (1960), December, 141-143.
10. Forest Products Laboratory, "Wood Finishing: Weathering of Wood," 4 pp., USDA For. Serv. Res. Note FPL-0135, For. Prod. Lab., Madison, Wis., 1975.
11. Browne, F. L., and Simonson, H. C., Forest Prod. J. (1957), 7 (10), 1-7.
12. Wengert, E. M., J. Paint Tech. (1966), 38 (493), 71-76.
13. Kuhne, H., Leukens, U., Sell, J., and Wälchli, O., Holz als Roh und Werkst. (1970), 28 (6), 223-229.
14. Sell, J., and Leukens, U., Holz als Roh und Werkst. (1971), 29 (1), 23-31.
15. Sell, J., and Leukens, U., Holz als Roh und Werkst. (1971), 29 (11), 415-424.
16. Leukens, U., Sell, J., and Wälchli, O., Holz als Roh und Werkst. (1973), 31, 45-51.
17. Sell, J., Muster, W. J., and Wälchli, O., Holz als Roh und Werkst. (1974), 32, 45-51.
18. Sell, J., and Wälchli, O., Holz als Roh und Werkst. (1974), 32, 463-465.
19. Tarkow, H., Southerland, C. F., Seborg, R. M., and Kalnins, M., USDA For. Serv. Res. Pap. FPL-57, 60 pp., For. Prod. Lab., Madison, Wis., 1966.
20. Laughnan, D. F., Forest Prod. J. (1959), 9 (2), 19A-21A.
21. Miniutti, V. P., The Microscope (1970), 18 (1), 61-72.
22. Miniutti, V. P., J. Paint Tech. (1969), 41 (531), 275-284.
23. Miniutti, V. P., USDA For. Serv. Res. Pap. FPL 74, For. Prod. Lab., Madison, Wis., 1967.
24. Miniutti, V. P., The Microscope (1967), 15 (3), 4-16.
25. Miniutti, V. P., J. Paint Tech. (1973), 45 (577), 27-34.
26. Black, J. M., USDA For. Serv. Res. Note FPL-0134, For. Prod. Lab., Madison, Wis., 1973.
27. Black, J. M., and Mraz, E. A., USDA For. Serv. Res. Pap. FPL-232, 40 pp., For. Prod. Lab., Madison, Wis., 1974.
28. Black, J. M., and Mraz, E. A., USDA For. Serv. Res. Pap. FPL 271, 7 pp., For. Prod. Lab., Madison, Wis., 1976.
29. Feist, W. C., Mraz, E. A., and Black, J. M., Forest Prod. J. (in press).
30. Feist, W. C., Forest Prod. J. (submitted for publication).
31. Desai, R. L., and Clarke, M. R., Canad. For. Indust. (1972), 92 (12), 47-49.

32. Desai, R. L., Dolenko, A. J., and Clarke, M. R., Canad. For. Indust. (1974), 94 (2), 39-40.
33. Shields, J. K., Desai, R. L., and Clarke, M. R., Forest Prod. J. (1974), 24 (2), 54-57.
34. Hulme, M. A., and Thomas, J. F., Forest Prod. J. (1975), 25 (6), 36-39.
35. Shields, J. K., Desai, R. L., and Clarke, M. R., Forest Prod. J. (1973), 23 (10), 28-30.

20

Wood-Polymer Composites and Their Industrial Applications

JOHN A. MEYER

Chemistry Department, S.U.N.Y. College of Environmental Science and Forestry, Syracuse, N.Y. 13210

Wood and man have coexisted on this planet from the beginning and wood, as a renewable resource, has provided man with tools, weapons and shelter. During the millennia of man's development he learned how to make it harder and stronger. This modification was accomplished by drying and heat tempering his wooden tools and weapons. As man increased his knowledge of the world he lived in he attempted other modifications of the basic resource to better fit his increased requirements. Over the years tars, pitches, creosote, resins and salts have been used to coat wood or fill its porous structure.

With the advent of the polymer or plastic age, scientists had yet another group of chemicals to coat and treat the ancient raw material, wood. During World War II, phenol-formaldehyde, based on the research of the Forest Products Laboratory, was used to treat wood veneer and to form the composite into airplane propellers. Today, this same "Compreg" is used for cutlery handles throughout the world.

Table I illustrates the range of new treatments introduced during the period of 1930 to 1960 (1). Some of the monomers are of the condensation type and react with the hydroxyl groups in the wood, while other chemicals react with the hydroxyl groups to form crosslinks. Another group of compounds simply bulk the wood by replacing the moisture content in the cell walls.

During the early 1960's a new class of chemicals containing one or more double bonds was used to treat wood; vinyl type monomers that could be polymerized into the solid polymer by means of free radicals (2). This vinyl polymerization was an improvement over the condensation polymerization reaction because the free radical catalyst was neither acidic nor basic, nor does the reaction leave behind a reaction product that must be removed from the final composite, such as water. The acid and base catalysts used with the other treatments degrade the cellulose chain and cause brittleness of the composite. Vinyl polymers have a large range of properties from soft rubber to hard brittle solids depending upon the groups attached to the carbon-carbon backbone.

Table I. Processes for Wood Modification

Acetylation	Hydroxyl groups reacted with acetic anhydride and pyridine catalyst to form esters. Capillaries empty. Anti-shrink Efficiency (ASE) about 70%.
Ammonia	Evacuated wood exposed to anhydrous ammonia vapor or liquid at 150psi. Bends in ½ inch stock up to 90°.
Compreg	Compressed wood-phenolic-formaldehyde composite. Dried treated wood compressed during curing to collapse cell structure Density 1.3 to 1.4. ASE about 95% Usually thin veneers for cutlery handles.
Crosslinking	Catalyst 2% zinc chloride in wood then exposed to paraformaldehyde heated to 120°C. for 20 mins. ASE 85%. Drastic loss in toughness and abrasion resistance
Cyanoethylation	Reaction with acrylonitrile (ACN) with NaOH catalyst at 80°C. Fungi resistant, impact strength loss.
Ethylene oxide	High pressure gas treatment, amine catalyst. ASE to 65%.
Impreg	Noncompressed wood-phenolic-formaldehyde composite. Thin veneers soaked, dried and cured under mild pressure. Swells cell wall, capillaries filled. ASE about 75%.
Irradiated Wood	Exposure to 10^6 rads gives slight increase in mechanical properties. Above this level cellulose is degraded and mechanical properties decrease rapidly. Low exposure used to temporarily inhibit growth of fungi.
Ozone	Gas phase treatment degrades cellulose and lignin, pulping action.
β-Propiolactone(β-P)	β-P diluted with acetone, wood loaded and heated. Grafted polyester side chains on swollen cell wall cellulose. Carboxyl end groups reacted with copper or zinc to decrease fungi attack. Compression strength increase.

Table I. Processes for Wood Modification (continued)

Staybwood	Heat stabilized wood. Wood heated to 150°C. to 300°C.
Staypak	Heat stabilized compressed wood. Wood heated to 320°C. then compressed; 400 to 4000 psi, then cooled and pressure released. Handles and desk legs.

A few examples of vinyl monomers are listed below.

$CH_2=CH_2$	$CH_2=CH-CH_3$	$CH_2=CH$–phenyl
ethylene	propylene	styrene
$CH_2=CH-Cl$	$CH_2=CH-CN$	$CH_2=C(CH_3)-C(=O)-O-CH_3$
vinyl chloride	acrylonitrile	methyl methacrylate

Since most vinyl monomers are non-polar, there is little if any interaction with the hydroxyl groups attached to the cellulose molecule. In general, vinyl polymers simply bulk the wood structure by filling the capillaries, vessels and other void spaces in the wood structure.

In the early 1960's, the Atomic Energy Commission sponsored research at the University of West Virginia for the development of wood-polymer composites using gamma radiation to generate free radicals, which in turn initiated the polymerization of the vinyl monomers (3). This support was expanded to other organizations such as, Lockheed-Georgia who supplied samples for industrial evaluation, the North Carolina State University (4,5) for evaluation of properties, and the Arthur D. Little Co. for economic evaluation. The first paper on the catalyst-heat process for making wood-polymer composites was presented at the 1965 Annual Forest Products Research Society meeting in N.Y. City. Research support for this development was provided by the SUNY College of Environmental Science and Forestry at Syracuse, N.Y. During the past ten years the industrial development and applications have been slow but steady for both processes.

This discussion is not intended to be an exhaustive review of the wood-polymer literature, but rather an overview of the processing procedures used today. In general, the free radicals used for the polymerization reaction come from two sources, temperature sensitive catalysts and Cobalt-60 gamma radiation. In each case a free radical is generated by the process, but from that point the vinyl polymerization mechanism is the same. Each process for generating free radicals has its own peculiarities.

Chemistry of the Process

"Vazo" or 2,2'-azobisisobutyronitrile catalyst is preferred over the peroxides because of its low decomposition temperature and its non-oxidizing nature. Vazo will not bleach dyes dissolved in the monomer during polymerization.

$$CH_3\text{-}\underset{CN}{\overset{CH_3}{\underset{|}{\overset{|}{C}}}}\text{-}N{=}N\text{—}\underset{CN}{\overset{CH_3}{\underset{|}{\overset{|}{C}}}}\text{-}CH_3 \longrightarrow 2\ CH_3\text{-}\underset{CN}{\overset{CH_3}{\underset{|}{\overset{|}{C}}}}\cdot + N_2$$

This first order reaction is independent of the concentration of Vazo and independent of the type of monomer (6).

Table II. Half Life of Vazo Catalyst vs. Temperature (7)

Temperature ^{o}C	$t_{1/2}$ minutes
0	4×10^7
7	1×10^7
18	1×10^6
30	1×10^5
46	1×10^4
70	270
100	5.5

The rapid decomposition of Vazo catalyst with an increase in temperature can be used to advantage in the bulk vinyl polymerization in wood. A moderate temperature of 60°C. can be used to initiate the reaction, and since the half life is more than 10,000,000 minutes or about 20 years at 0°C. the catalyzed monomer can be stored safely for months. Catalyzed monomers have been stored for over a year at 5°C. by the author without any deleterious effect. The nitrogen released during the Vazo catalyst decomposition is normally soluble in the monomer-polymer solution. At high temperatures the nitrogen forms gas bubbles in the highly viscous monomer-polymer solution and the final wood-polymer will contain void spaces. In the wood-polymer this is of little consequence, since the methyl methacrylate monomer (MMA) shrinks about 25% during polymerization creating additional void spaces in the solid polymer. The cost of Vazo catalyst is in the range of \$0.50 to \$5.00 a pound depending upon the amount ordered. One gram theoretically, will produce 7.4×10^{21} free radicals and at \$5.00 perpound this is 6.7×10^{23} free radicals per dollar, or about one cent per gram. When gamma radiation is used as a source of free radicals many complications arise immediately, the least of which is the chemistry of the process. Radiation today implies government regulations and the concern of the environmentalists.

Safety requirements must be satisfied before a Cobalt-60 source can be installed. Space is not available here to review the safety regulations and the facilities necessary for a radiation source large enough for a production facility. Radiation trained personnel must be on the staff before a license can be issued. At least 500,000 to 1,000,000 curies of Cobalt-60 are required for a production source for making wood-polymers, and at a dollar per curie, a considerable investment must be made before production can begin. Cost considerations aside, the Cobalt-60 radiation process does have some distinct advantages in making wood-polymer composites. Since the monomer is not catalyzed it can be stored at ambient conditions as long as the proper amount of inhibitor is maintained. The rate of free radical generation is constant for a given amount of Cobalt-60 and does not increase with temperature.

When gamma radiation passes through a material such as wood or a vinyl monomer it leaves behind a series of ions and excited states as the energy of the gamma ray is absorbed through photoelectric, compton and pair production collisions(*). The ions and excited states generated in the absorbing material immediately rearrange to form free radicals, which in turn initiate the polymerization process.

$$\underset{\text{Excited State}}{(\text{Monomer})^*} \longrightarrow \underset{\text{Free Radicals}}{R^\cdot + R^\cdot}$$

The free radicals usually consist of H˙ and the radical monomer˙. As mentioned above, once the free radical is generated, the polymerization reaction is the same as that of a normal catalyzed vinyl monomer bulk polymerization (8).

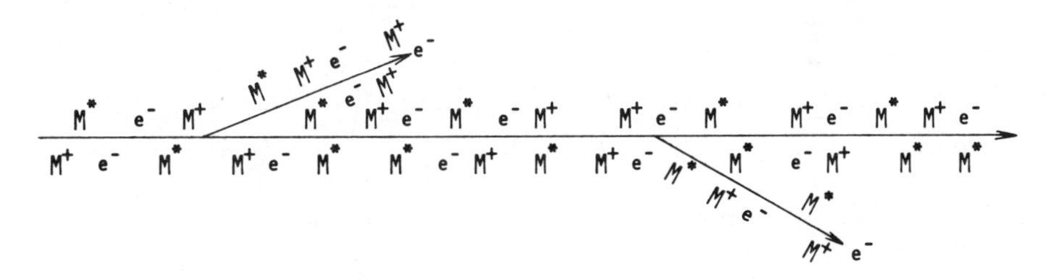

Figure 1. Ionized and excited molecules along the path of a gamma ray

* *Cobalt-60 produces two gamma rays of 1.17 and 1.33 million electron volts (MeV). Up to 30 eV are required to rupture covalent bonds and to cause ionization.*

The initiation step can be represented in general by

$$R^{\cdot} + M \text{ (monomer)} \longrightarrow R\text{-}M^{\cdot}$$

and the propagation step by

$$R\text{-}(M)_n\text{-}M^{\cdot} + M \longrightarrow R\text{-}(M)_{n+1}\text{-}M^{\cdot}$$

and the termination step by the recombination of growing radicals

$$R\text{-}(M)_n\text{-}M^{\cdot} + R\text{-}(M)_n\text{-}M^{\cdot} \longrightarrow R\text{-}(M)_n\text{-}M\text{-}M\text{-}(M)_n\text{-}R$$

Since the cell wall structure of the wood is not swollen by the vinyl monomer, there is little opportunity for the monomer to reach the free radical sites generated by the gamma radiation on the cellulose to form a vinyl polymer branch. From this short discussion, it is reasonable to conjecture that there should be little if any difference in the physical properties of catalyst-heat initiated or gamma radiation initiated in situ polymerization of vinyl monomers in wood.

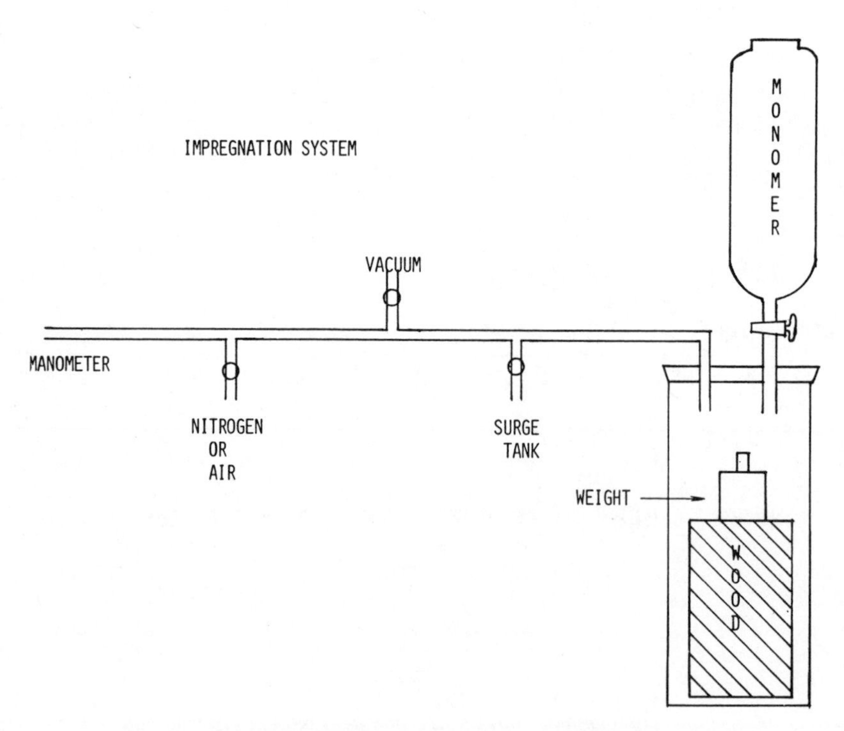

Figure 2. Apparatus used to impregnate wood

Impregnation Process

In both processes the impregnation of wood is carried out by first evacuating the air from the wood vessels and cell lumens(8). Figure 2 represents the components necessary for making wood-polymers on a laboratory as well as an industrial scale. Any type of mechanical vacuum pump is adequate, if it can reduce the pressure in the apparatus to 1 millimeter of mercury or less. Some industrial producers only reduce the pressure to about 28 inches of mercury. Experience has shown that the air in the cellular structure of most woods is removed as fast as the pressure in the evacuation vessel is reduced. A half hour pumping at 1 mm pressure is sufficient to remove the air. The vacuum pump is isolated from the system at this point. The catalyzed monomer containing crosslinkers, and on occasion dyes, is introduced into the evacuated chamber through a reservoir at atmospheric pressure. The wood must be weighted so that it does not float in the monomer solution. In the radiation process, the catalyst is omitted from the monomer. A surge tank ten times the volume of the treating vessel is included in the system to allow the air dissolved in the monomer to expand without greatly changing the pressure in the system. An alternate to the surge tank is to pump on the system as the monomer is admitted into the evacuated vessel. With this procedure much monomer is lost due to the high vapor pressure of MMA, which is 40 mm at room temperature. After the wood is covered with the monomer solution atmospheric air pressure is admitted, or dry nitrogen in the case of the radiation process. Immediately the monomer solution begins to flow into the evacuated wood structure to fill the void spaces. Care must be taken to maintain enough monomer solution above the wood so that air is not readmitted to the cell structure. The soaking period like the evacuation period depends upon the structure of the wood: maple, birch and other open celled woods fill in about a half hour, while other woods require longer periods of time. A block of 3 x 3 x 7 inch maple absorbed 300 ml of monomer in less than 10 minutes.

After the monomer impregnation is complete the wood-monomer is removed and placed in an explosion proof oven, or the Cobalt-60 source for curing. On a laboratory scale or small production unit the wood-monomer is wrapped in aluminum foil before placing in the curing oven at 60°C. In larger production operations the wood-monomer is placed directly into the curing oven, usually in the basket which held the wood during impregnation. In the radiation cured procedure the thin metal can, in which the wood was impregnated, is flushed with nitrogen and is lowered into a water pool next to the Cobalt-60 source. With high vapor pressure monomers, the wood surface is depleted to some extent by evaporation, but this depleted area is usually removed by machining to expose the wood-polymer surface. As already mentioned, MMA has a vapor pressure of 40 mm at room temperature; on the other hand t-butyl styrene has a vapor pressure of only 0.8 mm at room temperature.

Monomers For Wood-Polymer Composites

Many different vinyl monomers (9) have been used to make wood-polymers during the past ten years, but methyl methacrylate (MMA) appears to be the preferred monomer for both the catalyst-heat and radiation processes. In fact, MMA is the only monomer that can be economically polymerized with gamma radiation. On the other hand, all types of liquid vinyl monomers can be polymerized with Vazo or peroxide catalysts. In many countries styrene and styrene-MMA mixtures are used with the Vazo or peroxide catalysts.

All vinyl monomers contain inhibitors to prevent premature polymerization during transportation and storage. If these inhibitors, such as hydroquinone (HQ), monomethyl ether of hydroquinone (MEHQ), t-butyl catechol(TBC) and 2,4-dimethyl-6-t-butyl phenol (DMTBP) are not removed before use, the catalyst or radiation must generate enough free radicals to use up the inhibitor before polymerization will begin. This induction period depends on the amount and type of inhibitor present in the monomer. In the case of radiation, the inhibitor must be kept at a minimum for efficient use of the gamma rays. The production of commercial polymethyl methacrylate rod or sheet stock, sold as Lucite or Plexiglas, requires about 0.01% Vazo catalyst with the inhibitor removed. With 50 parts per million of DMTBP in MMA in wood, 0.25% Vazo catalyst is required to obtain complete polymerization. Wood contains natural inhibitors which is the reason for the high Vazo content. Again, the amount of natural inhibition will depend on the species of wood. Monomers extract the soluble fractions from the wood structure, and with repeated use, the extractives content builds up in the monomer to where excessive foaming is produced under vacuum and the polymerization reaction is completely inhibited. Additional catalyst must be added.

The polymerization of vinyl monomers is an exothermic reaction and a considerable amount of heat is released, about 18 kCal per mole. In both the catalyst-heat and gamma radiation processes the heat released during polymerization is the same for a given amount of monomer. The rate at which the heat is released is controlled by the rate at which the free radical initiating species is supplied and the rate at which the chains are growing. As pointed out above, the Vazo and peroxides are temperature dependent and the rate of decomposition, and thus the supply of free radicals, increases rapidly with an increase in temperature. Since wood is an insulator due to its cellular structure, heat flow into and out of the wood-monomer-polymer material is restricted. In the case of the catalyst-heat process heat must be introduced into the wood-monomer to start the polymerization, but once the exothermic reaction begins the heat flow is reversed. The temperature of the wood-monomer-polymer composite increases rapidly, because the heat flow out of the wood is much slower than the rate at which the heat is generated. Figure 3 illustrates the heat transfer process (10).

The time to t_0 is the time for the wood-monomer mass to reach oven or curing temperature at T_b. During the period of constant temperature, the induction period, the inhibitor is being removed by reaction with the free radicals. Once the inhibitor is eliminated from the monomer and wood, the temperature rises to a maximum which corresponds to the peak of the exothermic polymerization reaction. Polymerization continues to completion although at a decreased rate and the temperature returns to that of the curing chamber. The time to the peak temperature depends upon the amount of catalyst present, the type of monomer, the type of crosslinker, and the ratio of the mass of monomer to that of the wood. The wood mass acts as a heat sink. Figure 4 illustrates the effect of increased Vazo catalyst on the decrease in time to the peak temperature, and the increase in the peak temperature(10).

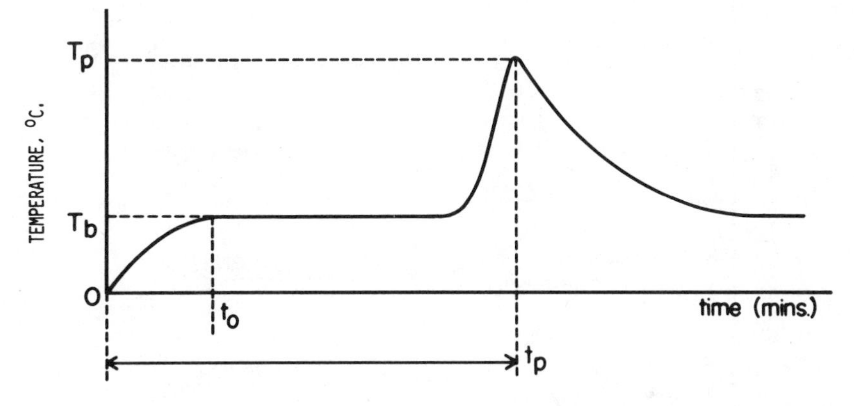

Figure 3. Idealized temperature–time exothermic curve

The addition of a crosslinker (1-5%) to a vinyl monomer will increase the peak exothermic temperature and decrease the time of polymerization. At the beginning of the wood-monomer polymerization, the initial heat of the curing oven and the polymerization reaction expands the volume of the monomer so that it bleeds out of the ends of the wood where it evaporates or polymerizes into a foam. This decreases the polymer loading in the wood and wastes monomer. By the addition of divinyl monomer, such as ethylene glycol dimethacrylate (EGDMA) which contains two double bonds, a gel forms initially and prevents the monomer from expanding out of the wood. Crosslinking also increases the molecular weight to the point where the now thermosetting polymer will decompose before melting. The non-melting characteristics of the wood-polymer composite is important in the machining and sanding of the final product.

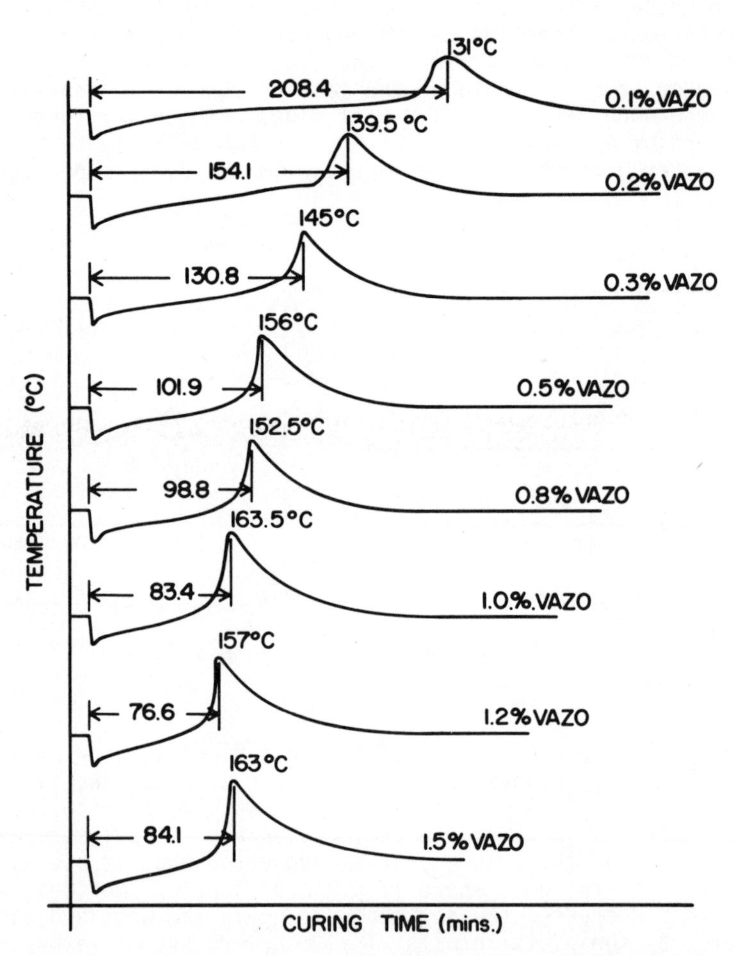

Figure 4. Temperature–time curves. Effects of varying concentrations of Vazo catalyst on the polymerization exotherm of basswood–MMA composite.

The gel or Trommsdorff effect (11) is the striking auto-acceleration of the vinyl polymerization reaction as the viscosity of the monomer-polymer solution increases. Chain termination involving the recombination of two free radicals becomes diffusion controlled and this results in a decrease in the rate of termination. The concentration of active free radicals therefore increases proportionally. To sum up the gel effect: the rate of Vazo catalyst initiation increases with temperature; the rate of propagation or polymerization increases with the viscosity; and the rate of termination of the growing polymer chains decreases with the viscosity. This of course also results in an increase in the molecular weight of linear polymers, but this has no practical significance when crosslinking is part of the reaction.

As mentioned before, a given gamma radiation source geometry will supply free radicals at a constant rate for vinyl monomer polymerization. An increase in temperature would only affect the propagation and termination rates. The exothermic heat from the vinyl monomer polymerization is still released in the wood-monomer composite, but the temperature is much lower because of the slow rate of initiation. Complete radiation curing of wood-monomer composite usually requires 8 to 10 hours depending upon the radiation source geometry; the Vazo initiated reaction is over in 30 to 40 minutes. In other words, all the catalytic heat of polymerization of a given monomer mass is released in 1/16 of the time it is in the radiation process. Since the wood-monomer material in a thin metal can is immersed into a water pool for irradiation the cooling by the water radiation shield also assists in lowering the temperature. Additional heat is added to the wood-monomer polymer composite by the absorption of the gamma rays by the wood although this is small compared to the exothermic heat from the polymerization.

When the heat of polymerization is released quickly in a wood-monomer composite the high temperature increases the vapor pressure of the moisture in the cell walls and drives it out of the wood. The change in volume of the cell wall causes changes in dimensions which is manifested by shrinkage and distortion of the original shape. Wood-polymer composites cured by the catalyst-heat process must be machined to the final shape after treatment. On the other hand, since the heat of polymerization by gamma rays is released over a longer period of time, the temperature of the wood-polymer remains low and not as much cell wall moisture is driven off. Therefore, the amount of distortion and dimensional change is somewhat less.

Soluble dyes can be added to the monomer solution to color the final wood-polymer composite. Any color of the visible spectrum can be added, browns to simulate black walnut, red and blues for the bicentennial theme. The color emphasizes the grain structure of the particular species and combines with the polymer to add a three demensional depth not present in surface finished wood. A dense black wood-polymer is difficult to obtain because of the wood's light color and the tendency of the micro structure of wood to chromotographically separate a dye of several components into its separate colors. Dyes have an inhibiting effect on the polymerization of wood-monomer composites, some more so than others. Additional catalyst can be added to overcome this inhibition but in the radiation process of a given geometry additional time must be allowed for complete curing.

Some research has been done on the addition of polar solvents to the nonpolar monomer in an attempt to swell the cell wall structure and anchor it in a swollen state (9). This can be done and the antishrink efficiency (ASE) does increase, but after the solvent evaporates, the wood is only partially loaded which in turn decreases the physical properties. Wood-polymer composites normally have about 10-15% ASE, which means that there is some penetration of the cell wall structure to reduce the swelling over that of untreated wood.

The mechanism of water absorption by dry wood proceeds in two steps. Water entering dry wood in vapor form is absorbed into the cell wall and hydrogen bonds to the cellulose. As a result, the cell wall swells, and the overall dimensions of the wood increase. After 25-28% is absorbed (based on the oven dry weight of wood) and the cell wall has swollen to its maximum, additional water will be condensed in the capillaries or other void spaces in the wood until it is filled. The fiber saturation is the point where the cell walls have absorbed the maximum amount of water and are swollen to the maximum extent, but no water has condensed in the capillaries. This fiber saturation point is surprisingly consistent, 26 to 30% considering the large number of species of wood on this planet. As pointed out before, normal wood-polymer material contains polymer only in the void spaces that are available and little if any in the cell walls. This loading of the capillary vessels reduces the rate of of water vapor diffusing into the cell walls. But, given enough time (a factor of 10 to 20), at high humidity, water will eventually reach the cell walls and cause substantially the same volume swelling as untreated wood. Figures 5 and 6 show the differences in water absorption by basswood-polymer composites (9).

Timmons (12) showed that the water in a never dried freshcut tree, when exchanged with a series of organic solvents could be replaced with radioactively tritium labeled MMA (hydrogen-3). After polymerization and the use of autoradiography the MMA was located in the cell wall structure. If the same wood was dried normally, and then treated in the usual manner with tritium labeled MMA, the polymer was located exclusively in the capillaries and none was found in the cell walls.

The flow into wood, especially liquids, is along the grain. Siau made (13) air permeability measurements and found the following.

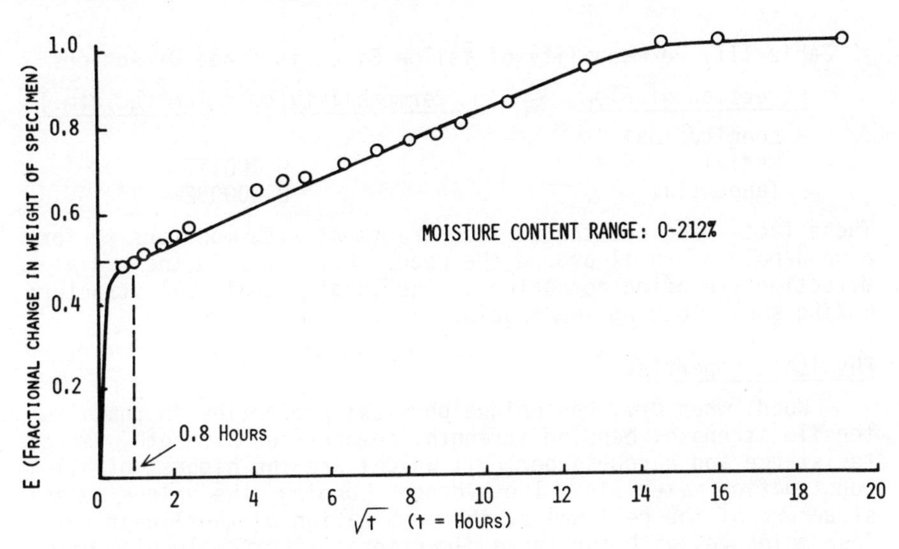

Figure 5. Fraction of total change with time for untreated basswood

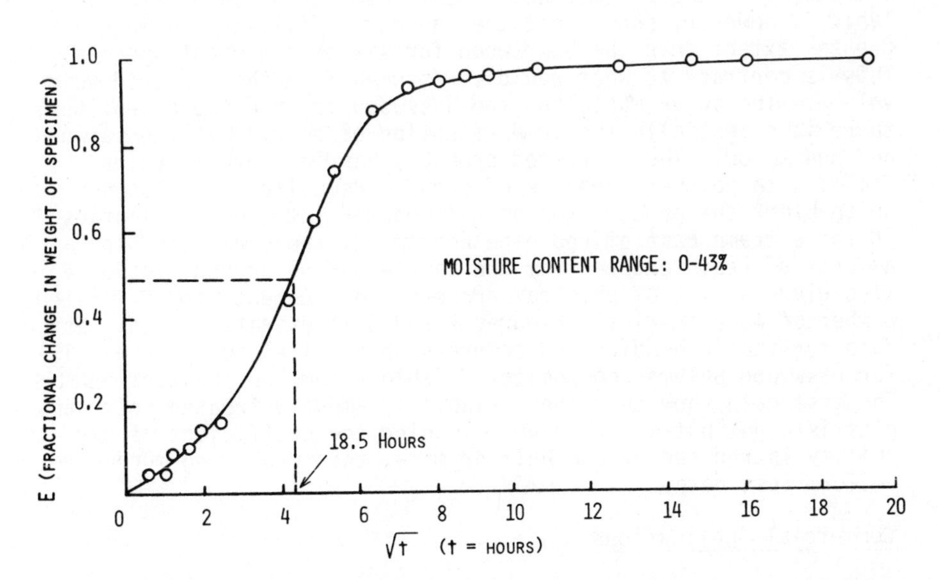

Figure 6. Fraction of total weight change with time for basswood treated with t-butyl styrene

Table III. Permeability of Yellow Birch in Three Directions

Direction of Flow	Permeability($cm^3 cm/cm^2 sec.atm.$)
Longitudinal	696
Radial	0.000177
Tangential	0.000092

These facts preclude the surface treatment with monomers to form a wood-polymer shell around the wood. Ray cells in the radial direction are often connected to the longitudinal cell structure making shell loading improbable.

Physical Properties

Wood, when dry, has unique physical properties in that its tensile strength, bending strength, compression strength, impact resistance and hardness per unit weight are the highest of all construction materials. The hydrogen bonding, the unique helical structure of the cell walls, the combination of the linear cellulose molecules with the three dimensional lignin molecules impregnated with low molecular weight extractives makes wood an infinitely variable resource. All the unusual features of wood are the reason for the "ART" of wood treatment.

The polymer loading of wood depends not only upon the permeability of wood species, but also on the particular piece of wood being treated (14). Since the void volume is approximately the same for the sapwood and heartwood for each species, it would be expected that the polymer would fill them to the same extent. Table IV however, shows that the sapwood is filled to a much greater extent than the heartwood for six of the eight species. This is contrary to what would be assumed from the measured void volume. The sugar maple and the basswood are the two exceptions; there is essentially the same retention of polymer in the sapwood and heartwood. The heartwood probably has less of the voids filled with polymer, because of organic deposits and tyloses, which block the penetration of the monomer into the capillaries. In the extreme case of red pine heartwood, there was visible amounts of resin exuded from the sample during drying. Table IV also gives a list of physical property enhancements for a limited number of wood species. Figures 7 and 8 illustrate typical test data for static bending and compression parallel to the grain(15) for basswood-polymer composites. Table V sums up the test results. The test data show that the variability among untreated test samples is high, but after polymer loading the coefficient of variability is reduced by one half or more, thus producing much more uniform test data.

Commercial Applications

Radiation Process. Commercial production of wood-polymer composites began in the mid 1960's using the radiation process.

TABLE IV

Species		Void Volume %	Voids Filled %	Polymer in WPC %	Density Increase %	Compression Strength Increase %	Tangential Hardness Increase %	Permeability Ratio Untreated/Treated
Acer rubrum	S	64	65	46	82	171	280	270
Red Maple	H	63	56	71	67	82	209	40
Acer saccharum Marsh.	S	60	61	40	65	160	229	225
Sugar Maple	H	58	60	38	58	125	200	186
Prunus serotina Ehrh.	S	64	63	45	78	202	289	717
Black Cherry	H	63	46	37	56	86	124	428
Tilla americana L.	S	80	61	63	168	435	626	107
Basswood	H	77	66	62	160	288	505	104
Betula lutea Michx. F.	S	55	67	37	58	146	215	1047
Yellow Birch	H	52	60	31	43	56	120	1896
Liquidambar styracifulua	S	68	58	48	88	175	243	2884
Red Gum	H	65	35	33	46	33	95	110
Pinus resinosa Ait.	S	68	65	51	100	636	523	1395
Red Pine	H	68	6	8	7	1	1	14
Fagus grandifolia Ehrh.	S	59	53	36	53	201	261	213
Beech	H	55	34	24	30	30	112	19

Sapwood = S
Heartwood = H

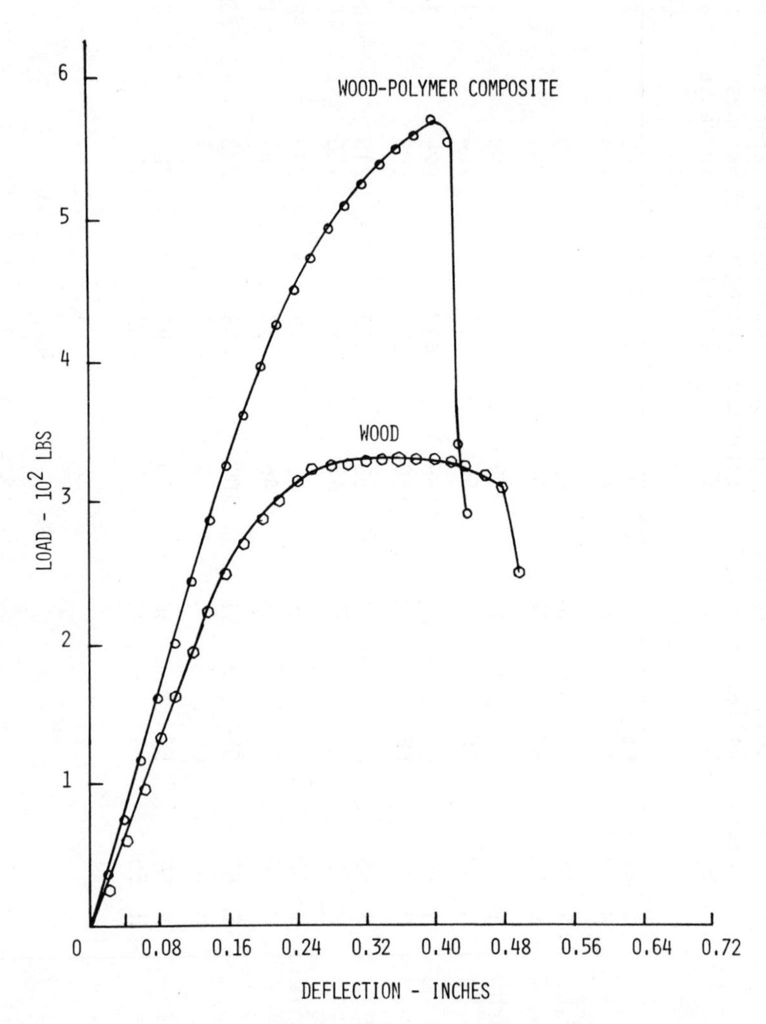

Figure 7. Example of typical bending test data

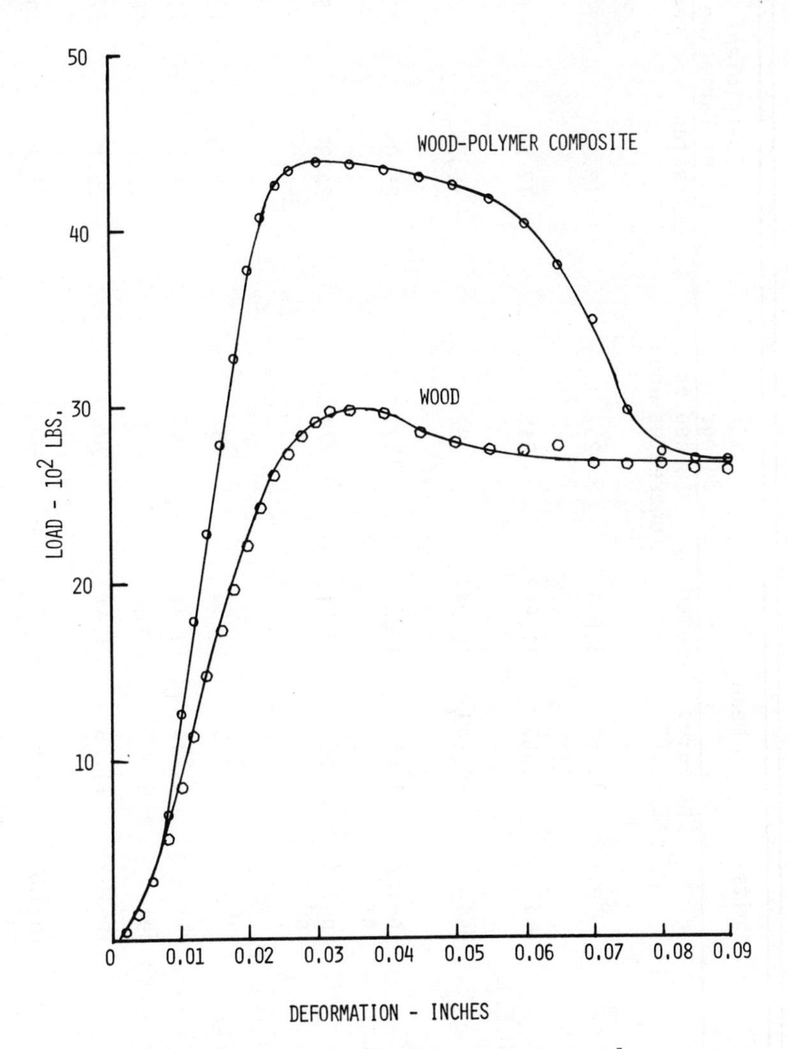

Figure 8. Example of typical compression test data

TABLE V. - Summary of Test Results For Treated and Untreated Basswood

Strength Property	Units	Mean		% Change (Based on untreated wood)	Coefficient of Variation	
		Untreated MC -7.2%	Treated		Untreated	Treated
Static Bending						
Modulus of Elasticity	10^6psi	1.356	1.691	25%	13.3%	10.4%
Fiber Stress at Proport. Limit(P.L.)	psi	6,387	11,587	81%	17.8%	6.3%
Modulus of Rupture	psi	10,649	18,944	78%	15.7%	5.4%
Work to Proportional Limit(P.L.)	in-lb/in^3	1.66	4.22	154%	25.0%	11.6%
Work to Maximum Load	in-lb/in^3	10.66	17.81	77%	24.3%	5.3%
Compression Parallel to Grain						
Mod. of Elasticity	10^6psi	1.113	1.650	48%	25.2%	13.8%
Fiber Stress at P.L.	psi	4,295	7,543	76%	22.5%	11.6%
Max. Crushing Strength	psi	6,505	9,864	51%	18.9%	14.1%
Work to P.L.	in-lb/in^3	11.28	21.41	90%	25.1%	17.5%
Toughness	in-lb/in^3	41.8	62.6	50%	17.2%	9.5%

Parquet flooring was the primary product and the increased hardness and abrasion resistance were used to advantage in high traffic commercial installations. The long life and ease of maintenance justified the increased cost over normal flooring. Of the several original companies who produced irradiated wood-plastic composites only one is a serious high volume producer today. The ARCO Chemical Company, a Division of Atlantic Richfield at Karthaus, Pa. started production in 1968 using the facilities built at this location by a former nuclear reactor company (16). Their production has increased from 1600 to 400,000 square feet of flooring per month. Over the last five years over 8,000,000 square feet of wood-acrylic flooring was installed in the United States and most of this was produced by ARCO under the trade name of PermaGrain (17). The original terminology wood-polymer, wood-plastic composites, and WPC has given way to acrylic-wood for commercial products.

Red Oak is the wood most used for the production of irradiated acrylic-wood parquet flooring, however acrylic Aspen chip board and acrylic particle board are finding their way into the flooring market. Smaller amounts of maple, ash and black walnut have been used. The polymer loading of red oak is about 40% which makes the acrylic-wood flooring quite expensive, approximately three times that of ordinary red oak parquet flooring. The enhanced beauty of acrylic-wood is one of its primary selling points along with its greater hardness and abrasion resistance. Maintenance consists of wet mopping and machine buffing to restore the original luster (ARCO supplies a special maintenance kit). Cigarette burns can be removed by simply steel wooling away the charred wood, and buffing, since the finish is throughout the thickness of the acrylic-wood. An untreated red oak floor must be stripped of its old finish and a new finish applied every year or even less in high traffic areas. When the total cost of installation and yearly maintenance over the life of the floor is considered, the cost of acrylic-wood flooring is one of the least expensive. The data in Table VI was compiled by ARCO in 1972. It is estimated that for red oak parquet flooring the cost of the wood is 50%, the cost of the MMA monomer, dyes, crosslinkers and fire retardants is 25%, and the labor and overhead is 25% (17).

ARCO also produces a smaller volume of acrylic-wood for other uses. Eight foot long sections of hand rails are produced using maple and ash. Again the superior abrasion resistance and hardness are important as well as the look and feel of the finished hand rail. During the past year more than a half million knife handles of acrylic black walnut have been produced for a cutlery company whose high quality products are sold through jewelry stores. In this case the esthetic properties of looks, feel and weight are the key to increased sales. Other small quantities of acrylic-wood for archery bows, stair treads, window sills and golf clubs have been produced. It must be under-

TABLE VI. Relative High Traffic Flooring Costs (¢/ft^2)

	Average Installed Cost	Refinishing Required		General Maintenance	Wear Resistance *	Life (Years)	Yearly[1] Cost
		Cost	Frequency Per Year				
PermaGrain	150	0		20	5 Mils	50	23
Terrazzo	150	7	4	22	SP	20	37
Nylon Carpet	150	2	2	25	SP	7	48
Synthetic Gym Floor	325	2	2	25	SP	20	44
Wood	60	25	5	25	SP	30	52
PVC Tile	90	7	4	22	SP	15	35
Tile	45	7	4	22	SP	12	33
Quarry Tile	225	7	4	20	SP	25	36

[1] Yearly Cost = $\frac{\text{Installed Cost}}{\text{Life}}$ + Refinishing Costs + General Maintenance

* Taber Abrader - 4,000 Cycles
SP - Surface Penetration - 1,000 Cycles

stood that most production information is proprietory and not made available for public distribution.

The ARCO process follows the procedure outlined earlier in this paper (16). The cannisters used for impregnation and irradiation of the wood-monomer composite are 1 x 2 x 13 feet in size and hold about 450 square feet of flooring. After the metal cannisters are loaded with wood they are inserted into a special treating tank for MMA monomer impregnation. Caution must be observed so that the thin metal wall of the cannister is not collapsed or blown out. After the monomer treatment, draining of the excess MMA and flushing with dry nitrogen, the cannisters are lowered into the 25 feet deep water pool next to the Co-60 source which contains about 800,000 curies. The cannister is programmed to proceed along one side of the Co-60 source and back along the other side so that the radiation enters both sides of the cannister during the 8 hours of curing time. After receiving from 1.0 to 2.0 megarads of radiation the cure is complete. The cannister is removed from the water pool and the acrylic-wood is processed into its final shape by machining, sanding and buffing. The acrylic-wood flooring is shipped from inventory to the installation site where it is installed and buffed to its final finish. The acrylic-wood floor can be used immediately after installation. Since acrylic-wood flooring production is only a small percentage of the total, it is expected that a larger share of the commercial flooring market will be acrylic-wood in the future.

Catalyst-heat Process. At the present time production using the catalyst-heat process for making wood-polymer composites is much smaller than the radiation process, but its use is much more widespread throughout the United States and the world. The simplicity of the catalyst-heat process and low initial cost to begin production is the key to the use by small companies who make high cost small volume items.

The first small production system was assembled by the Bowling Division of AMF in 1966 when the first wood-plastic billiard cues were produced. The initial impregnation unit cost two hundred dollars and the dry kiln associated with the saw mill was used as a heat source for curing the wood-monomer, which was enclosed in a capped pipe. After several successive size increases the production system consisted of a ten inch pipe six feet long with end caps as the impregnation tank and a similar pipe with an insulated copper steam coil wound along its length was used as a curing oven. The catalyzed monomer was stored in five gallon cans in a refrigerator at 34°F. The monomer solution was transfered to the impregnation chamber by hand pouring. A high volume exhaust fan at floor level removed the monomer vapors. The total volume of catalyzed monomer for each treatment was 35 gallons. Since each batch of fifty cues absorbed five gallons of catalyzed monomer, five gallons of new monomer solution (0.25% Vazo and 5%

crosslinker) was prepared for each run. By adding new catalyzed monomer to the thirty gallons from the previous run the proper inhibitor level was maintained and the level of extractives was kept at a low level. With this system the curing time was about four hours for each batch and three to four runs each day gave the required level of production. This level of production was maintained for several years until the market became saturated, and AMF discontinued their billiard line. The cost of making the wood-plastic cue at that time was fifty cents and this cost was recoverable by increasing the price of the cue by five dollars. Total cost of the treating system was less than two thousand dollars. In addition to the billiard cues, AMF Bowling Division produced hard maple wood-plastic bowling alley flooring for numerous test facilities, but the bowling alley boom died out before commercial installations were built. Other items were produced in limited quantities, such as, lacrosse sticks, dynamite tamping sticks, police night sticks and chair arms. At the present time the treating plant is in storage.

In 1968 Bear Archery installed a small production plant for making archery bows and in 1971 the Wing Archery Division of AMF began production. These plants use vertical autoclave type impregnation cylinders about four feet high and three feet in diameter with quick locking and release lids. The wood to be impregnated is stacked into cylindrical baskets which fit into the impregnation cylinder. The wood-monomer composite remains in the same basket during the curing cycle. Bear Archery uses a large room-like heated chamber with circulating air where many baskets of wood-monomer can be cured simultaneously. Wing Archery has a hot water heated radiator type of curing oven with circulating air for each basket of wood-monomer composite. In each case the catalyzed monomer is recycled after the addition of new monomer solution. Both companies use dyes dissolved in the catalyzed monomer solution to produce various colors. Combinations of colors are laminated together to give a variety of colors in the final archery bow.

There are several reasons for using wood-plastic composites for archery bows. By using domestic hardwoods, such as hard maple, birch and occasionally oak, hickory and ash, a reliable source of supply is at hand. Tropical woods must be imported from all over the world and the quality is often less than satisfactory. In many cases more than half of the tropical wood has to be rejected. By dying the hard or sugar maple in the process of making wood-plastics the color quality can be controlled and a great variety of colors can be produced which far exceeds the colors available in tropical hardwoods. The finishing of tropical hardwood bows is demanding in that many coats of sealer have to be applied to the open grain after which half a dozen coats of lacquer are applied. The hard maple wood-plastic only requires one coat of sealer and one coat of lacquer to seal the surface against moisture. The savings in finishing alone pays for the cost of making the wood-plastic composite. Production at Bear Archery has exceeded 1000 bows a day while at Wing Archery production has been as high as 350 bows per day. According to Wing Archery management the original cost of their installation was about four thousand dollars. This cost included a refrigerated storage room for the catalyzed monomer which is transfered to the impregnation vessel by vacuum and returned to storage by gravity.

Several other archery bow manufacturers have installed catalyst-heat systems for making wood-plastics but little is known about their production. A number of pilot plant operations have been installed in the United States but little of this proprietory information is available. Pilot plant quantities of cutlery handles have been produced for market studies, a laminated type of flooring was produced using a styrene-polyester composite cured in a hot press, parquet flooring using the catalyst-heat system is presently being produced for market studies. Many other small volume items, such as, jewelry, bird calls, transformer cores, etc. are finding their way into the market place.

Dunbar of Ontario, Canada and the Wee Piper of Vermont are producing bagpipes and chanters from catalyst-heat wood-plastics supplied by Wing Archery. The wood-plastic composite has replaced the black African wood usually used for bagpipes. Pipers claim that the tone of the new wood-plastic chanter is more mellow. Traditionally, bagpipe parts of ebony wood are black and this intense black has not been duplicated in wood-plastics. Generally, wood-plastic bagpipe parts are made in the natural maple color or dyed a deep brown.

Kelly Putters of Oregon produces a variety of putters made from catalyst-heat wood-plastics using myrtle wood and broad leaf maple which is native to the state. The light and dark wavy grains of these two woods are enhanced by the plastic treatment and have a three dimensional depth in their natural color. Several other golf club manufacturers have made small quantities of wood-plastic laminated birch and solid persimmon drivers for market studies. The dimensional stability of the wood-plastic composite is important during the manufacturing when the various face and base plates are inserted.

The exact number is not known, but many high school industrial arts departments have installed the catalyst-heat system for making wood-plastics. The students produce the composite material and then fabricate it into various items during their shop classes.

World Wide Production

Some sketchy information is available on the production of wood-plastics in countries around the world. During the 1960's support for making wood-plastics by the radiation process was available from the various Atomic Energy Agencies. This support resulted in a great amount of research in Canada, Finland, Sweden,

Japan, and South America, and commercial amounts of wood-plastic flooring were produced for airport terminals and office buildings. Today there is no known commercial gamma radiation production of wood-plastics outside of the United States.

The catalyst-heat system for making wood-plastics is used on a commercial scale in Japan, Germany and Italy. Few details are available on the German production of shuttle cocks for the textile industry. In Italy a plant in the Bologna area is producing about three cubic meters a day using styrene monomer. This polystyrene wood-plastic is used for buttons, desk sets and other high cost low volume items. The same organization is planning a future plant in Spain. The Japanese have been very active in research and the production of wood-plastics using the catalyst-heat system with MMA and styrene mixtures. Applications research at the Government Forest Experiment Station in Tokyo has been carried through to production by the Iwaso Company, Ltd. in Ishikawa, Japan. Iwaso Ltd. produces a range of wood-plastic colors for industry including the Pilot Pen Co. Pilot pen and pencil sets with solid dyed wood-plastic bodies are available in U.S.A. at most stationery stores. Iwaso Ltd. produces wood-plastics for vases, bowls, desk caddies, unusual paper weights, letter holders and the ancient abacus.

Fifteen years ago, when wood-plastic composites were first introduced many people predicted that this process would solve the problem of wood dimensional stability and great claims were made for its future use. Now that the physical properties of wood-polymer composites are better understood, specific commercial products are being produced which take advantage of the desirable aesthetic appearance, the high compression strength, increased hardness and abrasion resistance and improved dimensional stability. Future use of wood-polymer composites will depend upon the imagination of the producer and the market place.

Literature Cited

1. Meyer, J.A., Loos, W.E. Forest Products Journal (1972)19(12)
2. Siau, J.F., Meyer, J.A. and Skaar, C. Forest Products Journal (1965) 15 (4) 162-166
3. International Atomic Energy Agency "Impregnated Fibrous Materials, Report of a Study Group Bangkok, Thailand" Vienna (1968) STI/PUB/209
4. Ellwood, L., Gilmore, R., Merrill, J.A. and Poole, W.K. "An Investigation of Certain Physical and Mechanical Properties of Wood-Plastic Combinations" Report No. ORO-638 (RTI-2513-T13) Division of Isotope Development, U.S. Atomic Energy Commission Contract AT(40-1)-25-13, Task 20
5. Ellwood, L., Gilmore, R.C., Stamm, A.J. Wood Science (1972) 4 (3) 137-141
6. Riddle, E. H."Monomeric Acrylic Esters" p.29 Reinhold Publishing Co. 1954

7. DuPont Product Bulletin "Dupont Vazo", Industrial and Bio-chemicals Dept. Wilmington, Deleware 19898
8. Meyer, J.S. Forest Products Journal (1965) 15 (9) 362-364
9. Langwig, J.E., Meyer, J.A. and Davidson, R.W. Forest Products Journal (1969) 19 (11) 57-61
10. Duran, J.A. and Meyer, J.A. Wood Science and Technology (1972) 6, 59-66
11. Trommsdorff, E., Kohle, H. and Lagally, P. "Zur Polymerisation des Methatcrylatsaure Methylesters" Colloquium on high polymers Makromol Chem. 1 169-198 (1948)
12. Timmons, T.K., Meyer, J.A. and Cote, Jr., W.A. Wood Science (1971) 4 (1) 13-24
13. Siau, J.F. and Meyer, J.A. Forest Products Journal (1966) 16 (8) 47-56
14. Young, R.A. and Meyer, J.A. Forest Products Journal (1968) 18 (4) 66-68
15. Langwig, J.E., Meyer, J.A. and Davidson, R.W. Forest Products Journal (1968) 18 (7) 33-36
16. Witt, A.E. and Morrissey, J.E. Modern Plastics (1972) 49(1) 78-82
17. Witt, A.E. "Applications in Wood Plastics" Paper presented at First International Conference of Radiation Processing, Dorado Deach, Puerto Rico, May 1976

21

Wood Softening and Forming with Ammonia

M. BARISKA

Institut fur Microtechnologische Holzforschung, Eidgenossische Technische Hochschule, CH 8006 Zurich, Switzerland

C. SCHUERCH

Department of Chemistry, S.U.N.Y. College of Environmental Science and Forestry, Syracuse, N.Y. 13210

A number of years ago, anhydrous liquid ammonia was found to be a powerful plasticizing agent for wood, which softened wood strips so that they could be formed readily into dramatic and complex shapes and caused them to develop permanent set in the new form (1). About the same time, the plasticizing action of aqueous ammonia was under investigation in Latvia (2), and although the effects were much less extreme, the softening was sufficient to serve as a basis for extensive scientific and technological work that has led to some industrial developments (3-18). The use of gaseous ammonia also proved successful when the accelerating effect of moderate moisture content on absorption was recognized (19), and the versatility of the method can now probably be greatly extended by application of the results of investigations in the low pressure range (20). The complete definition of the system wood-water-ammonia as a function of temperature, pressure and composition is still far from complete, but general trends are apparent enough to be summarized at this time. Both the processing and final properties of wood are markedly affected by the conditions of the treatment, and ammonia forming of wood is thus, in a sense, a spectrum of processes from which one can be selected to produce minor softening in the cold or, at the other extreme, one that causes greatly enhanced flexibility and stable final permanent set.

During a period of concern for dwindling natural resources, a serious scientific and technological investigation of methods of wood forming could be of considerable social value. Wood has many advantages as a naturally renewable material of high strength to weight ratio, but it also has some properties which limit its utility and lead to waste. It has variable physical and mechanical properties within a single sample due to its complex microscopic structure. It is not dimensionally stable to moisture changes. Although it can be easily worked by tools and machines, it cannot be easily molded to a complex shape. As a result of the last, very large losses of material accrue during fabrication from tree to products with complicated forms.

Frequently excessive material is required in the product itself to allow for the weakness of crossgrained sections and joints. These disadvantages all have pertinence to the softening and forming of wood. To the extent that wood can be softened and bent, thinner stock may be used because joints and crossgrained sections can be avoided and design restrictions can be minimized. However, wood's natural rigidity and variable properties now limit its forming to rather high grade stock, and some methods -- including the ammonia treatment -- alter somewhat the physical and mechanical characteristics.

A number of hydrogen-bonding solvents, other than ammonia, have also been investigated for wood softening. These include amines (21-25), formamide (21), dimethyl sulfoxide (26), phenol and urea (27). These differ in the degree of flexibility produced in the wood and the ease with which they are removed. All are more expensive than ammonia and most are more difficult to remove.

Since the purpose of this report is primarily to interpret the behavior of the wood-ammonia system and to relate it to practical application, we will first briefly review current methods of forming wood. We will discuss in some detail methods of softening wood with water and heat and the molecular changes underlying them and then extend these concepts to explain the more complicated but similar ammonia-based systems, and their possible practical implications.

Alternative Forming Methods. Wood can be formed into complex shapes with least chemical or molecular change by cutting thin slices, assembling them with the grain parallel or perpendicular and gluing them in place over the desired form. These processes of lamination and plywood moulding depend on the fact that stiffness of any lamella-shaped material varies as the third power of its thickness, so flexibility is greatly increased as the thickness is decreased. The separation of individual slices by a glue line inhibits flaw propagation, and the crossed orientation of plywood also enhances dimensional stability to moisture. Reasonably complex curves can thus be produced in furniture manufacture, and the inherent expense of these methods is offset by using lower quality veneers for the interior layers.

Wood can also be softened for forming by plasticization with water. Wood shows colloid character: it is often defined as a gel, predominantly a matrix of microfibrils surrounded by a fluid medium, hydrate water. Wood substance is generated by the living cell in a water-saturated milieu and, therefore, has by nature a certain flexibility, which is altered even by drying and remoistening. If retained in a never-dried state, thin sections are extremely supple and can be bent and woven readily in the cold. Once dried, wood can be resoftened by raising its water content especially with increase in temperature. For larger dimensional stock, soaking in hot water or preferably steam treatment is virtually required (27). Most simple wood bending is done by treating wood with saturated steam. Usually, the steaming of wood is interrupted before wood reaches saturation or is highly plasticized. Normally, wood will be softened sufficiently for forming by steaming one minute per mm of thickness. Then the softened "work piece" is placed on a form and bent rapidly by hand or machine. The bent piece is restrained in its new form and put aside to set.

Steaming affects the compression strength of the wood to a greater extent than the tension strength. During the bending procedure one tries to take advantage of this fact. For bending, a flexible metal strap with two blocks attached to its ends is placed on the side of the wood which is to be convex. On bending, with the blocks tightly in place on the ends of the wood, the entire wood member is placed in compression while the steel strap is under tension. Under these conditions a minimum radius of curvature ratio can be obtained. On a microscopic level, buckling of the plasticized cell walls occurs towards the lumena under compression. At the weakest points layers and microfibrils of the cell walls are partially separated, with partial destruction of the cell wall, to form so-called slip lines and slip planes. As a result, the ultimate compressive strength of the wood decreases while the tension strength normally is not affected. Thereafter, the formed wood retains a higher flexibility, but also greater shrinkage and swelling behavior due to its looser structure caused by the slip regions of the cell walls.

Variations on the hot steam process have been reported. Several repeated treatments of steaming and bending can permit more extreme bends than a single operation. Wood is more plastic at non-equilibrium states of moisture content if it has been subjected to stress, e.g. to bending during water adsorption (28). Water adsorption sets up an additional stress which on the tension side of the wood beam shows the same direction as the bending stress. Thus, in this region of the wood, creep will be accelerated during softening while the compression stress has been lowered.

If the wood is beaten parallel to the grain during steaming, cells are partially separated from one another and the work piece can be bent to various forms without spring back. Some deciduous species of wood can be steamed under compression, retained in the wet condition and later bent in the cold (29). High frequency heating of wood provides a rapid plasticization process comparable to conventional steam processing. While moist fiberboard tends to defiber on forming, dry fiberboard heated to 400° for five seconds can be bent to complex shapes and rapidly cooled with little diminution of strength (30).

It is thus clear that both the success in processing and the final properties of the formed wood can depend on the particular combination of plasticizer, heat, and mechanical treatment used. In the more complex ammonia-water-wood system, much more effort is required to determine optimum conditions for particular results, since a much wider range of results is possible.

Steam bending has severe limitations in the quality and number of species of wood that can be bent (27). In general straight grained high density northern hardwoods usually give a minimum radius of curvature ratio or a minimum failure rate. These variations between woods reflect differences in microscopic structure and chemical organization of the material, for phase geometry can be as important as molecular structure in determining the properties of both natural and synthetic multiphase systems (31). Therefore, it is clear that the mechanical behavior of the wood-water system cannot be explained entirely at the molecular level or as interaction of macromolecules with solvent. Nevertheless, the general trends observed do follow general principles of solvent-polymer interaction and can be so explained.

Fundamentals of Wood-water Interactions. Goring has lucidly described the influence of heat and water on wood components (32): As temperature is raised, solid polymers absorb energy and the chains develop more violent motion until a rather narrow temperature range is reached at which intermolecular bonds are broken and the macromolecules become capable of large scale displacement with respect to each other. In this range the polymer properties change rapidly and amorphous polymers generally undergo a change from a glassy to a rubbery or plastic state. Similar softening points are observed for both lignin and hemicellulose. Furthermore, the presence of water acts as a typical low molecular weight diluent, lowering the softening point or tack temperature of lignin from about 190° to 70-116° C. Very similar behavior is observed with isolated hemicellulose. In contrast, water cannot penetrate the crystal lattice of cellulose and cellulose softens at about 230° whether wet or dry. The softening of wood for forming depends directly on these polymer-solvent interactions and are a dramatic indication that much of the wood stiffness is due to intermolecular association forces, predominantly hydrogen bonds.

The extra suppleness of never-dried wood can be related to the fact that never-dried cellulose has a much higher equilibrium moisture content at all relative humidities than cellulose after drying (33). The multiple hydrogen bonds that are formed on drying form partially ordered regions that cannot entirely again be loosened with water alone. Possibly the same phenomenon occurs with lignin and hemicellulose.

Effect of Applied Force. In addition to considering the influence of water on the wood, one must consider the effect of an applied force in conjunction with water as plasticizer.

If a force is applied to wood within the proportionality limits, the wood will bend and if the force is released, the wood returns to its original form with an elastic recovery. In contrast, if the wood is dried under stress, a substantial superposition of stresses occurs in conjunction with the drying and shrinking process. Since the ordering of macromolecules or larger structural elements under tension is different from those under compression, as the water molecules are removed, new hydrogen bonds can form between different subunits of the structure to support the distorted structure in its new form. In that case one would expect that internal stresses would be present in the wood and on resteaming the wood could recover its original shape.

On the other hand, there may be during drying some true plastic deformation, some amount of irreversible displacement of macromolecules, fibrils or fibers relative to one another. For example, if a beam is dried during bending under load, an additional tension stress will be set up in the outer zones while in inner regions of the beam compression stresses will be introduced. The resultant stresses from the drying and bending can surpass the proportionality region between load and corresponding deformation on the tension side of the beam and creep will occur. In that case one would expect permanent set not recovered on resteaming. Thus some degree of permanent set is not limited to ammonia forming, and even small quantities of ammonia in aqueous systems probably facilitate creep.

Plasticization with NH_3 — Processes on Molecular Level

Differences Between Ammonia and Water. There are significant differences in physical properties between wood *fully* plasticized by ammonia and wood plasticized by steam. The ammonia-saturated wood shows comparatively little elastic deformation with stress and undergoes a large time-dependent plastic deformation and creep. Therefore, most extreme results are obtained if forming is carried out slowly or in some case perhaps intermittently. When the bending force is released, the wood does not return to its original shape. If the wood is dried and then wet with water it swells more on the compression side than the tension side and straightens, but on drying returns essentially to its formed shape.

These macroscopic differences can be related to the molecular interactions between wood components and the two solvents. Nayer and Hossfeld (34) have shown that wood swelling in a series of solvents increases with increase in hydrogen bonding capacity of the solvent and decreases with increase in molecular size. Ammonia as a solvent with a similar molecular size but a greater hydrogen bonding capacity than water would be expected to swell and soften wood more and such is the case (1b). Anhydrous ammonia can even penetrate the cellulose crystal lattice in wood and relax inter-crystalline forces. The solubility and swelling of lignin also increase with the hydrogen bonding capacity of solvents and are at a maximum in solvents of intermediate cohesive energy density (δ = 10 - 12) (35). Ammonia is much closer to this optimum value than is water. These factors undoubtedly are predominant in making ammonia a superior softening agent for wood, allowing creep during forming and permanent set in the final product.

Effect of Water. Wood is usually treated with ammonia in the presence of some amount of water. The effect of water depends not only on the amount of water but also somewhat on the history of the wood sample and the method of treatment. Thus, when oven-dried veneer strips were treated with cold liquid ammonia-water mixtures at ambient pressure, the flexibility of the treated wood was substantially decreased when the moisture content of the ammonia was much above 10% (26). Other protonic solvents act similarly (26). In apparent contrast, the rate of sorption of ammonia from the gas phase by wood is markedly enhanced by moisture in the wood (19). Bone dry wood absorbs ammonia quite slowly at ambient temperatures but if the wood has ten to twenty percent moisture content, sorption and plasticization occur much more rapidly. Presumably the moisture opens the pore structure of the wood and also dissolves ammonia much more readily than bone-dry wood. On continued treatment, the water is presumably displaced from the wood by the ammonia since the x-ray diffraction pattern of the wood is usually Cellulose III, a modification which can not be formed unless most of the water is displaced into the vapor phase (26). The reverse phenomenon, displacement of ammonia by water has been proven by chemical analysis as well as by physico-chemical methods (36). (Fig. 1). Non-protonic solvents can be used in mixtures with liquid ammonia, allowing relaxation, and in some cases inhibiting checking and shrinkage (Carbowax 400) (26).

Sorption, Kinetics and Transport. If wood is immersed directly in liquid ammonia at ambient pressure, convection of the liquid is inhibited by the presence of air, and if the wood is at ambient temperature, it must be cooled to less than -30° C before liquid can flow into the pore structure. Nevertheless, under the best of circumstances much of the wood substance must be reached by diffusion rather than convection because a variety of physical restrictions inhibit liquid flow in wood (37,38).

The kinetics and thermodynamics of gaseous ammonia sorption and diffusion have been studied in detail (36,39,40). In general, two stages of the gaseous ammonia plasticization process can be distinguished. In the initial phase the hydrate envelope of the wood substance interacts with ammonia, causing the formation of ammonia-complexes, $NH_4^+OH^-$, NH_4OH etc. These

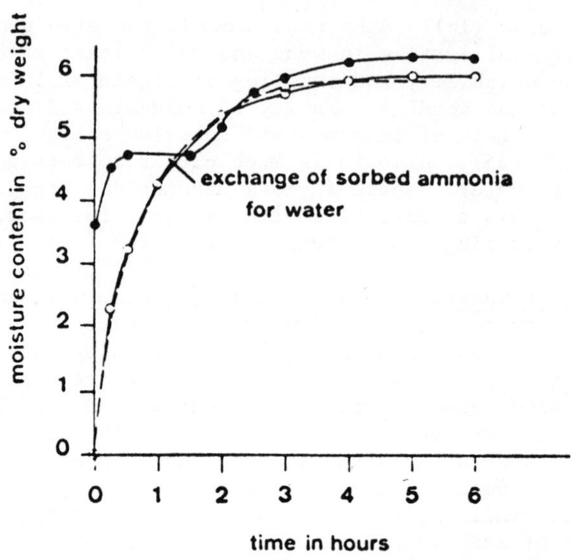

Figure 1. Water vapor adsorption kinetics of ammonia-treated ramie cellulose (● first exposure to water vapor after drying from ammonia; ○ repeated exposures thereafter to water vapor after drying from 6% moisture content) (36)

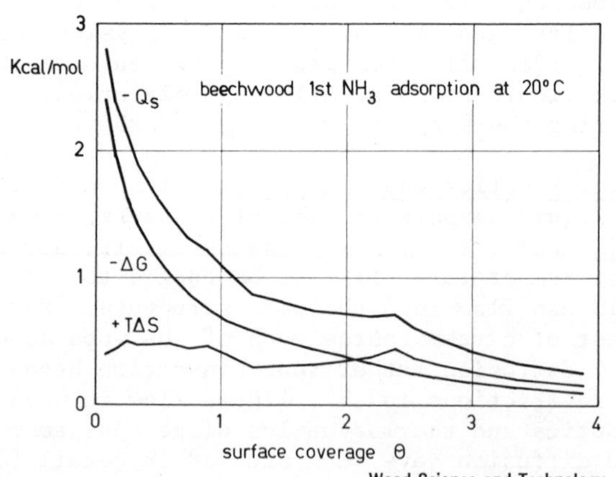

Figure 2. Variation of the integral net heat of sorption ($-Q_s$), change of free energy ($-\Delta G$), and change of the integral entropy ($T\Delta S$) with surface coverage (θ) of beechwood (39)

are also sorptively active and accelerate ammonia adsorption relative to the specific surface area of the wood. Therefore, at very low vapor pressures, the rate of diffusion of ammonia vapor surpasses that of the water vapor by two orders of magnitude, (Table I) and at higher vapor pressures, by at least one order of magnitude. In the course of the diffusion NH_3 is also sorbed on all wood surfaces. During sorption of NH_3 by cellulosic material in the low vapor pressure range, the isotherms show a steep rise which indicates that strong bonding forces act between sorbate and adsorbent. The integral sorption heats of ammonia calculated from the measured isotherms are the same, whether the adsorbent is beech wood, birch wood or cellulose. The calculated integral net heat of sorption shows that the bond between sorbate and adsorbent must be stronger than between the NH_3 molecules in the liquid phase (Fig. 2), and corresponds (about 3 Kcal/mole) to the formation of hydrogen bonds between wood and ammonia. The calculation of the change of entropy as a function of the coverage of the sorption area, reveals that there is in addition at least a second type of NH_3 uptake. In the middle range of vapor pressure ($p/p_o = 0.5 - 0.7$) all cellulosic adsorbents are characterized by a more intense ammonia uptake, and in this respect, the shape of the isotherm differs from the type II described by Brunauer (41) and displays a double S-form (39) (Fig. 3). The sorbate concentration in the higher range of vapor pressure increases asymptotically, indicating capillary condensation with fiber saturation capacities considerably higher than in water vapor. (Table II). Therefore, it is clear that the ammonia sorbate layer directly adhering to the wood is bound chemosorptively, and that the overlaying sorbate layers are accumulated by physisorption followed by capillary condensation. The isotherms discussed are not reproducible for subsequent complete ad- and desorption treatments lower the ammonia sorption capacity of the material as is shown in Table II. The values presented in Table II indicate that with ammonia the specific surface area of wood is on the average, twice or three times larger than with water (cf. refs. 42,43 et al. for data on water sorption) and that the ammonia sorption area of the samples decreases from one sorption cycle to the next. Thus the substance of the adsorbent apparently undergoes a process of densification with each ammonia contact (44). Along with this process of densification, there is also a decrease in degree of crystallinity. The extent of change in crystallinity is highly dependent on the species of the specimen and on the previous NH_3-treatment but usually lies between 5 and 18% (45-47).

The displacement of ammonia by water is of practical significance, since some difficulty has been noted in removing ammonia from thick work pieces. Clearly a drying cycle after treatment with a reasonably high humidity might be advantageous in difficult cases.

Table I Diffusion coefficients of ammonia vapor (D_{NH_3}) and of water vapor (D_{H_2O}) in wood specimen with equal dimensions (Yellow birch).

Vapor pressure steps (p/po)	$D_{NH_3} \cdot 10^{-7}$ (cm^2/sec)	$D_{H_2O} \cdot 10^{-7}$ * (cm^2/sec)
0. - .1	390	3.4
.1 - .2	18	2.0
.2 - .3	14	
.3 - .4	18	2.3
.4 - .5	13	
.5 - .6		1.4
.6 - .7	20	
.7 - .8	12	0.7

*According to G. N. Christensen (1960), quoted only for qualitative comparison.

Table II Some results of the analysis of the NH_3 sorption isotherms on beechwood.

Specimen	Treatment	Treating Temp. °C	Specific Surface Area m^2g	Fiber Saturation Point %
Beechwood	1. NH_3-ads.	20	872	69*
	1. NH_3-des.		772	
	2. NH_3-ads.		780	58
	2. NH_3-des.		659	
	3. NH_3-ads.			

*Compared to approximately 35% in water.

Processes on Supramolecular and Microscopic Levels

If ammonia pressure is sufficiently high, the hydrate envelope of the wood can be exchanged for ammonia. Thereafter slower processes result in the penetration and loosening of wood substance during the first hour of ammonia contact. As a result the density of the wood substance decreases temporarily by approximately 10% (Fig. 4), at full tank pressure, accompanied by an excessive swelling of the cell walls, and the mechanical properties of the bulk wood change sharply (48). If then the ammonia is removed from the saturated wood, the wood substance will be densified to an increasing degree depending on the length of treatment. The pore volume, especially that of pores with radii falling into the range from the cell luminae down to the tori, is thereby reduced by more than one-third (Fig. 5) by the following mechanism. Because of the decrease in capillary radius, the menisci of the receding NH_3 - fluid remain active in the cell wall capillary tubes at still lower relative vapor pressures of NH_3 than at that of water. These menisci exercise strong transverse tensile stresses on the capillary tube walls. At the same time there is an accumulation of drying stresses, as in the case of evaporating water. The vectors of these forces are mostly in the radial direction, and easily overstress the plasticized cell wall. Consequently, first the pores with larger diameters, then the smaller ones can partly or entirely collapse (44).

After the complete removal of the ammonia from the wood, the extent of the collapse can be measured. This collapse is entirely caused by the decrease in void volume, and is made up to a larger extent by the partial closure of the cell luminae and to a smaller extent by the reduction of the pore volume of the cell walls. At the molecular level, wood substance does not seem to remain loosened after the NH_3-treatment except for an increase in amorphous areas.

It is clear from the preceding discussion that collapse, densification and increased amorphous character can be minimized by single gaseous treatments of short duration. The degree of relaxation is less than that under full swelling conditions, but can be in the range of practical working conditions as has been demonstrated in Zurich and is discussed later.

It should also be remembered that the structure of wood substance is not homogeneous. There are physical discontinuities: well ordered regions of microfibrils, fibrillar surfaces, varying orientations of fibrils in the layers of the cell walls, differences in lignin and pectin content between tracheids, middle lamellae and parenchyma cells and differences in density between spring wood and summer wood. At the present time it is not known in detail at what sites ammonia sorption begins or the minimum pressure of ammonia required to induce a particular level of softening, (although

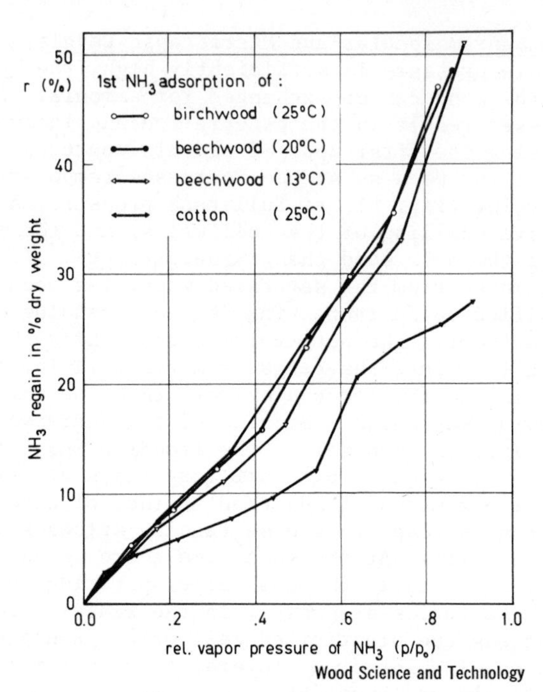

Figure 3. Adsorption isotherms of wood and cotton on first exposure to ammonia vapor (39)

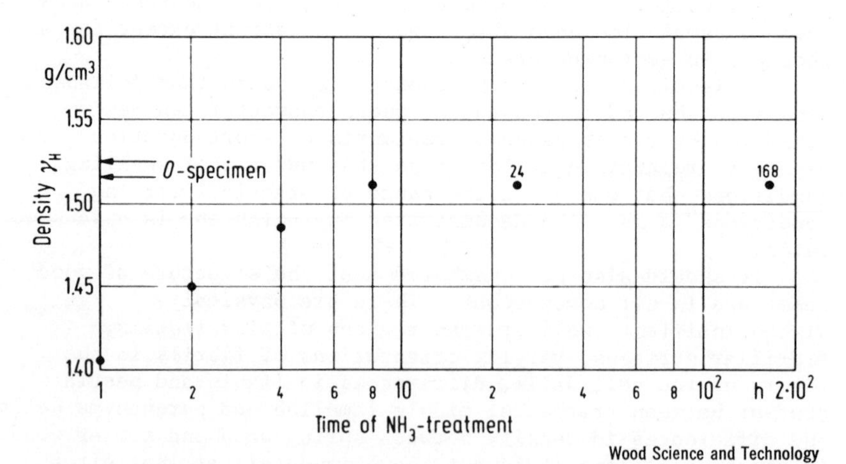

Figure 4. Density (γ_H) of dry beechwood cell wall substance (initial value 1.52 g/cm^3) after ammonia saturation for various periods of time (in hours), followed by vacuum drying (44)

there have been some analyses of the preferential uptake of ammonia in particular tissues.) (49) It is quite likely that the inter-crystalline uptake of ammonia which occurs above 0.5 relative vapor pressure (50), and the excessive swelling characteristic of the use of full tank pressure are unnecessary for most forming operations. Translation between supermolecular structural units may occur under stress as has, for example, been observed for aqueous ammoniacal systems (16a). Therefore, it may be possible at intermediate ammonia pressures also, to obtain much greater relaxation and creep than can be produced in steam forming.

At the present time, there has been too little experimental forming done at intermediate ammonia pressures.

Processes on Macroscopic Level

On ammonia treatment, the gross swelling behavior and changes in the strength and the structure of wood are very similar to those observed during and after steaming of wood with water vapor. The changes described below refer to wood saturated with ammonia.

Swelling and Shrinking: The rate of swelling of wood of all species in ammonia is faster than in water (43,51). This is understandable since the rate of diffusion of ammonia surpasses that of water. At equilibrium, almost all species show superswelling in tangential direction during ammonia-soaking. There are, however, a few exceptions where superswelling occurs in the radial direction (e.g. Douglas fir (43), hard maple (51). The superswelling is associated with loosened wood structure, with the slip regions of the cell walls. Upon removal of ammonia from full vapor pressure, all species undergo an excessive shrinkage (44,45,52) mostly followed by collapse of the cell structure. Repeated water soaking of ammonia-treated material enhances this dimensional collapse (44) (see Fig. 6).

The swelling and shrinkage of ammonia-treated species in water is higher than those of untreated material, but water penetrates ammonia-treated material more slowly than untreated wood (43,51). Under most circumstances, wood will not be subjected to repeated water treatment after forming and the influence of moisture vapor is much less severe. If wood samples are subjected to ammonia treatment, then vacuum dried and finally subjected to atmospheres of 50 to 98% relative humidity, the treated wood and untreated controls both absorb an excess of moisture and then on further standing in the same atmosphere lose moisture (50). The initial "overshoot" is slightly higher with ammonia treated samples but the final equilibrium values appear somewhat lower with the treated samples (53). Swelling measurements show similar trends. Similar phenomena have been observed with other polymer solvent

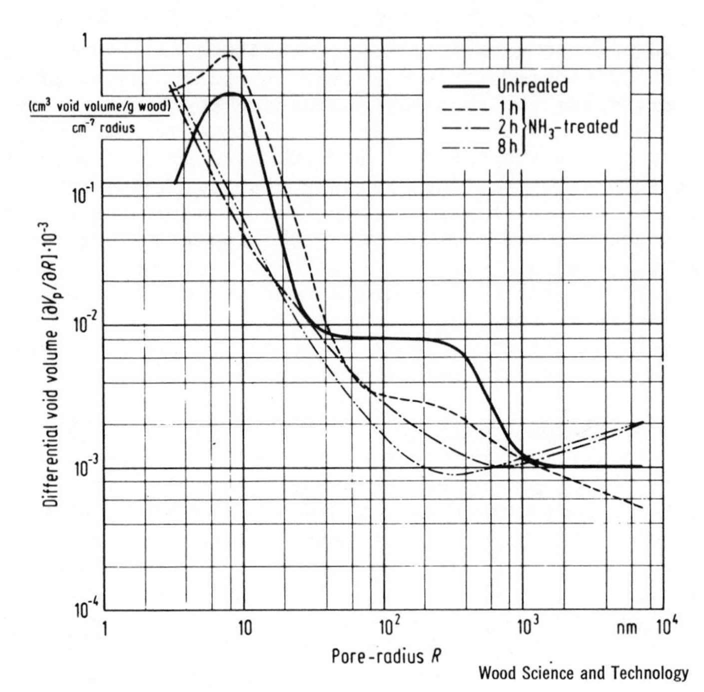

Figure 5. Volumes of pores with radii ranging from 7.5–7,500 nm in correlation with the entire void volume of 1 g of beechwood before and after treatments of 1, 2, and 8 hr, followed by vacuum drying (Compare *Fig. 4*) (44)

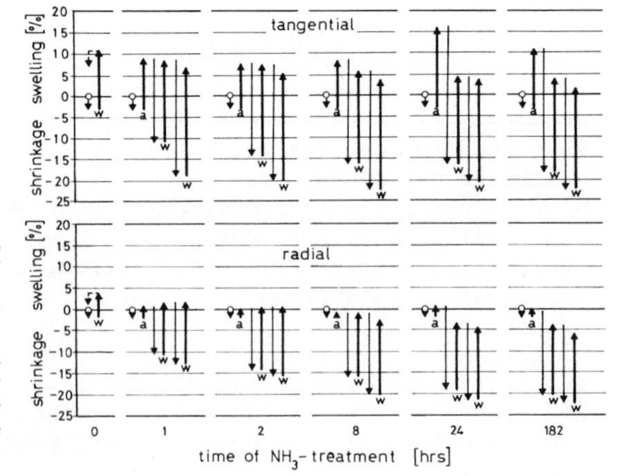

Figure 6. Linear shrinkage and swelling measurements of beechwood after alternating treatments with ammonia and water. Arrow ↑ symbolizes swelling: a *in ammonia,* w *after two days of watering. Arrow ↓ symbolizes shrinkage. Reference values are dimensions in air-dry state.*

systems and can be related to recrystallization phenomena (53). For practical purposes, the sensitivity of ammonia treated wood to moisture vapor can probably be considered about the same as untreated wood.

Specific Gravity: The specific gravity of wood can be raised about 10-40% as a consequence of a single ammonia treatment over an extended period (44,51). Repeated treatment enhances the densification.

Color: The degree of color change in wood is determined by the time and temperature of treatment. Often the color change is advantageous. One method of control has been described (54) and apparently others are in use in the Soviet Union.

Mechanical Properties: Bending strength is increased after ammonia treatment and drying, but the amount is dependent upon the species (51) (3% for beech compared to 30% for poplar) (5). The modulus of elasticity is lowered to about 1/5 - 1/10 of the initial value (36), but after treatment it regains approximately its original value (45,51). Compression and tensile strengths are enhanced after treatment and drying, by about 10-40%, again depending on the species (51). Much of this increase reflects the increased amount of material per unit cross section (51). The toughness of wood is greatly reduced by 30-40% as a result of treatment (51). All mechanical properties are reported to be improved by treatment with aqueous ammonia, but it is questionable that this is true of toughness.

The overall effect of ammonia on wood results from processes, some of which lower and others enhance the strength properties of wood. Among the former are the following: A decrease in the degree of crystallinity which causes the introduction of weaker secondary bonds into the substance. Dislocation of components in the cell wall making the material more loose. Crimpings and microfailures in the cell wall which eliminate primary and secondary bonds. Among the latter are: The increase of specific gravity by the collapse of cells and cell walls, giving more substance per unit volume to withstand forces; elimination of local stress maxima in the tissue by relaxation or creep. Reduced anisotropy of the properties in the tangential and radial directions (44) as the substance becomes denser and by collapse tangential and radial cell wall parts are mixed. Often the most prominent effect is the increase in specific gravity. It should be emphasized that maximum effects are observed on extended treatment times at high relative vapor pressures.

Rheological Properties. Rheological studies show that ammonia affects the compressive mechanical behavior of wood to a much higher extent than the tensile behavior (55,16). Since the compression strength is highly dependent on lignin content, this effect probably suggests lignin plasticization early in the process.

The question whether ammonia treated wood shows linear or non-linear viscoelastic behavior has not been answered so far. The measurements reported by Bach (56) apparently were not under constant temperature conditions. Strain recovery after loading in the plasticized state is small. The longer the loading period the smaller the recoverable strain. This suggests plastic flow under load and a conversion of delayed elastic strain into an irreversible deformation.

Practical and Potential Uses of Ammonia in Wood Technology

The use of ammonia for softening wood has not been limited to bending, compressing (56) and forming whole wood. For example, two different applications in pulping have been reported (57,58). In one proposal, lignocellulosic material is defibered by an explosive decompression of wood chips which have been plasticized with ammonia at elevated temperatures and pressures. In a second investigation wood chips were plasticized with ammonia in an Asplund pressurized refiner, during defibration. Low energy requirements appear to be characteristic of the latter process. Pulp quality and potential applications were described in both cases.

There are also reports of the production of boards and moldings from plasticized wood particles, pulp or sawdust without adhesives (5,8,59,60). By compressing and heating to high temperatures, particle boards can be produced which have mechanical properties comparable to conventional resin bonded boards. However, their specific gravity is on the average about twice that of ordinary commercial products (59).

The use of aqueous ammonia in the preparation of wood fiber filled phenolic plastic molding has been investigated. In this case ammonia both plasticizes the wood filler and catalyzes the phenolic methylol condensation. With proper formulations and treatments, it is possible to maximize the quantity of wood fiber that can be used and minimize the resin without deterioration of properties of the molded product (61).

Preliminary investigations suggest that wood slicing with knives can be applied successfully to ammonia plasticized wood and that thicker boards and veneers can be cut than by conventional methods. Savings of material and energy are envisioned over the use of sawing (62).

Some wood species suffer severe checking on kiln drying due to internal stresses. It has been suggested that the use of ammonia gas in the kiln might lead to stress relaxation and lower losses during drying. It is, however, certain that the discovery of appropriate conditions would require a systematic investigation for in the drying of large dimensional stock fully plasticized with ammonia, checking is more severe under conventional drying conditions than with water-wet green wood. Investigations in the low partial pressure range would be indicated.

The most extensive research that has been undertaken on wood softening and forming is that at the Institute of Wood Chemistry in Riga, Latvian S.S.R.. Their studies have been primarily upon the influence of aqueous ammonia on wood and have included fundamental scientific and engineering research, and process and product development. Some of their publications also discuss investigations of gaseous treatments. Although some of their technology is directed toward bending and forming operations, the main thrust of the research appears to be directed toward the improvement of physical properties of woods by compression. Their technology depends upon a more detailed study of the rheology of wood under varied conditions of temperature and ammonia concentration than has been attempted in the West (2-18). Probably their forming operations, especially that of compression, do not require as complete relaxation as is possible with pure ammonia and they may use longer forming times than has been customary in Western experimentation. Long forming times are indicated in their American patent (18), and these would allow, presumably, more creep than would be expected from some results reported in the West on the influence of water in the forming process.

They have prepared some compressed wood samples which show, in addition to increased density and surface hardness, a lower moisture regain than untreated wood up to 80% relative humidity. Using materials such as these, they have experimented with the manufacture of parquet flooring, a rather severe test of dimensional stability. A variety of other finished products have been prepared, some on pilot plant scale. However, it is not known to what extent they have appeared as products in the open market. Their technology is available through licensure.

The original suggestion that ammonia could be used to produce extreme flexibility and permanent set in wood was based on observations of the effect of liquid ammonia at low temperature and atmospheric pressure on well dried wood samples (1,63). These conditions produce maximum swelling and relaxation, and fibrillation and discoloration tend to be less than with equally lengthy treatments at higher temperatures and pressures. However, safety hazards are severe and the method is extremely wasteful of chemical, since the liquid in the luminae does not contribute to the softening process. Most work since that time has been carried out in pressure vessels on wood with moisture content about ten percent. This technique has been explored at a number of centers of industrial arts and by several artists, some of whose work has appeared in public exhibitions. Greater use of the process in arts and crafts is clearly indicated, for no other technique allows the formation of such extreme shapes or fluid lines in wood so easily and without loss of strength.

In the many examples produced to date, dimensional stability has not been a problem.

A recent process development from Zurich by M. Bariska (20) has a number of important and interesting features. The method is specifically designed to test the applicability of ammonia forming to commercial practice. Secondly it constitutes a lower extremum, an investigation of conditions of minimum useful plasticity. Furthermore it is based on extensive investigations of kinetics, thermodynamics, sorption processes, and structural changes characteristics of the ammonia wood system.

Some of the pertinent concepts underlying the method are the following: Much of the stiffness of wood is contributed by lignin, and a high concentration of lignin exists in the middle lamella. Ammonia has a high thermodynamic activity at room temperature and an extremely high rate of diffusion. It, therefore, appeared likely that a rapid brief impregnation of wood with ammonia might cause sufficient plasticization of the middle lamella to permit wood forming. This conjecture was tested as follows:-

Specimens of air-dried European beech (ca 12% moisture content, 1/2 in. x 1 in. x 36 in.) were plasticized in saturated steam and in ammonia vapor at room temperature (ca 23° C). The treatment time for the steam treatment controls was that typical of industrial practice, about 1 minute per millimeter of thickness. The treatment period for ammonia softening varied from the same time down to the shortest possible blow, which consisted of filling a three gallon treatment tank with ammonia gas at tank pressure and then decompressing, a total period of about 30 seconds. After softening, the individual work pieces were bent to a form with varying radii of curvature (30, 15, 7.5, 3.8, and 1.9 cm) and kept in a mold for thirty minutes. Regardless of treatment conditions, compression failures occurred in the range of less than 7.5 cm., radius of curvature. Spring back of the treated and rehardened pieces that were bent to a radius of 30 cm. was measured at different times (Fig. 7). A statistical analysis of the data showed no difference between the two softening methods at the 95% confidence limit. Color change due to ammonia treatment was indistinguishable from that of steam plasticization to the unaided eye.

Judged by these results, rapid low temperature ammonia softening of wood is at least as effective as conventional hot steam plasticization and the effects of the two methods are essentially indistinguishable. Ammonia softening consumes less energy on the site and does not require setting time. However protection against corrosion and against irritation of working personnel is required. Under some local conditions and for specific applications this method may be preferable to conventional practice.

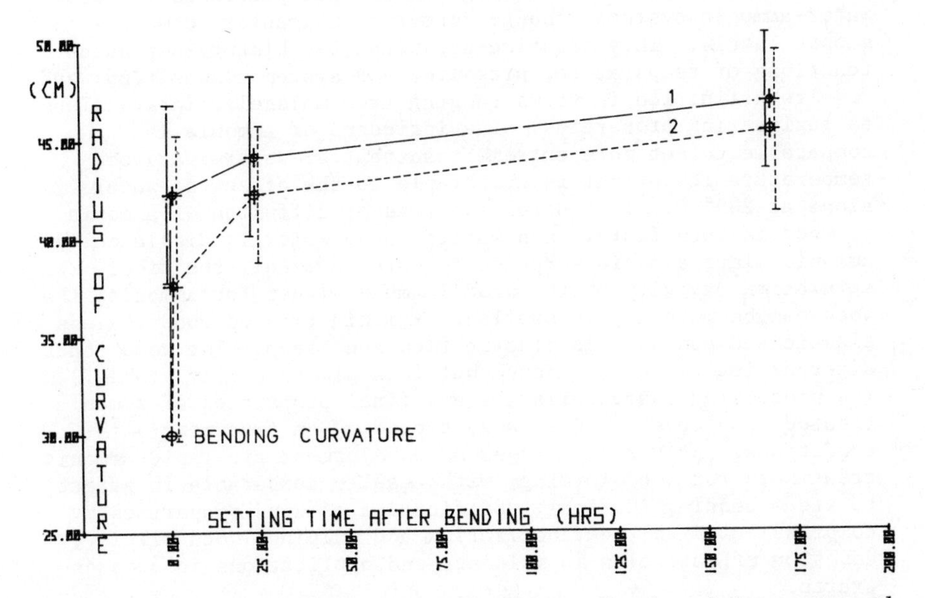

Figure 7. Spring-back of treated and rehardened pieces of beechwood after treating with ammonia (1) or steam (2), bending to a 30-cm radius, and setting for various times

It is of interest that the minimum practical relaxation of wood by ammonia approximates the maximum obtainable with steam. Specific effects and applications which cannot be achieved by means of steam forming may, therefore, be obtained under the wide range of conditions possible in the wood-water-ammonia system.

Abstract

During sorption, softening and forming processes in wood-water-ammonia systems, changes occur on molecular to macroscopic levels. They are time-dependent and history-dependent functions of temperature, pressure, and system composition, and require definition in terms ranging from molecular interactions to engineering properties. The influence of ammonia is comparable to but more extreme than that of water. At room temperature its effect is comparable to the effect of water alone at 200° C. In general the rate of diffusion of ammonia in wood is much faster than water; bound water is displaced by ammonia since ammonia sorption is more powerful; the fiber saturation capacity of the wood is much higher for ammonia; the wood can be more highly swollen. Ammonia-treated wood is less elastic and can undergo plastic flow and creep. The wood tends after drying to be more dense but less dimensionally stable. The processing characteristics and final properties of ammonia-treated wood cover a wide range depending on treatment conditions. At various stages of development are rapid ammonia treatments for wood forming, with results comparable in effect to steam bending (Zurich), modification of wood properties by compression (Riga), defiberization and pulping, wood slicing, adhesion of particles in molding, and applications in arts and crafts.

Literature Cited

(1) Schuerch, C. (a) Ind. Eng. Chem., (1963), 55, 39.
(b) Forest Prod. J., (1964), 14, 377-381.
(2) Kalnins, A. J. and Darzins, T. A, Latv. Lauksaimn. Akad. Raksti, XI, (1962))421-2.
(3) "Modification of Wood" [Modifikatsiya Drevesiny], Academy of Science Latvian SSR, Institute of Wood Chemistry, "Zinatne," Riga (1967) (Russian). This book contains thirty one articles by the Institute's staff. Abstracted by article in A.B.I.P.C. (Jan. 1970), 40 #7, 563 ff.
(4) Kalnins, A. J., Darzins, T. A., Jukna, A. D., Berzins, G. V., Holztechnologie, (1967), 8, 23. (German)
(5) Kalnins, A. J., Cellulose Chemistry and Technology, (1969), 3, 199. (Russian)
(6) Kalnins, A. J., Berzins, G., Skrupsis, W. and Rumba, A., Holztechnologie, (1969), 10, 17. (German)
(7) Berzins, V. G. and Doronin, J. G., Holztechnologie, (1970),

11, 125. (German)
(8) Vejina, L., Latv. P.S.R., Zinat. Akad. Vestis.(1970), 2, 52-54. (Russian)
(9) Berzins, G., Latv. P.S.R., Zinat. Akad. Vestis., (1970), 10, 130. (Russian)
(10) Berzins, G., Latv. P.S.R., Zinat. Akad. Vestis., (1970), 2, 61-69. (Russian)
Doronin, Y. G., Latv.P.S.R., Zinat. Akad. Vestis., (1970), 2, 55-60. (Russian)
(11) Onisko, W., Matejak, M., Silwan, (1971), 115, (2), 39-50. (Polish)
(12) Erins, D., Karklins, J., Odincous, P., Veveris, G., Khimya Drevesiny, (1971), 7, 159-69. (Russian)
(13) Erinsh, P. P., Odintsov, P. N., Alksne, I. M., Khimya Drevesiny, (1971), 9, 19-28. (Russian)
(14) Erinsh, P. P., Cinite, V. A., Khimy Drevesiny, (1971), 9, 29-38. (Russian)
(15) Lielpeteris, U. U., Ziedinsh, I. O., Khimy Drevesiny, (1971), 9, 167-171. (Russian)
(16) Rocens, K., Holztechnologie, (1976), 17, 40-45. (German)
(a) Work of Erins, P. P., and Odinkov, P. N., herein quoted.
(17) For other contributions of the Institute of Wood Chemistry Riga, see Chemical Abstracts (1971), 76, 101409q, 101410h; (1972), 77, 76925r, 166446q, 166442k; 78, 5605w, 5606x, 5607y, 17847q.
(18) Various patents issued on the basis of work at the Institute of Wood Chemistry, include the following to Berzins, G. V. et al.
Ger. Offen 2,020,810 (Cl. B27k) 07 Sept 1972.
U.S. Patent 3,646,687 Mar. 7 (1972).
U.S.S.R. Patents No. 208,923; 316,309; 299,364; 3,646,687.
(19) Davidson, R. W., "Plasticizing Wood with Anhydrous Ammonia," Technical Bulletin, Dept. Wood Products Engineering, SUNY College of Environmental Science and Forestry, Syracuse, NY 13210.
(20) Bariska, M., herein reported, to be published in entirety elsewhere.
(21) Sadoh, T., Yamaguchi, E., Bull. Kyoto Univ. Forests, (1968), 40, 276-283. (Japanese)
(22) Sadoh, T., Journ. Japan Wood Research Soc., (1969), 15 (1), 29-34. (Japanese)
(23) Sadoh, T., Journ. Japan Wood Research Soc., (1970), 16 (7), 334-338. (Japanese)
(24) Sadoh, T., Journ. Japan Wood Research Soc., (1968), 14 (3), 175. (Japanese)
(25) Beall, F. C., "Wood Forming Method" U.S. Pat. 3,717,187 (Feb. 20, 1973).
(26) Schuerch, C., Burdick, M. P., Mahdalik, M., I. and E.C. Product Research and Development, (1966), 5, 101-105.

(27) Peck, E. C., "Bending Solid Wood to Form" Agriculture Handbook No. 125, U.S. Dept. of Agriculture Forest Service (1968).
(28) Takemura, T., Memoirs College of Agriculture, Kyoto Univ., (1966), 88, 31-48. (English)
(29) Stevens, W. C. and Turner, N., Technical Brochure, "Experiments with Flexible Wood." Forest Products Research Laboratory, Princess Risborough, April 1966.
(30) Back, E. L., Didriksson, E. I. E., Johanson, F., Norberg, K. G., Forest Products Journal (1971), 21 #9, 96-100.
(31) Alfrey, T., Angew. Chem. Internat. Ed., (1974), 13 (2), 105-107.
(32) Goring, D. A. I., Pulp and Paper Mag. of Canada, (1963), 12, T-518-527.
(33) Urquhart, A. R., Eckevall, N., J. Text. Inst., (1930), 21, 499-510.
(34) Nayer, A. N., Hossfeld, R. L., J. Amer. Chem. Soc., (1949), 71, 2852-2855.
(35) Schuerch, C., J. Amer. Chem. Soc., (1962), 74, 5061-5067.
(36) Bariska, M. and Popper, R., J. Polymer Sci., (1971), C 36, 199-212.
(37) Schuerch, C., I. and E. C. Product Research and Development, (1965), 4, 61-66.
(38) Schuerch, C., Forest Products Journal, (1968), 18, 47-53.
(39) Bariska, M. and Popper, R., Wood Science and Technology (1975), 9, 153-163.
(40) Bariska, M., Habilitationsschrift: "Physikalische and Physikalisch - Chemische Aenderungen im Holz waehrend und nach NH_3-Behandlung." Eidgenossische Technische Hochschule, Zurich, October 1974 p. 93 ff. (German)
(41) Brunauer, S., "The Adsorption of Gases and Vapors, (1945), Vol. 1, University Press, Princeton, N.J.
(42) Spalt, H.A., Forest Products Journal, (1958), 8, 288-295.
(43) Stamm, A. J.,"Wood and Cellulose Science," The Ronald Press Co., New York, 1964.
(44) Bariska, M., Wood Science and Technology, (1975), 9, 293-306.
(45) Pentoney, R. E., I. and E. C. Product Research and Development, (1966), 5, 105-110.
(46) Fukada, E., Wood Science and Technology, (1968), 2, 299-307.
(47) Lewin, M., Roland, L. G., J. Polymer Sci., (1971), C 36, 213-229.
(48) Bariska, M., Strasser, Ch., J. Polymer Sci., (1976), 41, in press.
(49) Bariska, M., Bulletin, International Association of Wood Anatomists (1969) No. 2, 3-8.
(50) Bariska, M., Skaar, C., Davidson, R. W., Wood Science (1969), 2 (2) 65-72.

(51) Pollisco, F. S., Dissertation, SUNY College of Environmental Science and Forestry, Syracuse 1968.
(52) Schuerch, C., U.S. Patent 3,282,313, (Nov. 1, 1966).
(53) Pollisco, F. S., Skaar, C., Davidson, R. W., Wood Science, (1971), 4 (2) 65-70.
(54) Davidson, R. W., Schuerch, C., J. Polymer Sci. (1971), C, 36, 231-239; U. S. Patent 3,642,042 (Feb. 15, 1972).
(55) Bach, L. Materials Science and Engineering, (1974), 15, 211-220.
(56) Bach, L., Hastrup, K., Materiaux et Constructions, (1973), 6, (32), 137-139.
(57) O'Connor, J. J., Tappi, (1972), 55 (3) 353-358.
(58) Peterson, R. C., Strauss, R. W., J. Polymer Sci., (1971), C 36, 241-250.
(59) Graf, G., et al., Holztechologie, (1971), 12, 235-238; (1972), 13, 152-155.
(60) Shaines, A., U.S. Pat. 3,514,353 (May 26, 1970).
(61) Jukna, A. D., Inst. Wood Chem., Riga, personal communication.
(62) Davidson, R. W., Baumgardt, W. G., Forest Products Journal (1970), 20 (3) 19-24.
(63) The first observation of the plasticization of wood by ammonia was by Stamm, A. J., Forest Products Journal, (1955), 413-416.

22

Wastewater Management in the Solid Wood and Board Products Industries: A Review

WARREN S. THOMPSON

Mississippi Forest Product Laboratory, P.O. Box 552,
Mississippi State University, Miss. 39762

Concern over the deteriorating quality of the environment has focused world-wide attention on pollution in its various forms and on industry's contribution to this problem. Because of stringent legislation, the incentive to develop pollution abatement and control technology has perhaps been greater in the United States than in other countries in North America. Over 250 pieces of pollution legislation have been passed at state and federal levels since the early 1960's.

Effluent guidelines for the timber products processing industry have been developed by the United States Environmental Protection Agency (EPA) as part of an over-all pollution control program ordered by Congress in its 1972 Amendments to the Federal Water Pollution Control Act of 1965. This legislation has significance to the wood products industry in general, since the pollution problems that it addresses are essentially the same worldwide. This paper describes the pollution problems related to water in wood-based industries in the United States, summarizes current effluent standards as they apply to the industry, and discusses the technology available to meet these standards.

Federal Effluent Standards

Under the provisions of the 1972 Amendments to the Federal Water Quality Control Act, the U. S. Environmental Protection Agency is required to develop effluent limitations for all industrial point sources of discharge. Guidelines for the wood products processing industry have been promulgated. They identify in terms of the amounts of constituents and the characteristics of pollutants the degree of discharge reduction that must be achieved through the application of two levels of control and treatment designated as Level I Technology and Level II Technology.

Level I Technology is defined as the best practicable control technology currently available and must be achieved by all point sources of discharge by July 1, 1977. It is based on the average existing performance by exemplary plants of various sizes, ages, and types of processes within an industry. Level II Technology is the best available technology economically achievable and must be attained by all point sources by July 1, 1983. Performance under this level of technology is based on the best control and treatment technology existing in an industry or which is transferable from another industry.

In the development of effluent guidelines, each industry was subdivided into discrete categories based on processes employed. The range of control and treatment technology used by or indicated for each category was evaluated to determine what constituted "best practicable" and "best available" technology. A summary of the current discharge limitations for various facets of the timber products industry is given in Table I.

As shown in this table, a zero discharge of process wastewater pollutants is required by EPA for all facets of the timber products industry except wet storage of logs, wet-process fiber products manufacture, and one segment of the wood preserving industry. This requirement, by definition, holds that a zero discharge is "the best practicable technology currently available" by which the affected industries can control pollution. The technology indicated in most cases is in-plant process changes and modifications which will permit recycling of process water.

Specific effluent limitations have been established for those segments of the timber products industry for which a zero discharge was not judged to be economically or technically feasible. These limitations set forth the average daily discharge value for each pollutant, based on 30 consecutive days, as well as the maximum one-day value. Expressed in terms of units of production, the limitations set maximum discharge values for each point source of pollution and, by subtraction, indicate the amount of each pollutant that must be removed from process wastewater before it can be discharged. Typical effluent limitations are shown in Table II for wood preserving, hydraulic debarking, and steaming of veneer logs.

Technology for Zero Discharge

The exact technology used to meet the effluent limitations for an industry is left to the initiative of the individual companies. However, the development document on which the effluent limitations are based includes treatment and control technology. Plants that must meet a zero discharge requirement are more limited in their choice of abatement measures than those that are permitted a discharge, however small. Where the volume is large and discharge to publicly owned treatment works is not possible, recycling is the only economically viable method of achieving zero discharge. Basically, this involves installing the equipment needed to clean up process water sufficiently so that it can

Table I. Summary of discharge requirements for timber products industry

Industry Subcategory or Process	Discharge of Process Water[a]		Principal Pollutants
	1977	1983	
Wet Storage of Logs	Yes[b]	Yes	Floating solids
Log Washing	No	No	SS, BOD
Sawmills	No	No	SS, BOD
Particleboard	No	No	Phenols, BOD
Insulation Board	Yes	Yes	SS, BOD
Debarking	No[c]	No	SS, BOD
Veneer Mills			
(Hot-water vats)	No	No	BOD (phenols)
(Steam chambers)	Yes	No	BOD (phenols)
Plywood Mills	No	No	BOD, Phenols
Finishing	No	No	BOD, Metals
Wood Preserving			
(Steaming)	Yes	Yes	Phenols, Oils, COD
(Boulton)	No	No	Phenols, Oils, COD
(Salts)	No	No	Heavy Metals, COD
Hardboard			
(Dry process)	No	No	BOD, SS
(Wet process)	Yes	Yes	BOD, SS
Paper	Yes	Yes	BOD, SS

[a]Cooling water, yard runoff, and boiler blowdown are excluded from definition of "process water."

[b]Effluent guidelines limiting the amount of specific pollutants that can be discharged have been established for each process for which a discharge is permitted.

[c]Hydraulic debarking operations are permitted a discharge under 1977 guidelines, but not under 1983 guidelines.

Table II. Summary of effluent standards for selected segments of the wood processing industry

Industry Subcategory	Pollutant	Effluent Limitation 1977 Daily Max.	1977 30-Day Avg.	1983 Daily Max.	1983 30-Day Avg.
		(kg/1000m^3)			
Wood Preserving					
Boulton Plants		No discharge of pollutants permitted			
Steaming Plants	COD	1100	550	220	110
	Phenol	2.18	.65	.21	.064
	Oil & Grease	24.0	12.0	6.9	3.4
	pH		6.0-9.0		
Salt Plants		No discharge of pollutants permitted			
		(kg/m^3)			
Hydraulic De-barking	BOD	1.5	0.5	No discharge of pollutants permitted	
	TSS	6.9	2.3		
Steam Conditioning of Veneer Logs	BOD	0.72	.24	No discharge of pollutants permitted	

be reused in the manufacturing process along with initiating strict water conservation programs.

Sources of contaminated water in the manufacture of products for which no wastewater discharge is allowed include wet-type debarking operations, steam or hot-water conditioning of logs, veneer drier washwater, glue equipment washwater, and certain types of preservative treatments. Only limited data are available on the quantity and characteristics of these effluents (Table III). Each source of wastewater is discussed below in terms of the problem that it presents and the technology that is available to deal with it.

Debarking Operations. Debarking is not ordinarily a significant source of wastewater, except where hydraulic debarking is employed. Where this is the case, water discharges in the amount of 50,000 to 120,000 liters/M^3 of wood are common. Since this water has a BOD content in the range of 50 to 250 mg/liter, as well as a relatively high content of suspended solids, the effluent from a hydraulic debarking operation can be the most significant part of the waste stream at plants employing this technique.

Because of the large volume of wastewater generated in hydraulic debarking and the problems associated with recycling it, waste treatment and disposal techniques used by the pulp and paper industry are applicable. This involves the employment of heavy-duty clarifiers to remove suspended solids, followed by biological treatment to remove oxygen-demanding substances.

Hydraulic debarking is being phased out by the industry -- in part due to the pollution problems associated with its use and in part because of the decrease in the number of over-sized logs for which the technique was originally developed. Because other wet-type debarking methods generate relatively little wastewater, all of which can be recycled, it is not anticipated that debarking will contribute significantly to the waste load from plants in the future.

Log Conditioning. Veneer and plywood plants in the United States use either steam vats or hot water vats to condition logs preparatory to peeling or slicing. Steam vats are the more common nationally, but both types are employed extensively in the South and Southeast.

Condensate from steam vats is the major source of wastewater at plants that use this method of conditioning. The volume varies with the number and size of vats but amounts to an estimated 114 liters/minute at plants with annual productions of 9.31 million M^2. Effluent from the vats of a typical softwood plywood mill has a BOD loading of 410 kg/day at 2500 mg/liter concentration and a total solids loading of 325 kg/day at 2,000 mg/liter concentration. Unlike wastewater from other sources, that from steam vats is difficult to eliminate because the contaminated

Table III. Volume and characteristics of wastewater from selected wood products manufacturing operations for which no discharge of pollutants is permitted

Source of Wastewater	COD	Phenols	Suspended Solids	Total Solids	Heavy Metals[b]	Volume[c]
	mg/l					(l/wk.)
Debarking	-	-	1,450	-	-	-
Log Conditioning						
Steam vat	4,900	0.44	661	3,388	-	815,000[d]
Water vat	7,293	<.70	935	2,660	-	-
Dryer Washwater	3,140	2.06	1,855	2,883	-	57,000
Glue Washwater[a]						
Phenolic	32,650	25.7	15,250	19,850)		
Protein	8,850	90.5	5,900	8,850)	-	53,500
Urea	21,050	-	10,200	27,500)		
Wood Preserving	50-8,000	10-300	200	1,200	1-10	2,000-60,000

[a]Values are based on an assumed water-to-glue dilution ratio of 20:1.

[b]Metals are major pollutants only at wood preserving plants that treat with water-borne salts of copper, chromium, arsenic, and zinc.

[c]Estimates for plants producing 9.3 million M^2 annually on 9.53 mm basis.

[d]Based on use of steam vat without condensate recovery and reuse.

condensate cannot be reused for steam production.

Hot water vats may be heated either directly with live steam or indirectly by means of heating coils. Where the former method is used, the volume of discharge and its composition are similar to that of steam vats. When the vat is heated indirectly, the volume of wastewater generated is small, there being no constant discharge. The oxygen demand of this wastewater is normally much higher than that from steam vats because of its longer contact time with the wood. Hot water vats are normally emptied periodically, regardless of the heating method used, and refilled with clean water. A few plants settle spent vat water and reuse it.

Effective wastewater management in this phase of veneer and plywood production can be accomplished in part by in-plant operational changes and in part by equipment modification. For plants with hot water vats that are heated indirectly with steam coils, the discharge results from spills in loading and unloading the vat and from periodic cleaning. Plants operating in this manner need only to collect the water in small ponds or tanks and reuse it for vat makeup water. Adjustments in pH by the addition of lime or caustic soda may be required to control acidity and prevent the development of corrosion problems.

Plants equipped with hot water vats that are heated by direct steam impingement generate a volume of water too large to be reused for vat makeup water. This method of heating is increasing in use because it permits a more rapid attainment of the desired water temperature. The discharge from this method of log conditioning can be reduced to a manageable level by installing sufficient steam coils to provide the desired rate of temperature rise in the vat. This modification would eliminate the continuous discharge from vats now heated by direct steam impingement and permit the wastewater resulting from spillage and cleaning operations to be recycled, as described above.

Wastewater discharged from steam vats is more difficult to eliminate because it can be recycled only with difficulty. Either modification of existing equipment or changes in steaming technique is necessary to obtain a completely closed system.

Conversion of steam vats to hot water spray tunnels has been effected by a number of plants. Although this system does not heat logs as rapidly as direct steaming, it is a practical alternative for most plants. The water used in this system is passed through heat exchange coils prior to being sprayed over the logs. Hot water sprays can be placed in existing steam vats with only minor modifications of existing equipment. Alternatively, the steam vats can be equipped with a water reservoir fitted with coils and steam for conditioning generated by heating the water. The condensate formed in the vat returns to the reservoir where it is revaporized. This procedure has been successfully adopted by the wood preserving industry to reduce the volume of effluent generated by direct steam conditioning of green stock.

Dryer Washwater. Veneer dryers accumulate wood particles and resin deposits which necessitate periodic cleaning with water or water and detergents. The characteristics of this water vary with the amount of washwater used, the extent to which the dryer is scraped prior to application of water, operation of the dryer, and species of wood, among other factors. Flow volume from washing operations varies among plants and with individual operators, but generally ranges from less than 2,000 to over 24,000 liters per dryer per week. Analysis of washwater from two softwood plywood plants differing greatly in efficiency of water use revealed average BOD, suspended solids, and phenol contents of 292, 1411, and 2.71 mg/l, respectively. Expressed on a unit of production basis, these values were 31.13, 161.04, and .016 kg/million M^2.

In addition to washwater, most dryers are equipped with deluge systems to extinguish fires. Because fires are common in dryers, especially those that are poorly maintained, this source of water can add significantly to the waste disposal problem at some plants.

Wastewater management in this, as in other operational phases of veneer and plywood production, is based principally on water conservation. Plants with annual production of 9.3 million M^2 (9.53 mm basis) currently use between 50,000 and 60,000 liters of dryer washwater weekly. It has been shown that this volume can be reduced by 75 percent or more by the simple expediency of scraping the dryer and blowing it out with air prior to the application of cleaning water. The small volume of water required to clean the dryer after the scraping operation can be stored on company property, or disposed of by land irrigation. No discharge of wastewater from the plant is necessary.

Glue Mixing Systems. Adhesives used in the board and laminated products industries in the United States are virtually all of the phenolic, urea, and protein types. Protein glues are gradually being phased out of the industry, while phenolic glues are being used more extensively. All three types have the potential of greatly increasing the BOD loading of the discharge from the industry. The chemical and biochemical properties of this waste vary with the type of glue used and the degree of dilution that occurs in the washing operation. For a dilution ratio of 20:1, COD, suspended solids, and phenol values for phenolic adhesives are on the order of 32,000, 15,000, and 25.7 mg/liter, respectively. Comparable values for protein glue are, in order, 8,800, 6,000, and 90.5 mg/liter.

The main source of wastewater from gluing operations is cleaning of glue-handling equipment, primarily glue spreaders and mixing tanks. Glue washwater may be reused for glue makeup without adversely affecting the quality of the resulting glue mix. However, at plants that do not practice water conservation in clean-up operations, the volume of washwater generated exceeds the requirements for glue makeup by a factor of two or more.

Pollution control measures currently used in the industry to control discharge of glue washwater include lagooning, evaporation, incineration, and reuse of washwater. Water conservation, including the recycling of glue washwater for glue makeup, is an accepted technology in the industry. A high proportion of the larger plants in the industry have achieved zero discharge by reducing the volume of washwater to an amount equal to the volume of water required for glue mixing.

Wood Preserving. Two of the three segments of the wood preserving subcategory have a zero discharge limitation. These segments are composed of the plants that employ inorganic salt formulations as preservatives and those that condition stock by the Boulton process. Most of the estimated 200 plants in the former industry segment have achieved zero discharge by water conservation measures and by reusing wastewater resulting from spillage, equipment washing, and other sources as makeup water in preparing fresh batches of treating solution.

Boulton plants have largely met the zero discharge requirement by reusing preuse water for cooling purposes. Excess water is either evaporated by adding heat to a cooling tower using a small heat exchanger or is discharged to a publicly owned treatment works.

Technology for End-of-Pipe Treatments

In addition to recycling, evaporation, and incineration, plants that are allowed a discharge have the option of employing end-of-pipe treatments to upgrade the quality of the water to the point that it will meet applicable effluent limitations and can be discharged. The most common treatments are physical or physiochemical processes to remove floating and settleable solids and emulsified materials, followed by biological or chemical oxidation to remove oxygen-demanding substances and to break down toxic organic chemicals. A typical treatment regime is as follows:

Oil Separation
Equalization
Chemical Coagulation
Sand Filtration
pH Control
Biological Oxidation
Secondary Clarification

The technology actually employed in treating a given waste depends upon waste characteristics primarily, but it is also influenced by such other factors as available land area, volume of discharge, and disposition of treated waste. Oil separation, for example, is necessary only for oily wastewater and, among timber products industries, is employed only in the pretreatment of wood preserving wastes. Adjustment of pH is required only if the pH value falls outside the range of 6.0 to 9.0.

Biological treatment is the most practical "end-of-the-pipe" method by which industries can meet effluent limitation requirements. Oxygen demand and suspended solids are the only two wastewater pollutants of any consequence in most wood industry effluents, and, with the notable exception of the pulp and paper and wet-process hardboard categories, the discharge volume is small. It follows that a conventional treatment consisting of primary settling and biological oxidation is the indicated technology. Indeed, several segments of the industry are successfully using this treatment regime. Wood preserving effluents present a somewhat more difficult problem because they contain phenolic compounds, principally cresols and pentachlorophenol, in addition to a high oxygen demand. However, biological treatments, when preceded by appropriate primary treatments, have been shown to provide satisfactory results (Table IV).

There are several biological treating techniques widely used both for industrial and municipal wastes. These are activated sludge, trickling filtration, oxidation ponds, and aerated lagoons. Among these, the activated sludge process is probably the best over-all in terms of efficiency of operation, resistance to upset, and land area required. Small package units are available that are suitable for use by many wood-based industries. However, the most common treating methods used in this industry are oxidation ponds and aerated lagoons. Oxidation ponds are appealing because the initial capital investment, exclusive of land, is relatively small. The problem is that they are frequently treated as sumps instead of the delicately-balanced, biotic systems that they are. When so used, they operate inefficiently, if at all. Aerated lagoons tolerate higher organic loadings and, therefore, require less land area than oxidation ponds for the same volume of waste. Like oxidation ponds, their main appeal is a relatively low initial capital investment and absence of complicated equipment. However, they, too, require a substantial land area and are subject to upset by shock loading.

One of the most promising treatment techniques for wastewater is soil irrigation. Although used off and on in the U. S. since the 1920's, it has been thoroughly researched in recent years and is now being promoted for use with both municipal and industrial wastes. The process consists simply of spraying wastewater on a prepared field. Soil microorganisms break down the organics and the filtering effect of the soil removes solids and much of the color. It is more efficient than other biological systems in use, in terms of percent reduction of oxygen demand and organic toxicants, provides a means for a plant to achieve a zero discharge, and is relatively inexpensive if land that can be devoted to that use is available. It is currently being used successfully by a number of wood preserving plants in the South and is the indicated technology for other segments of the wood-based industry. Typical data obtained using this process to treat wood preserving wastewater containing pentachloro-

Table IV. Quality of effluent from each stage of a multiphase waste treatment system of a phenol-bearing wastewater

Parameter	Raw Waste Load	Treatment Phase[a]				
		A	B	C-1	C-2	C-3
				(mg/l)		
COD	21,670	6,500	3,250	700	700	700
Total Phenols	-	135	127	<12.5	<12.5	<12.5
Penta-chlorophenol	100	25	17	<2.5	<2.5	<2.5
Oil and Grease	2,985	895	<100	<10	<10	<10
				(kg/day)[b]		
COD	654 (1,439)[b]	306 (674)	153 (337)	33 (73)	(73)	(73)
Total Phenols	-	6.37 (14.01)	5.99 (13.18)	<.59 (<1.30)	<.59 (<1.30)	<.59 (<1.30)
Penta-chlorophenol	2.99 (6.58)	1.18 (2.59)	.80 (1.76)	<.12 (<0.26)	<.12 (<0.26)	<.12 (<0.26)
Oil and Grease	90 (198)	42 (93)	<45 (<100)	<.45 (<1.0)	<.45 (<1.0)	<.45 (<1.0)
pH	4.7	4.7	7.0	7.0	7.0	7.0

[a]Treatment phase A = oil separation; B = flocculation, filtration, and pH adjustment; C-1,2,3 = biological treatment by activated sludge, aerated lagoon, and soil irrigation, respectively.

[b]Values in parentheses are waste loadings in pounds per day.

phenol are shown in Table V.

Table V. Efficiency of COD and pentachlorophenol removal from wood preserving wastewater by soil irrigation

Month	COD[a] Removal Efficiency (%)	PCP[a] Removal Efficiency (%)
Apr.	99	99
June	99	98
Aug.	99	99
Oct.	98	97
Dec.	95	95
Feb.	92	94
Annual Average	97.9	98.31

[a]COD annual average: 19,420 mg/liter. PCP annual average: 75.45 mg/liter.

In addition to biological treatment, phenol-bearing wastes are also amenable to treatment by chemical oxidation, activated-carbon filtration, and certain other techniques that are effective in removing phenols. This methodology is of primary interest to the wood preserving segment, since among timber industry sub-categories that are permitted a discharge it is the one that has the most serious problem with phenols.

Chemical-oxidation treatments of wood preserving wastewaters containing phenols have been successfully conducted on both a laboratory and commercial scale using either chlorine or a chlorine compound, principally calcium hypochlorite. Its effectiveness varies with the type of phenolic compound in the effluent, either cresols from creosote treatments or pentachlorophenol from treatments employing that chemical. Also influential in this regard are effluent pH, the effectiveness of pretreatment, particularly flocculation and filtration, and the amount and type of organic materials other than phenols present in the wastewater.

The theoretical ratio of chlorine to phenol required for complete oxidation is about 6:1. For m-cresol the ratio is 3.84:1 (1). However, because of the presence of other chlorine-consuming compounds, such as oils from the preservatives employed and carbohydrates leached from the wood, much higher ratios are usually required. Thompson and Dust (2) found that the minimum concentration of calcium hypochlorite required to oxidize all phenols in creosote wastewater was equivalent to a chlorine:phenol ratio of 14:1 to 65:1, depending upon pH, oxygen demand, and source of the wastewater. Comparable ratios for wastewaters containing pentachlorophenol ranged as high as 700:1. As shown by the data in Table VI, even massive chlorine dosages were at times ineffective in removing the last traces of phenolic compounds from wastewater, thus indicating that they were resistant to oxidation.

Table VI. Effect of chlorination with calcium hypochlorite on the pentachlorophenol content of wastewater from a wood preserving plant

	Pentachlorophenol (mg/liter)					
$Ca(OCl)_2$ as Chlorine (gm/liter)	Unflocculated pH 4.5	Unflocculated pH 7.0	Unflocculated pH 9.5	Flocculated pH 4.5	Flocculated pH 7.0	Flocculated pH 9.5
0	21.5	19.0	20.5	12.0	12.0	14.0
0.5	10.0	14.0	10.0	6.0	9.0	11.0
1.0	8.0	10.0	8.0	4.0	8.0	9.0
1.5	6.0	8.0	8.0	2.0	5.0	6.0
2.0	6.0	7.5	8.0	0	3.6	7.0
3.0	3.5	6.0	5.0	0	0	4.0
4.0	2.0	6.0	4.0	0	0	0
5.0	2.0	5.8	4.0	0	0	0

A large proportion of the chlorine added to wastewater is consumed in oxidizing organic materials other than phenolic compounds. Thus, for example, in the work of Thompson and Dust cited above, the wastewater COD content averaged 20,400 mg/liter before and 10,400 mg/l after treatment with chlorine at a rate of 2 g/liter. However, the COD was further reduced to only 10,250 mg/liter upon the addition of 10 g/liter of additional chlorine. This result suggests that a portion of the organic content of the wastewater was resistant to chemical oxidation, as indicated above for phenolic compounds.

Results of commercial employment of chlorination to treat wood preserving effluents have been mixed. Treatments applied to wastewater as part of a regime that included only oil separation and flocculation-filtration generally have been of questionable value. Conversely, its use as a polishing treatment following biological oxidation has been more successful. For example, White, et al., (3) achieved no detectable reduction in either phenols or pentachlorophenol in full-scale prechlorination treatments of a waste scheduled to receive a biological treatment.

However, a 50 percent reduction in both chemicals was obtained in post-chlorination treatments of the same waste.

The presence in wood preserving wastewater of oils, carbohydrates, and other organic substances also limit the effective use of activated carbon adsorption for removal of phenolic compounds. These materials are also adsorbed by the carbon, and since their concentrations exceed that of phenols in most wastewaters, the useful life of activated carbon is determined by the rate at which they are adsorbed. Recent laboratory studies indicate that this adsorption rate is unacceptably high (Figure 1). Consequently, the economics of carbon adsorption in the treatment of wastewater containing a high COD content are not favorable. However, the process does have an indicated use as a polishing treatment for waste that has received a biological treatment.

In actual practice, pollution abatement and control activities in the timber products industries consist in part of in-plant process changes to reduce the volume of wastewater that must be treated and treatments of the residual waste to meet state and federal guidelines and standards. Separation of contaminated waste streams, recycling of cooling water, reuse of process water, and changes in processing techniques are methods in common use to reduce the total volume of water that must be treated. For example, some plants in the wood preserving industry have reduced their discharge volume by 75 percent, while improving effluent quality, by relatively simple changes in processing methodology (Figure 2). Similar improvement is possible in the veneer industry by converting hot water vats heated by direct steam impingement to systems heated with steam coils.

Many wood-based industries with discharges have not inaugurated waste management programs. Instead, they are storing their wastewater in lagoons on company property. This practice is not a viable, long-range solution to pollution control in the South because of the excess of rainfall over evaporation. Monitoring procedures proposed by EPA, when approved and issued in final form, will require that test wells be sunk around such lagoons and sampled periodically to ensure that no groundwater contamination results from these storage facilities.

In addition to its effluent guidelines, EPA will soon issue an additional regulation related to water pollution that directly affects the timber products industry. Entitled "Pretreatment Standards for Incompatible Pollutants for the Timber Products Processing Point Source Category," its purpose is to outline pretreatment standards for incompatible pollutants that are discharged to publicly owned treatment works. As such, it is of primary interest to plants that are currently discharging process wastewater to the treatment works or that plan to do so in the future. By "incompatible pollutant" is meant those substances in a plant's discharge that may adversely affect the treatment works or pass through such works unchanged and pollute the receiving stream.

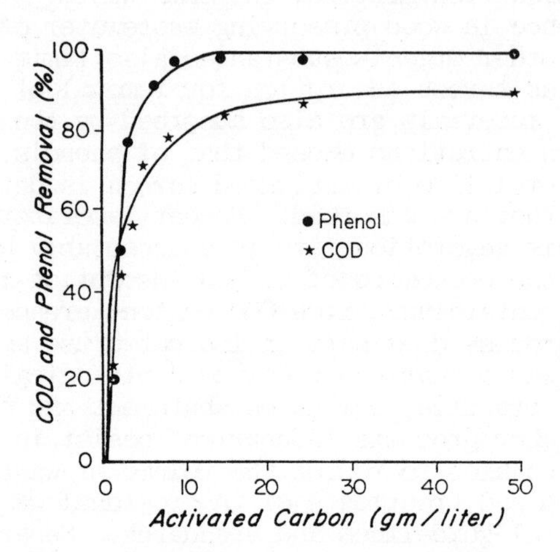

Figure 1. Relationship between weight of activated carbon and removal of COD and phenols from a creosote wastewater

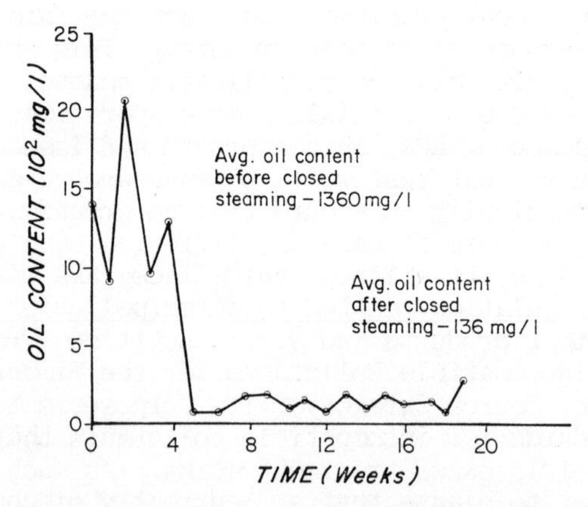

Figure 2. Variation in oil content of effluent with time before and after initiating closed steaming at a wood preserving plant

Among the various facets of the timber products industry, only the wood preserving industry has pollutants in its discharge that have been classified as "incompatible" by EPA. The field work required to develop pretreatment standards for this industry has been completed and the issuance of standards is expected during 1976. Of primary concern here are the phenolic compounds in oily preservatives and heavy metals in salt-type preservatives.

Summary

Federal effluent guidelines and standards have been prepared and promulgated for all subcategories of the timber products industry. Under these standards, industry must install the best practicable treatment and control technology currently available by July 1, 1977, and the best available treatment and control technology economically achievable by July 1, 1983. A zero discharge has been judged by the Environmental Protection Agency to be both practicable and economically achievable for most subcategories of the timber products industry.

Industry subcategories that are permitted a discharge under existing standards are wet-fiber processing and one segment of the wood preserving industry. Control of pollutant discharge is being accomplished by in-plant process modifications that reduce the volume of wastewater generated and by treatments that include flocculation and filtration, followed by biological treatment.

Literature Cited

1. Manufacturing Chemists Association. (1972). The effect of chlorination on selected organic chemicals. U. S. Environmental Protection Agency, Water Pollution Control Series, Project 12020 EXE. 104 pages.
2. Thompson, W. S. and J. V. Dust. (1972). Pollution control in the wood preserving industry. Part 3. Chemical and physical methods of treating wastewater. Forest Prod. J. 22(12): 25-30.
3. White, J. T., T. A. Bursztynsky, J. D. Crane and R. H. Jones. (1976). Treating wood preserving plant wastewater by chemical and biological methods. Grant No. 12100 HIG. Office of Research and Development, U. S. Environmental Protection Agency, Corvallis, Oregon. 97 pages.

INDEX

INDEX

D

E

F